QUANTUM FIELD THEORY, AS SIMPLY AS POSSIBLE

QUANTUM FIELD THEORY, AS SIMPLY AS POSSIBLE

A. ZEE

PRINCETON UNIVERSITY PRESS   PRINCETON AND OXFORD

Published by Princeton University Press
41 William Street, Princeton, New Jersey 08540
99 Banbury Road, Oxford OX2 6JX

press.princeton.edu

All Rights Reserved

ISBN 978-0-691-17429-7
ISBN (e-book) 978-0-691-23927-9

Library of Congress Control Number: 2022944272

British Library Cataloging-in-Publication Data is available

Editorial: Ingrid Gnerlich and Whitney Rauenhorst
Production Editorial: Karen Carter
Jacket/Cover Design: Karl Spurzem
Production: Danielle Amatucci
Publicity: Kate Farquhar-Thomson and Sara Henning-Stout
Copyeditor: Cyd Westmoreland

This book has been composed in Sabon LTStd and Jost

Printed on acid-free paper. ∞

Printed in the United States of America

10   9   8   7   6   5   4   3

Dedication

To all those who taught me quantum field theory, the theory of not quite everything, directly and indirectly[1]

Contents

Preface

What is quantum field theory?

By now, there are numerous books introducing quantum mechanics[2] to the general public, but I am not aware of any popular book on quantum field theory.[3] When I told a distinguished theoretical physicist that I was working on a popular book on quantum field theory, he exclaimed, "Your book is really going to fill a gap. By now, everybody and his grandmother has heard about quantum mechanics, but nobody knows anything about quantum field theory." I replied, "Exactly, but even more strangely, by now everybody and his grandmother has heard about string theory." Readers of popular books on physics have jumped directly from quantum mechanics to string theory, it would appear.

Einstein showed, in 1905, that to describe particles moving close to the speed of light, we need to modify our everyday notions about space and time and unify the two into a spacetime described by special relativity. Around the same time, it became apparent that the world of atoms and subatomic particles, such as the electron, is governed not by classical mechanics but by a hitherto unknown quantum mechanics.

Consider the electron orbiting in a hydrogen atom, for example. While it behaves like a quantum particle for sure, it moves around rather slowly. Calculations show that the electron is moving at less than 1% of the speed of light. Hence, special relativity is not needed[4] to study how it behaves, for example, when absorbing and radiating light. Indeed, quantum mechanics was pieced together slowly through meticulous studies of how atoms absorb and radiate light.

Back in the 19th century, Faraday and Maxwell introduced the electromagnetic field, which led to the understanding that light is a form of electromagnetic wave. Thus, the electromagnetic field propagates through space at the speed of light, and so, by definition, is relativistic. Hmm, the nonrelativistic electron is interacting with a relativistic field.

So, in elementary treatments of atoms radiating light based on quantum mechanics, as it was formulated around 1925 and fed to unsuspecting undergraduates, the electron is treated as a nonrelativistic quantum point particle, while the light it radiates is treated as a relativistic, but classical, field. I tell students that this half-assed treatment, even though it produces results in agreement with experiment, is intellectually unbalanced and unappealing. Theoretical physics is more than a bunch of calculations.

Soon and sure enough, in 1930, Paul Dirac proposed that the electron is also described by a relativistic field, and together with others, he launched quantum field theory. Meanwhile, it was also understood that the electromagnetic field should also be governed by quantum rules, whereupon the field was found to consist of quanta subsequently named photons.

Quantum field theory was developed, in the late 1940s, by the likes of Feynman[5] and Schwinger, and nurtured by a number of greats. (Know ye that Feynman's main contribution to theoretical physics is in quantum field theory, not quantum mechanics, as is often indicated on the web.) It culminated during the 1970s in a triumphant rebirth, with a victory parade that filled the awestruck spectators with wonder and joy, as one of my professors told me.

In short, quantum field theory emerged out of the union of quantum mechanics and special relativity. It is also the most accurately tested theory in physics, far exceeding the limits set on Newtonian physics and on quantum mechanics.

Genesis of this book

Some twenty years ago, I wrote a textbook on quantum field theory and later, a textbook on Einstein gravity. I wanted to address what I regard as the most beautiful subject in theoretical physics first, then the second most beautiful.[6] At some point, Ingrid Gnerlich, my editor at Princeton University Press, and I talked about the possibility of two more accessible books to go with my first two textbooks. So I signed a two-book deal back in 2016.

I took the easier path first. In spite of curved spacetime and all that, Einstein gravity is probably more concrete and easier to understand than quantum field theory. And thus *On Gravity* came out in 2018. See figure 1. But still I put off writing a popular book on quantum field theory, and worked on my 2020 book *Fly by Night Physics* instead. Subconsciously, I knew that a popular book on quantum field theory would be difficult to write.

The reason is obvious. To even broach quantum field theory, I will have to get through both special relativity and quantum mechanics, which I cannot expect the reader to know intimately. Thus, parts I and II of this book are devoted to a lightning overview of these two pillars of 20th century physics. By necessity, I have to be almost absurdly brief, unless you want to be holding a thousand page book in your hands. Much as I would like to introduce this beautiful and enchanting quantum field theory to the largest possible audience, the reality is that those who have never heard of special relativity and quantum

Figure 1. Thus far, *On Gravity* has been translated into Chinese, Czech, Turkish, Italian, and Spanish. The Spanish publisher translated the cover design also! Notice that a "la" was added to the title but omitted from the design. Also, the "gravedad" failed to fall all the way to the bottom but was held up by an unknown force.

mechanics are rather unlikely to read this book anyway. At the other end of the spectrum, some readers should be able to zip through these two introductory parts.

Before parts I and II, I present a prologue introducing quantum field theory, sketching the road map for our quest. This is meant to provide an overview, not a detailed explanation, of what quantum field theory is all about. So, please do not be alarmed if you fail to understand every word in the prologue.

Since the publication of my textbook on quantum field theory in 2003, I have given general talks around the world.[7] When I spoke with Ingrid, I thought, somewhat naively, that I could convert the slides of my talk into a book. But I soon realized that for the slides to make sense, I would have to add a great deal of connective and explanatory tissues. Still, you might notice that a few of the figures in this book are in the form of slides.

Who is this book for: two ends of the spectrum

The young and the old, that's who.

I would love to reach the next generation of theoretical physicists, the bright college students, and even high school students inclined to major in physics.

For them, I throw in tidbits to entice and mystify. These are mainly to be found in the endnotes. Some of the chapters in part V are written with this particular group in mind.

I sometimes think of myself when young. I was told by a professor to read, during the summer of my junior year, an enormously thick book on quantum field theory, perhaps one of the worst textbooks ever written on any subject in physics. While I struggled and suffered, swimming in the fearful humidity of central New Jersey, a book like the one you now hold would have helped me greatly, showing me the forest instead of the trees.[8] (The punchline of the story is that in the fall, when the professor returned, he told me to read the book again, starting from page 1. I sometimes do the same to a student. It is actually good advice: read the same book twice.)

From the mail I have received, I gather that many readers of my popular books are scientists, engineers, medical doctors, lawyers, and other professionals, many retired, some not. Quite a few are brave enough to tackle my textbooks. I applaud these older readers, and address them as I write.

For both groups, and for others, don't be cowed. Understand that quantum mysteries are mysterious and not understandable in terms of everyday experience. If there is something you don't quite grasp, simply move on. If you didn't understand some concept or another, it is because I do not have the space to explain it in detail, nor the language in some cases. This book is meant to give you a flavor of quantum field theory, not to bestow mastery. When I write a popular physics book, I have the habit of sending the manuscript to a bunch of friends who are not physicists. One friend who has read a number of popular books likened them to fast food. He said that this book was like a more nutritious and substantial meal, which signaled to him that a fine cuisine, to be served up by textbooks, awaited him.

Princeton University Press routinely sends book manuscripts to expert readers to referee. One reader wrote: "[This book] offers insight that are not available in any other book, and in return it expects a lot from its readers. So this is not your average popular science book .... The point here is that Zee is pushing what can be done in the popular science format. Quantum field theory is not an easy subject to describe to the lay person. And Zee is forced to present this at a higher level than is normally done." Yes! This reader understands what I have in mind.

While I need to satisfy a whole spectrum of readers, I am particularly solicitous of the young, the future physicists of the world. For their benefit, I have put in a bit more math than I have to, but by and large I have put it into the endnotes.[9] Needless to say, I would be delighted beyond words if some college students, or even a few high school students, are inspired by this book to move on to a textbook on quantum field theory.

I daresay that this book might be helpful even to those readers who have already decided to go into theoretical physics. A friend to whom I sent this manuscript wrote, "This book would have been an excellent primer for me to read before studying Bjorken and Drell[10] when I attended Caltech in 1984."

I was once musing about the relative merits of writing a popular book and writing a textbook with Rob Phillips, a distinguished physics professor at Caltech. Citing himself as an example,[11] he opined that a popular book has the potential of drastically changing the trajectory of a young person's life. He is of course right; textbooks are for those who have already decided to learn physics.

Scope of this book

In my popular books, I prefer not to mouth a breathless account of the latest and the hottest. You can find that on the web, but caveat emptor! I would rather write about the oldest and the coldest, the established and the cherished. Yes, I am aware that some people like wild speculations, the more tenuous any conceivable connection with reality the better, the kind of stuff that even physicists would forget soon enough, or prefer not to know about in the first place. In this book, we will stay within the one and only universe we know and love. Be clear that this is not a book on string theory, a speculative theory that may well be correct, but a book on quantum field theory, a well established theory that in some form has been around for almost a hundred years.

As readers of all my other books know, I like to put in all sorts of tidbits, some more relevant than others, into endnotes. Readers have written to me saying that these endnotes, besides being informative, are even occasionally entertaining. On the other hand, some readers might find them distracting. If so, then simply ignore the endnotes and return to them later. Each to his or her own taste!

How much math do you need?

> To those who do not know mathematics, it is difficult to get across a real feeling as to the beauty, the deepest beauty, of nature .... If you want to learn about nature, to appreciate nature, it is necessary to understand the language that she speaks in.
> R. P. Feynman

I couldn't have said it better than The Man himself, or with more authority. Analogies of trying to appreciate theoretical physics without math abound: reading an essay about a piece of music without ever listening to it, watching a serious foreign movie without subtitles. Here is mine. Imagine yourself blindfolded, wearing thick winter mittens and trying to appreciate the beauty of a flower in bloom.

Some years ago, I wrote the preface to Feynman's classic book on quantum electrodynamics.[12] While reading that book, I kept muttering, "Oh Feynman, if you would write down an equation at this point, everything would become perfectly clear." But of course he was not allowed to do that. Neither am I. Yet, I dare to be delinquent and break some rules in this book.

If you flip through this book, you would see a sprinkling of mathematical symbols, even a few equations here and there. Relax! You are not entitled, let alone required, to take a final exam on quantum field theory.

In many cases, the math is not much more than a convenient notation replacing what would otherwise be a mouthful. And often it is just standard mathematical terminology. Would you rather that I keep saying "variation in space" instead of "spatial derivative"? Similarly, in the chapter on the Dirac-Feynman path integral, I started out saying "a fancy sum," but finally gave up and wrote "integral" and then the mathematical symbol $\int$, because that was what Dirac and Feynman were talking about, an integral, not a fancy sum! Clearly, you are not being asked to integrate anything, and if you prefer, you can keep on thinking of the integral as a "fancy sum." But from my visualization of the two types of readers I am addressing, I believe that many, both young and old, know some calculus, without insisting too much on the legal definition of "know." I also provide a list of mathematical symbols and notation, which some readers might find helpful.

As another example, it is manifestly impossible to discuss quantum mechanics seriously without complex numbers: it was formulated in terms of complex numbers from day one. It would be like a book on biology without words like "cells" and "DNA," or a book on accounting without numbers. My advice to you, no charge: Be suspicious of popular books on quantum mechanics that do not mention complex numbers.

In several of my books, I cited Einstein's dictum that "Physics should be made as simple as possible, but not any simpler." Here too, I try to make a rather abstruse subject as simple as possible but am often aware that if I make it any simpler, I would be at risk of descending into vapid generalities that characterize some popular physics books (excluding those which amount to mere breathless catalogues of astounding facts.) Still, it should be clear to discerning readers that at some places, if I were to go into a more detailed explanation, the book would easily grow to ten times its present size.

Choice of topics

Quantum field theory is an exceedingly rich subject, so that I necessarily have to leave out many interesting topics and skim over the ones I do include. Although the past several decades have seen a flowering of quantum field theory when applied to other areas of physics, notably condensed matter theory, it grew out of particle theory and reached its greatest glory there. Thus, in part V, I am obliged to go through the fundamental interactions, the strong, the weak and electroweak, grand unified, and gravity, limiting myself to one chapter on each. (One reader said that he would like more on particle physics, but this is a book on quantum field theory, not particle theory.) Eventually, at the urging of another reader, the rather long chapter on Einstein gravity was split into two: classical and quantum. Clearly, it is unconscionable to cram all this material into several chapters, but I have no choice.

On the other hand, I originally thought that a single chapter on spin and statistics, comprising perhaps five or six pages, would suffice. I could have left it at that, but then I realize that by merely saying things like "You cannot tell who is who in the quantum world," I would make this book even more incomprehensible than it need be.[13] So I lavish many pages on this topic, which I consider central to our understanding of the physical world, and expanded the material into part VI.

Preview of an exciting result: I devote one chapter (chapter V.1) exclusively to the crowning glory of relativistic quantum physics, namely, the existence of antimatter. The argument is incredibly and elegantly simple!

The ecological niche for this book

This book sits somewhere between a textbook and a popular book as traditionally understood. But still, it is a popular book, perhaps somewhat more advanced than most, meant to provide an overview, and so here and there I have to gloss over some technical details or say things that are not exactly right. A note to nitpickers: Yes, I actually know what I am talking about. For instance, I know the difference between invariance and covariance, between gauge field and gauge potential, and so on. After all, I did write a textbook on quantum field theory. But if I switch back and forth, always using the correct word, that would only serve to confuse most readers.

Several kinds of popular physics books are possible. The more descriptive kind focuses on processes and phenomena; this nucleon collides with that nucleon, this star dies off and explodes. Another kind is filled with wild speculations, talking about the beginning of time (or even more brilliantly, what happened before there was time, smile), a multitude of universes, and such like. Such books are easy to understand, but they come at a price: They do not offer a deeper understanding. In this book, I chose to emphasize the conceptual foundations, what makes quantum field theory quantum field theory. In other words, the difficult parts. But I could also sense the frustration of some readers.[14] Allow me to offer you some advice. You could read ten popular books but you will still understand less than what you would understand by reading a single textbook on quantum field theory. If you have the background necessary to tackle a textbook, then by all means go for it. Even struggling through a few introductory chapters would be worthwhile.

A disclaimer about history

Finally, a disclaimer. This is not a scholarly treatise, but a popular book intended to convey a flavor of what quantum field theory is about. Given a choice between historical accuracy and a livelier narrative, I have generally opted for the latter. No, Paul Dirac did not in a flash of insight realize the existence of antimatter. But if I were to mention all those who straightened him out, this book would be significantly thicker. No, Werner Heisenberg did not propose isospin as we know it, but instead suggested interchanging the proton and neutron. History is convoluted. Unfortunately, and also unhappily, popular physics book writers almost by necessity have to promote the Matthew principle[15] and perpetuate the myth that a mere handful of greats were responsible for advances in physics. I have included some historical material in endnotes.

Acknowledgments

I thank those who commented on various pieces of the manuscript: Linda Robbins Coleman, Joshua Feinberg, Andrew Greenwood, John Hart, Greg Huber, Brian Kent, Nadie LiTenn, Lewis Robinson, Richard Scalettar, Steve Weinberg, Mark Weitzman, Andrew Zee, and Peter Zee. (This group comprises an attorney, a composer and conductor of classical music and concert pianist, a biophysicist, a neurologist, a judge on the Federal Court of Australia, an evolutionary biologist, a computer scientist, a retired professional poker player, three physicists, and two undergraduate physics majors.)

At Princeton University Press, Ingrid Gnerlich has been enthusiastically supportive of this book since its conception. She has once again entrusted the copyediting to my longtime collaborator Cyd Westmoreland and the actual production to Karen Carter. I also thank Christie Henry, the director of Princeton University Press, for her vision and good humor. Craig Kunimoto, Alina Gutierrez, and David Reiss have been indispensable for help with the computer. As always, I have enjoyed the presence and encouragement of Janice and Max.

Notes

[1] Directly through courses and face-to-face instruction, Sidney Coleman, Julian Schwinger, Arthur Wightman, Sam Treiman, and a number of others, including but certainly not limited to James Hartle. Indirectly through their textbooks and popular books, Steve Weinberg, Richard Feynman, John Jun Sakurai, James Bjorken, and many others.

[2] Some tending to emphasize the more bizarre aspects perhaps a bit more than I would like.

[3] To some extent, my book *Fearful Symmetry* touches on several aspects of quantum field theory.

[4] For the innermost electrons in the heavy elements, such as uranium, the corrections due to special relativity are no longer negligible.

[5]The reader now holding this book in hand is likely an admirer of Feynman, as am I, but I do not worship him as a peerless god, as many do.

[6]Later, I wrote a textbook on group theory, to me the third most beautiful subject in theoretical physics.

[7]Most recently, Bangladesh, China, India, Sweden, and Brazil, and to a group of students at Cambridge University.

[8]By the way, I am now convinced that the author of the quantum field theory book I read never even saw the forest; he might have tripped over a log and banged his head. Indeed, he showed me more bark than tree.

[9]For instance, in chapter I.3, I dare to mention cosh and sinh in an endnote.

[10]Bjorken and Drell, *Relativistic Quantum Fields* is a celebrated textbook I also studied from. Note that Jim Bjorken, whom everyone in my community calls "Bj", is also mentioned in endnote 1.

[11]Rob chose not to go to college, but worked as an electrician, read books (including popular physics books), and got admitted to an elite graduate school.

[12]R. P. Feynman, *QED: The Strange Theory of Light and Matter,* Princeton University Press, 2014.

[13]Some readers may complain that it is incomprehensible anyway. The more sophisticated would realize that in a popular book, the best I can do is to give you a flavor of the subject. Of course you are warmly invited to move on to a textbook on quantum field theory, preferably mine.

[14]A few on the Amazon write that some of my books are too difficult, others complain that they are too easy. Well, simple: if you want less, go read a popular book, such as *Fearful,* and if you want more, read *QFT Nut.*

[15]The Matthew principle, coined by the sociologist R. K. Merton, operates in full force in theoretical physics. For a few examples in gravity, see *G Nut,* footnote on page 169, endnote 18 on page 376, etc.

Prologue: The greatest monument and a road map for a quest ████████

The greatest monument to the human intellect

Quantum field theory is arguably the greatest monument to the human intellect.

Wait for the indignant howls to die down. Anyhow, quantum field theory is the most accurately[1] tested piece of physics ever. Far more than Einstein gravity, far more than quantum mechanics itself.*

By now, the global intelligentsia has largely heard of quantum mechanics (or quantum physics, as it is often called). But surprisingly, to me at least, not many have heard of quantum field theory,[†] the logical outgrowth of quantum mechanics. Quantum field theory was started around the 1930s, a few short years after quantum mechanics itself was definitively formulated, and by some of the same greats[2] involved in developing quantum mechanics. Shortly afterward, it ran into serious obstacles, not surmounted until after World War II, thanks to a brilliant new generation of theoretical physicists, Julian Schwinger and Richard Feynman among them.

The most accurate in physics, really?

How accurate is quantum field theory?

---

*Of course, since quantum field theory is a generalization of quantum mechanics, a pedant might insist that a test of quantum field theory is also a test of quantum mechanics.

†Furthermore, strangely enough, probably more people have heard of string theory than quantum field theory. Perhaps something for the sociologists to study.

Our understanding of how fast a spinning electron precesses in a magnetic field surely ranks among the top ten hits of 20th century physics. A bit of background first for completeness, but to appreciate the marvel of quantum field theory, you do not need to follow the rest of this paragraph in detail. The electron has an intrinsic magnetic moment, a quantity that measures how the electron reacts to a magnetic field, just as the electron's charge measures how it reacts to an electric field. In classical physics, a spinning charged body has a magnetic moment proportional to its angular momentum—makes sense that the faster it spins, the bigger its magnetic moment. The ratio of its magnetic moment to its angular momentum is known as its[3] "g factor." For classical objects, $g = 1$.

Physicists were hugely surprised by experiments showing that $g = 2$ for the electron, indicating that it was not a tiny spinning ball. This mystery was subsequently explained by Dirac in a brilliant flash of insight.

But then the plot thickened. Eventually, with improvements in experimental techniques, the electron's g was measured to be slightly more than 2. As of 2011, experimentalists had determined the electron's g to a mind boggling accuracy of 1 part in 1 trillion (that is, an accuracy of[*] 10^{-12}), to be[4]

$$g_e/2 = 1.001\ 159\ 652\ 180\ 73(28)$$

Compare this with the theoretical prediction[5] from quantum field theory:

$$g_e/2 = 1.001\ 159\ 652\ 181\ 643(764)$$

The field theory prediction agrees with the measured value to 12 significant figures, which makes it by far the most accurately verified prediction in the entire history of physics.[6] (As I mentioned above, Dirac calculated the leading digit, namely, 1, before quantum field theory.)

To put this insane degree of accuracy in rough pictorial perspective, visualize a place about 1,000 kilometers (for those readers who cannot get over the king's foot, that is about 600 miles) from wherever you are.[7] Then the present discrepancy (if you could call it that!) of 10^{-12} between quantum field theory and experiment corresponds to an uncertainty of 10^{-3} millimeter in that distance. (The diameter of a human hair ranges from 10^{-2} millimeter to about 10 times that.)

You might think that Newtonian mechanics has been tested to a higher degree of accuracy. But a moment's reflection would convince you that cannot be true. How a spinning electron responds to a magnetic field is a fundamental property of the universe, in contrast to, say, how the large rock we call home spins. Consider the vast numbers of electrons in the universe. Every single one

---

[*]Here I am using scientific notation for numbers. The number described in words as 1 followed by n zeroes is denoted by 10^n. Thus, $10^3 = 1,000$. In words, $10^6 = $ a million, $10^9 = $ a billion, and $10^{12} = $ a trillion, etc. And 10^{-n} denotes the number obtained by dividing 1 by 10^n.

is expected to spin at the same rate and to have the same magnetic moment. Why? You are not amazed? Quite a mystery, left to quantum field theory to answer!

An inessential and digressive remark: I keep talking about the electron, but in the interest of total accuracy, I should mention that, for various technical reasons, some experiments are performed with the electron, while others are performed with the muon, an elementary particle that you could think of as the electron's "pleasantly plump" cousin. The muon behaves like the electron in many respects, except that it is about 200 times more massive than the electron. This need not concern the reader at this point. I will come back to it in chapter V.3.

Allow me a rant about the misleading poverty of our languages. Exhibit A: the word "theory." Philosophers might debate until they are red in the face about what constitutes a theory.[8] Happily, we are free to form our own opinion.[9] And fortunately for physicists, whether quantum field theory is merely a theory has no impact on society, unlike Darwin's theory of evolution.

Note added in proof

When I started writing these words almost five years ago, the gyromagnetic ratio was talked about only among quantum field theorists, and only occasionally. But then in April 2021, after the manuscript for this book was first completed and submitted to Princeton University Press, the popular press practically exploded about the latest measurement of the gyromagnetic ratio of the muon. See the addendum to chapter V.3.

What is it? Who needs it?

What is quantum field theory?

Quantum field theory arose out of our need to describe the ephemeral nature of life.

Indeed! My textbook *Quantum Field Theory in a Nutshell** opens as follows:

> Quantum field theory arose out of our need to describe the ephemeral nature of life. Birth and death, with some semblance of existence in between.

A physics in joke? Well, sort of. To appreciate this nerdy joke, come with me on a quest.

---

*Hereafter referred to as *QFT Nut*, and which I already alluded to in an endnote in the preface. As indicated in the bibliography, I will refer to my previous books as *QFT Nut, GNut, Group Nut, Fearful, Toy, G,* and *FbN*.

	Big	Small
Fast	Rocketship near lightspeed, no need for QM	The marriage of quantum mechanics and special relativity
Slow	Classical physics	Slow moving electron orbiting a proton in the hydrogen atom, no need for special relativity

In the peculiar confluence of special relativity and quantum mechanics a new set of phenomena arises: particles can be born and particles can die.

A new subject in physics, quantum field theory, is needed to describe birth and death, and some kind of life in between.

Figure 1. The square of physics: the map for our quest. In the upper right corner, we meet the confluence of quantum mechanics and special relativity.

A map for our quest

Our quest will take us over a strange landscape. So, first an overview before we embark. Dear reader, this prologue offers something like a travel brochure or video that a tour agency might show you before the trip. The rest of this book describes the actual trip, with more detailed explanations of the exotic scenery. So, do not worry if the brochure does not make complete sense. Things should become clearer as you get into the book. For now, sit back and savor what we will encounter.

The land of our quest is called the "square of physics," and figure 1 is a map of it. (You understand that the lines on the map are there to merely guide the eye. There is of course no abrupt transition between "slow" and "fast." Similarly, for "big" and "small.")

We will start in the southwest, in Newton's kingdom, where large lumbering objects, such as ourselves, planets, planktons, and jet planes, moving at speed far less than that of light, reside. Our goal is the northeast, where teeny bits of matter, such as photons, electrons, and quarks, might move at speeds close to the speed of light. On the way, we will encounter all sorts of enchanting phenomena, unknown and undreamed of in our "home village."

Starting in Newton's kingdom

In Newton's kingdom, we learned that force produces acceleration: $F = ma$; acceleration a is equal to the force F acting on the object divided by the mass

m of the object. Now hear the curse of angry frustration and derisive laughter from a medieval peasant pushing a heavy cart along a muddy road. Not only to him, but to medieval scholastics, this claim was utter lunacy. Aristotle sounded much more plausible: force produces velocity.* No force, no velocity. To keep the cart moving, you got to push.

The educated among us now understand that everyday life, alas, is dominated by friction, pain, and suffering. Aristotle appears to be right, and Newton wrong. But in fact Newton is right, and the venerable Greek, now banished from reputable physics departments everywhere, wrong. The fundamental laws of physics know squat about friction, pain, and suffering.

I believe that modern physics could only have begun in icy climes, such as England. Surely, Newton had seen skaters gliding along effortlessly on a frozen pond. This $F = ma$ stuff could never have occurred to a lad lazing under a coconut tree.[10]

Galilean relativity is as obvious as common sense; Einsteinian, not so much

We now move north on our map and forward in time. As the golden age of rail travel dawned, Einstein's contemporaries worried about synchronizing station clocks in nearby towns.[11] Let's follow one of Einstein's thought experiments. Suppose that Ms. Unprime,[12] sitting on a train gliding smoothly through a station at 3 meters per second, rolls a ball forward down the aisle at 2 meters per second. Mr. Prime, the stationmaster (with X-ray vision) standing on the platform, sees the ball moving forward at $2 + 3 = 5$ meters per second. Velocities add.

This is as obvious as common sense. After 1 second passes, Ms. Unprime sees the ball moving through 2 meters. In that 1 second, Mr. Prime sees the train moving through 3 meters, and hence the ball moving through 5 meters.

Now Ms. Unprime takes out her phone, and snaps a flash photo of her travel companion sitting across from her. The light in the flash moves forward at $c = 300,000,000$ meters per second. The Newtonian stationmaster thinks that the light is moving $300,000,000 + 3 = 300,000,003$ meters per second.

Einstein tells Mr. Prime, "No, you added wrong! You saw the flash moving forward at $c = 300,000,000$ meters per second also."

Actually, it was Maxwell who first said no. In Maxwell's electromagnetic theory, an electric field varying in time and space produces a magnetic field varying in space and time, which in turn produces an electric field varying in time and space, and so on, thus generating an electromagnetic wave, which light is. The speed with which the wave propagates has to do with how a

---

*Surveys have shown that most of the proverbial guys and gals on our streets, and also not a few university philosophers, are Aristotelian, not Newtonian, on this question.

varying electric field produces a magnetic field and vice versa, and has nothing to do with the person observing the light.

And thus Einstein shocked the physics world with his equation[13] $c = c$.

I already cautioned you that the two lines dividing the square of physics into four squares are merely to guide the eye. Newtonian mechanics does not give way to Einsteinian mechanics (more popularly known as special relativity) abruptly, but gradually, as the speed v of the objects involved approaches the speed of light c. Typically, the corrections to Newton's results are of order $(v/c)^2$. For example, the electron in a hydrogen atom moves at about one hundredth the speed of light, and the relativistic correction amounts to $(v/c)^2 \simeq (1/100)^2 = 1/10{,}000 = 1/10^4 = 10^{-4}$, that is, one part in ten thousand. Thus, in much of atomic physics, we could ignore special relativity.

Entering the restless quantum world

Next, let us move east from Newton's kingdom. See figure 1. Unless you are a Papuan head hunter, you have probably heard that we actually live in a quantum world, in which everything is constantly jiggling.[14] Hence the Heisenberg uncertainty principle: You can never know exactly where anything is. The quantum world is like a daycare center: Kids are zipzapping all over the place. In contrast to classical physics, quantum physics does not allow you to locate a particle and measure its momentum to arbitrary accuracy, no matter how much you refine your instruments.

Physicists use the Greek letter delta Δ (as in the Mississippi delta and in Delta Airlines, for example) to denote uncertainty. So, more precisely, Heisenberg tells us that the uncertainty in a particle's position,[15] denoted by Δq, multiplied by the uncertainty in its momentum, denoted by Δp, is equal to[16] a fundamental constant, introduced by Max Planck[17] and written as[18] $\hbar$. The uncertainty principle[19] states that* Δp multiplied by Δq equals roughly $\hbar$, that is,

$$\Delta p \times \Delta q \sim \hbar$$

Less uncertainty in one leads to more uncertainty in the other. If you know an electron's momentum accurately (less uncertainty in momentum), you won't know where it is (more uncertainty in position): $\Delta q \sim \hbar/\Delta p$. The smaller Δp, the larger Δq is.

And vice versa: If you try to locate an electron, you end up not knowing how fast it is moving. The smaller Δq, the larger Δp is.

Planck's constant $\hbar$ provides a measure for quantum uncertainty. A hypothetical world with $\hbar$ equal to zero would be entirely classical.

---

*I offer a heuristic derivation of the uncertainty principle in chapter I.5.

A word of encouragement. Don't worry too much about mastering the uncertainty principle with any degree of certainty. In this prologue, I am just giving you a bird's eye view of the land of our quest.

Uncertainty in time versus uncertainty in energy

Position and momentum are known as a complementary pair[20] in quantum physics. Time and energy form another complementary pair.[21]

The uncertainty in energy ΔE multiplied by the uncertainty in time Δt is equal to $\hbar$:

$$\Delta E \times \Delta t \sim \hbar$$

What this means is that if you narrow the time interval during which you observe a system, you won't know its precise energy. And vice versa: If you know the energy of a happening precisely, you won't know when it is happening.

I dare say that this energy time uncertainty[22] is much less often mentioned in the popular media than the position momentum uncertainty principle. Ironically, physicists use it much more often than the better known version: The uncertainty (technically known as the width) in the measured mass (which according to Einstein is the same as energy up to that famous factor of c^2) of a newly discovered unstable particle tells us about its lifetime, that is, the uncertainty in how long it has lived.

The Grand Old Man of Physics did not like it

An amusing digression. Let's break from the physics exposition to mention that the energy time uncertainty principle disturbed Einstein greatly. When the 26 year old Heisenberg introduced the uncertainty principle in 1927, the 48 year old "Grand Old Man" dominating theoretical physics worked furiously (but ultimately, fruitlessly) to disprove it, proposing a gadget (true to his patent clerk past) involving an unstable nucleus disintegrating, a clock, a spring, etc., to show that the precise time the nucleus disintegrated, and the amount of energy emitted, could be determined (see figure 2).

When Einstein presented the 42 year old Niels Bohr with a drawing of the gadget, allegedly showing the collapse of a pillar of quantum mechanics, the younger man was devastated. According to legend, Bohr stayed up all night and managed to come up with a counterargument to convince Einstein the

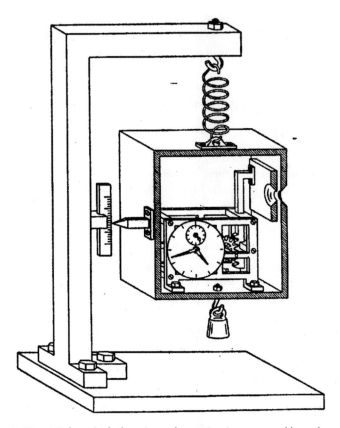

Figure 2. Einstein's box: A clock registers the precise time an unstable nucleus disintegrates and ejects a particle, which escapes through the hole shown. By weighing the box (as indicated by the ruler to the left of the drawing) before and after this event, Einstein claimed that he could determine how much energy was emitted. Never mind the details, since the contraption does not work, as was shown by Bohr.

next morning that his gadget failed to work. Ironically, the counterargument invokes physics that the Grand Old Man himself had invented. See figure 3.

As I mentioned above, not only do physicists now all believe in the energy time uncertainty principle, but they routinely invoke it in high energy experiments.

Two great advances distilled into two gold-plated equations

And thus, by the late 1930s, theoretical physicists had understood, separately, the realm of the very fast and the realm of the very small. They had ventured

Figure 3. Bohr and Einstein deep in discussion.
Photograph by Paul Ehrenfest, courtesy AIP Emilio Segrè Visual Archives.

forth from our home village, first north into special relativity, then east, into quantum mechanics. Shaking with trepidation, we now steel ourselves before heading to the wild frontier of the northeast (see figure 1).

Before doing that, let us distill the two great advances of 20th century physics into two gold plated equations, one for each advance, with an easy to remember "advertising slogan" to go with it. First, let us deal with quantum physics, and then turn to special relativity a short while later.

The gold plated equation of quantum mechanics
Uncertainty principle: $\Delta E \sim \hbar / \Delta t$
Advertising slogan: "Accounting errors could be tolerated for a short time!"

An accounting error of ΔE could be tolerated only for the short time $\Delta t \sim \hbar / \Delta E$. The larger the accounting error, the sooner it will be detected and set right. In contrast, a tiny accounting error might last for a long time. In this respect, the quantum world actually accords with the garden variety world: A sure fire embezzling scheme that might not be detected for a long time is to skim off a penny at a time.[23]

Students of quantum physics learn to deal with these fluctuating uncertainties. But what can these fluctuating uncertainties in energy do over a short duration? Actually, nothing all that much. Imagine having the students in a quantum mechanics exam calculate the behavior of two electrons in a box. They could calculate until they are blue in the face, but there will still be two electrons in the box, not one more, not one less. The ΔE can't be turned into electrons.

The other great advance is special relativity, with perhaps the most famous physics equation of all time.

The gold plated equation of special relativity
Energy and matter are interchangeable: $E = mc^2$
Advertising slogan: "Accounting errors could be turned into stuff!"

According to Einstein's famous equation $E = mc^2$, energy can be converted into mass and hence particles. Our proverbial embezzler could turn an accounting error into a Bugatti[24]—but only in his dreams if the world is classical. In that quantum-less world, even if relativistic, accounting errors are impossible.

Two separate strange worlds, but not strange enough

Two fascinating worlds, both strangely remote from our comfortable classical world in the southwest governed by nonrelativistic Newtonian physics! Indeed, each is bizarre in its own way,* and as such, has been dramatically described in popular physics books.

In a quantum world without relativity (a world governed by what is known in the jargon as nonrelativistic quantum mechanics), nothing much happens to the quantum fluctuations. The accounting errors get noticed after time Δt and are rectified.

In a relativistic world without the quantum (governed by what is known as relativistic classical physics), also nothing much happens. Yes, $E = mc^2$, and an energy fluctuation could be converted into particles, but there is no energy fluctuation in the first place.

To recap, at the beginning of the last century, physicists uncovered two bizarre worlds, the relativistic classical world and the nonrelativistic quantum world. Each ludicrously strange, but not strange enough!

The fun really begins when physics tried to combine the two.

---

*While the relativistic classical world is quite well understood in spite of mind bending happenings such as time dilation, the nonrelativistic quantum world still represents a fog of mystery to physicists after almost a century.

When Doctor Heisenberg met Professor Einstein

With both quantum mechanics and special relativity, something new could happen! Now, accounting errors abound, and they can be turned into stuff.

When physicists combined quantum mechanics and special relativity, around the middle of the last century, an exciting new subject, known as quantum field theory, emerged. With it came profound and novel concepts, one of which is nothingness.

In nothingness is everything

In quantum field theory, a state of nothingness is known as the vacuum.[25] But in quantum field theory, nothingness does not merely contain nothing; to the contrary, nothingness contains everything. The vacuum is a roiling sea of quantum fluctuations, boiling with particles and their corresponding antiparticles, coming into existence from nothing, and annihilating back into nothing after a short while. How short is determined by the energy of the particle antiparticle pair, in accordance with the uncertainty principle.

More precisely, when an energy fluctuation in the vacuum ΔE exceeds $2mc^2$, with m the electron's mass, then it could produce an electron and an antielectron (known as a positron), written as e^- and e^+, respectively.[26] With both quantum mechanics and special relativity, particles could magically appear!

But this magic can last for only a short time* Δt, before the carriage, aka the Bugatti, turns into a pumpkin, so to speak. Poof, the electron and the positron vanish into thin air! Physicists say that the electron and the positron annihilate each other.

Indeed, there is nothing special about the electron in this discussion. That is why physicists think of nothingness as a roiling sea of pairs of particles and antiparticles of every imaginable description, popping in and out of existence. The more massive the particle, the more ephemeral its existence.

But now we can take this argument one step farther. Instead of starting with nothingness, let us set two electrons crashing into each other with a huge amount of energy, call it $\mathcal{E}$, way more than $2mc^2$. Again, in the vicinity of the two colliding electrons, a quantum fluctuation could produce an electron and a positron. But now we don't need an accounting error: plenty of dough in the energy account with which to turn into stuff. It is all legit.

---

*Of order $\hbar/(2mc^2)$, as some readers might realize, which, given the known values of $\hbar$, m, and c, comes out to be about 10^{-21} sec. Pretty far from human experiences!

In contrast to our earlier story, there is no longer any restriction on the time duration that the pair could exist; the energy needed could be simply taken out of $\mathcal{E}$. The vacuum produces an electron positron pair costing at least $2mc^2$, taking the energy needed out of the two colliding electrons, which end up with some energy less than $\mathcal{E} - 2mc^2$. In the presence of two energetic electrons, we could produce an actual electron and positron pair, thus ending up with three electrons and a positron. Experimentalists would see, and in fact see quite often, two energetic electrons colliding and becoming three electrons and a positron:

$$e^- + e^- \rightarrow e^- + e^- + e^- + e^+$$

The marriage of quantum mechanics and special relativity led to quantum field theory

Indeed, as long as there is enough energy, nothing says that the pair produced has to consist of an electron and a positron. It could be a monster particle some theorist dreamed up last night and its antiparticle. Two electrons colliding with enough energy could well produce some hitherto unknown particles.

This explains, in a nutshell, why physicists are constantly clamoring for money to build ever more energetic accelerators to collide particles[27] with, thus producing more particles. The hope is of course that among these produced particles there might be some that nobody has ever seen before, thus resulting in a free trip to Stockholm.

The marriage of quantum mechanics and special relativity gives birth to a marvelously beautiful subject—music please!—known as quantum field theory.[28] It exhibits qualitatively new physics found neither in quantum mechanics nor in special relativity.

A case of the child being vastly more scintillating than the two parents!

Dear reader, the next time you meet a theoretical physicist, say, a captive sitting next to you on a plane, you could ask her which quadrant on the map of our quest she comes from.

Notes

[1]Please! Distinguish accuracy from precision. See N. Silver, *The Signal and the Noise: Why So Many Predictions Fail—but Some Don't*, page 46.

[2]Including, but certainly not limited to, Paul Dirac, Werner Heisenberg, and Wolfgang Pauli.

[3]For those who must know, g stands for "gyromagnetic."

[4]The (28) represents experimental uncertainty. See G. Gabrielse et al., arXiv: 1904.06174.

[5]Known as of 2016 and quoted also with uncertainty in the computation.

[6]I am glossing over the decades of arduous struggles both experimentalists and theorists dedicated their lives to, and the revolutionary technological advances that made the measurements possible.

[7]I was tempted to put in the "New Age" completion here: Wherever you are, there you are. (Deep! And even true.)

[8]The trouble stems from the distinctively different meaning of the word "theory" in everyday and in scientific usage. The detective has a theory that the blonde seen leaving the house on that fateful stormy night was actually the butler in disguise.

[9]A friend reminded me of a quote attributed to Daniel Patrick Moynihan, "You are entitled to your own opinion, but not to your own facts!" Sadly, the Zeitgeist in the United States has deteriorated considerably since his time.

[10]I know I know, I am confounding gravity and motion.

[11]P. Galison, *Einstein's Clocks and Poincaré's Maps: Empires of Time,* Norton, 2004.

[12]These names were used in *GNut.*

[13]Or, if you prefer, $c = c + v$, with v the velocity of the train. The meaning of the symbol $+$ would then have to be modified.

[14]This must be distinguished from "thermal jiggling" in the presence of a heat bath, which some readers might be familiar with.

[15]I use the letter q to denote the position of the particle instead of the more familiar x, for reasons to be explained in chapter I.2.

[16]Strictly speaking, greater than or equal to. The uncertainty principle tells us the best we could do, no matter how hard we try to reduce both Δq and Δp.

[17]For a fascinating biography of Max Planck, see B. Brown, *Planck: Driven by Vision, Broken by War.*

[18]Called "h bar," the letter h with a bar through it.

[19]The uncertainty principle is often stated and misinterpreted outside physics. Fashionable thinkers in other fields often attempt to display their brilliance by borrowing terms from physics. Some years ago, at an ultra elite east coast university (that I am too ashamed to name), a professor of theoretical architecture (believe it, all kinds of subjects are talked about in academia) waxed eloquent about the application of Heisenberg's uncertainty principle to the location of doors in avant garde designs. Fortunately, none of his designs were ever actually built as far as I know. In this connection, see also https://luysii.wordpress.com/2011/10/24/the-higher-drivel/.

[20]For those who know some math, the members of a pair are related by a Fourier transform.

[21]The two are not independent. We could derive one from the other. Consider a moving particle with $E = p^2/2m$. Then $\Delta E \sim p\Delta p/m \sim v\Delta p \sim v\hbar/\Delta q \sim \hbar/\Delta t$, since $\Delta q \sim v\Delta t$.

[22]See, for example, J. J. Sakurai and J. Napolitano, pages 78–80.

[23]It has been done; the key is of course to repeat this for millions of accounts. Would you miss a penny from your monthly credit card statement?

[24]While writing this, I learned from a *New York Times* article on March 3, 2020, that low level Saudi princes buy Porsche and BMW, mid level, Ferrari and Maserati, but the truly high level guys buy Bugatti.

[25]The use of that particular word is consonant with its use in everyday parlance. But the everyday vacuum, in spite of fantastic advances in pump technology, is still far from the quantum field theory vacuum.

[26]An electron cannot be produced by itself, since electric charge is conserved. A negative charge must be accompanied by a positive charge.

[27]In our story, I talk about colliding electrons. For technical reasons, it is easier to collide two protons, such as at the much celebrated and sadly disappointing Large Hadron Collider.

[28]For the interested reader who already knows some quantum mechanics and special relativity, many textbooks stand ready to teach you quantum field theory.

Our physical world

Preview of part I

Quantum field theory emerges from the union of special relativity and quantum mechanics, and so I am obliged to start by telling you a bit about both.

First, a lightning overview of the physical universe consisting of particles controlled by four fundamental interactions, the electromagnetic, the strong and the weak, and gravity. Then the notion of field in classical physics, as in the electromagnetic field and the gravitational field.

Special relativity is explained in chapters I.3 and I.4. The relevant mathematics does not exceed high school algebra. The "modern" approach to special relativity is through the geometry of spacetime, generalized from the familiar geometry of space with the help of a "valiant piece of chalk." From there we proceed to witness the birth of quantum mechanics.

Matter and the forces that move it ▬

The prologue to the book gave you a preview of our quest, something like the video a tour agency might show you. Now we embark on the actual trip.

Where do forces come from?

In just about any physics course, the professor would be talking about forces, the force of gravity, the electric force, so on and so forth. I am here to tell you that, until quantum field theory was invented, physicists did not really know where these forces came from. Sure, they could describe the forces, but that was about it.

So, that was a fairly big deal: quantum field theory could explain how forces arise.

Matter

First, I have to remind you that matter consists of molecules, and molecules are built out of atoms. An atom consists of electrons whirling around a nucleus, which in turn consists of protons and neutrons, collectively known as nucleons. The nucleons are made of quarks. That's what we know.[1]

The universe also contains dark matter and dark energy. Indeed, by mass, the composition of the universe is 27% dark matter, 68% dark energy, and only 5% ordinary matter. To first approximation, the universe may be regarded as one epic cosmic struggle between dark matter and dark energy.[2] The matter we know and love and of which we are made hardly matters. Unhappily, at present we know little about the dark side. Nevertheless, essentially all reputable speculations about the dark side are based on quantum field theory.

Forces

We know of four fundamental forces between these particles. When particles come into the vicinity of each other, they interact, that is, influence each other. Here is a handy summary of the four forces, known as gravity, electromagnetism, the strong interaction, and the weak interaction:

G: Gravity keeps you from flying up[3] to bang your head on the ceiling or from floating off like a space cadet.

E: Electromagnetism prevents you from falling through the floor and dropping in on your neighbors if you live in an apartment.*

S: The strong interaction causes the sun to provide us light and energy free of charge.

W: The weak interaction stops the sun from blowing up in our faces.

While we all have to come to terms with gravity, we know electromagnetism best, as our entire lifestyle is based on enslaving electrons.

Only four forces!

The world appears to be full of mysterious forces and interactions. Only four?

As you toddled, you banged your head against a hard object. What is the theory behind that? Well, the theory of solids can get pretty complicated, given the large variety of solids. But a simple cartoon picture suffices here: the nuclei of the atoms comprising the solid are locked in a regular lattice, while the electrons cruise between them as a quantum cloud. A collective society in which all individuality is lost! The atoms no longer exist as separate entities. The arrangement is highly favorable energetically; that is jargon for saying that enormous energy is required to disturb that arrangement. Revolution is costly. It takes quite a tough guy to crack a rock into halves.

So, the myriad interactions we witness in the world, such as solid banging on solid, could all be reduced to electromagnetism. What we see in everyday life is by and large due to some residual effect of the electromagnetic force: since common everyday objects are all electrically neutral, consisting of equal numbers of protons and electrons, the electromagnetic force between these objects almost all cancel out. Even the steel blade of a jackhammer smashing into rock is but a pale shadow of the real strength of the electromagnetic force.[4]

When you first emerged into this world, you might have thought that there must be thousands, if not millions, of forces in the world. Thus, to be able

---

*Plus a lot of other good deeds. Electromagnetism holds atoms together, governs the propagation of light and radio waves, causes chemical reactions, and last but not least, stops us from walking through walls.

to state that there are only four fundamental forces is totally awesome, a feat summarizing centuries of painstaking investigations. For example, realizing that light is due to electromagnetism stands as a towering achievement.

No contact necessary

Our common everyday understanding of force involves contact: we can exert a force on an object only if we are in contact with it. In a contact sport such as American football, without tackling the ball carrier, a linebacker could hardly exert anything on him. And in the movies, a slap is not a slap until the leading lady's palm makes contact with the leading cad's cheek. At the supermarket, you can push the shopping cart only if you grip the handle. If you could just hold out your hands and command the shopping cart to move, a crowd would gather and honor you as a wizard.

Everyday forces, except for gravity, are short ranged, indeed zero ranged on the length scales of common experience. These forces are but pale vestiges of the electromagnetic force, as I've just said. The palm molecules have to be practically on top of the cheek molecules before the latter could acquire any carnal knowledge of the former.

Gravity is the glaring exception. When the earth pulls Newton's apple down, no hand comes out of the earth grabbing the apple as in a horror movie. Gravity is invisible, thus all the more horrifying as we age.

Just about the only commonplace example of a force acting without contact is the refrigerator magnet: You can feel the refrigerator pulling on the magnet before the magnet makes contact with the refrigerator. This shows that the electromagnetic interaction, like gravity, is also long ranged.

Hence, in quantum physics, the word "interaction" is preferred rather than the word "force." No contact is necessary for particles to interact with each other. Indeed, the very concept of "contact" is problematical in the quantum world.

The universe as a finely choreographed dance

While the proverbial guy and gal on the street are plenty acquainted with gravity and electromagnetism, they have no personal experience with the strong and the weak interactions. But in fact, the physical universe is a finely choreographed dance starring all four interactions.

Consider a typical star, starting out in life as a gas of protons and electrons. Gravity gradually kneads this nebulous mass into a spherical blob, in which the strong and the electromagnetic forces stage a mighty contest.

The electric force causes like charges to repel each other. Thus, the protons are kept apart from each other by their mutual electric repulsion. In contrast, the strong force, also known as nuclear attraction, between the protons tries to bring them together. In this struggle the electric force has a slight edge, a fact of prime importance to us.[5] If the nuclear attraction between protons were a tiny bit stronger, two protons could get stuck together, thus releasing energy. Nuclear reactions would then occur very rapidly, burning out the nuclear fuel of stars in a short time, thereby making steady stellar evolution, let alone civilization, impossible.

In fact, the nuclear force is barely strong enough to glue a proton and a neutron together, but not strong enough to glue two protons together. Roughly speaking, before a proton can interact with another proton, it first has to transform itself into a neutron. This transformation necessitates the intervention of the weak interaction. Processes effected by the weak interaction occur extremely slowly, as the term "weak" suggests. As a result, nuclear burning in a typical star like the sun occurs at a stately pace, bathing us in a steady, warm glow.

Short and long ranged

The reason that the proverbial guy and gal in the street do not feel the strong and the weak interactions is because these two interactions are short ranged. The strong attraction between two protons falls abruptly to zero as soon as they move away from each other. The weak interaction operates over an even shorter range. Thus, the strong and weak interactions do not support propagating waves.

In contrast, the gravitational force between two masses and the electric force between two charges both fall off with the separation r between the two objects like $1/r^2$, the famous inverse square law of Newton. Gravity and electromagnetism are long ranged, as was mentioned earlier, and thus can and do support propagating waves. We will see how quantum field theory could explain this curious state of affairs in chapter III.2.

For r large, these forces still go to zero, but slowly enough that we can feel the tug of the sun, literally an astronomical distance away.[6] For that matter, our entire galaxy, the Milky Way, is falling toward our neighbor, the Andromeda galaxy.

Thus, in the contest between the four interactions, brute strength is not the only thing that counts: many phenomena depend on an interplay between range and strength. A case in point is fusion versus fission in nuclear physics. When two small nuclei get together, each consisting of a few protons and some neutrons, the strong attraction easily overwhelms the electric repulsion and they want to fuse. In contrast, in a large atomic nucleus, famously, the uranium nucleus, the electric repulsion wins over the strong attraction. Each proton only feels the strong attraction of the protons or neutrons right next to it, but each proton feels the electric repulsion from all the other protons in the nucleus. The nucleus wants to split into two smaller pieces, accompanied by the release of energy.

Notes

[1]Whether or not quarks and electrons are tiny bitty strings is an intriguing, but at the moment purely speculative, possibility.

[2]See *GNut*, chapter VIII.2.

[3]You know how fast the earth is spinning to cover about 24,000 miles in 24 hours. Anybody who has studied some physics could calculate what the centrifugal acceleration would be.

[4]Just about the only time the true fury of electromagnetism shakes us is when thunder and lightning fill the sky. While we modern dudes have totally enslaved electromagnetism, all ancient people attribute its occasional bursts of temper to the gods. We still devote one day a week to electromagnetism: Thursday is Thor's day.

[5]Quantum mechanics enters crucially here. The protons are not energetic enough to climb over the repulsive barrier set up by the electric force but have to tunnel through. See the discussion about Gamow tunneling in my book *Fly by Night Physics* to be abbreviated henceforth as *FbN*. See the bibliography.

[6]Of course, the feebleness of gravity compared to the other three interactions is also compensated for by the enormous number of particles contained in the sun and in the earth.

The rise of the classical field

Bizarre physics in the time of Newton

"So great an absurdity!" All right, class, who said that?

School children learn that the moon is attracted to the earth across the vastness of empty space. In contrast to their experience of pushing and shoving on the playground, no contact is necessary for a force to act. The earth is incessantly moving around the sun, as they know, and any change in the position of the earth is instantaneously communicated to the moon. In Newtonian gravity, the moon is slavishly yoked to the earth. In turn, the earth is yoked to the sun, and the entire galaxy moves as a collective entity.

That this sounds bizarre was already apparent to Newton, who complained in a letter to his friend Richard Bentley: "That ... one body may act upon another at a distance through a vacuum without the mediation of anything else by and through which their action or force may be conveyed from one to another is to me so great an absurdity that I believe no man who has in philosophical matters any competent faculty of thinking can ever fall into it."

Do you recall when you first learned about Newtonian gravity? Did you wonder[1] how a moon could know instantly that its planet had moved? Were you lacking in "competent faculty of thinking"? Ooh oh.

Faraday and our mother's milk

Look up at the night sky and admire the serenity of the moon. It is impossible to imagine, let alone to feel, the earth pulling on the moon, trying to bring that giant rock in the sky down to earth. But hold a tourist souvenir magnet close to your refrigerator, and you can feel the the magnetic force reaching across

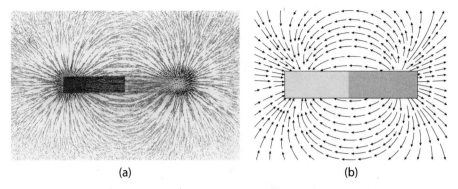

Figure 1. (a) Iron filings around a bar magnet. (b) The magnetic field of force around a bar magnet.

space. The great 19th century experimentalist Michael Faraday[2] introduced the term "field of force," or field[3] for short, in his study of magnetism.

Place a piece of cardboard on a magnet. Sprinkle iron filings (as if you usually have that around) on the cardboard. The iron filings eagerly, almost magically, line up to form a characteristic pattern. See figure 1. Faraday visualized a field of force around the magnet. When iron filings are introduced into this field, they are acted upon by the field.

Instead of the magnet acting directly on the filings, physicists think of the magnet creating a magnetic field around it, which in turn acts on the filings. The key point is that, even in the absence of the filings, the magnetic field still exists, just sitting around shooting the breeze, so to speak.

Analogously, physicists say that an electric field surrounds a charged sphere, with the field of force pointing radially outward like the spines of a sea urchin, outward because a test charge* with the same sign (that is, positive or negative) of charge as that on the sphere would be repelled, feeling a force in the direction of the arrow. See figure 2. In contrast, a test charge with the opposite sign of charge as that on the sphere would be attracted toward the sphere, and the arrows would point inward. Like and like repel, like and unlike attract, as was mentioned in chapter I.1 and as the reader surely already knows, if not in studying electrostatics, then perhaps in other contexts.

Gravity famously does not know about yin and yang, in sharp contrast to the electric force, as was also mentioned in chapter I.1. All masses attract each other, a fact responsible for some of the most salient features of the universe. The gravitational field looks just like the electric field around a charged ball, except that the force field is always pointing radially inward.

---

*A test charge is simply an infinitesimal charge imagined by physicists to test or measure an electric field, infinitesimal in order not to disturb or add significantly to the electric field.

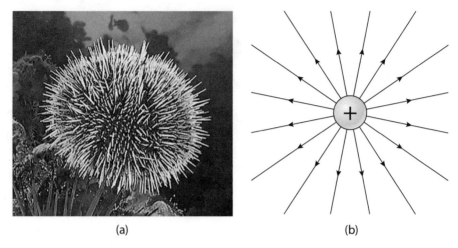

Figure 2. (a) A sea urchin, a leading California export to Asia. (b) The electric field of force around a positive charge.

That Einstein sure had a way with words. Listen to him:[4] "For us, who took in Faraday's ideas so to speak with our mother's milk,[5] it is hard to appreciate their greatness and audacity." Yes, some physicists have indeed forgotten[6] how audacious this concept of a field truly is.

Take-home message: Physicists need fields to make physics local. Following Newton, they run screaming away from the horror of action at a distance!

Able to leave home

Then physicists discovered that a moving magnet generates an electric field. At least in hindsight, the next question almost suggests itself: what does a moving charge generate? Moving charges are manifested most conveniently in the form of an electric current in a wire. Surprise! Electric currents do generate magnetic fields.

The important conclusion for physics is that an electric field changing in time could generate a magnetic field. Dualistically, a magnetic field changing in time could generate an electric field. This suggests that the electric field and the magnetic field would henceforth lose their separate identities, merging into one entity known as the electromagnetic field.

In Faraday's work, the electric field and the magnetic field served mostly as descriptive devices. But later, James Clerk Maxwell had the fantastic insight that an electromagnetic field changing in time could generate itself, moving across space as an electromagnetic wave. And thus, the electromagnetic field was able to leave home and take on a life of its own, bidding farewell to the charges, magnets, and currents that begot it in the first place.

Lo, our telecommunicating civilization was born!

As real as a rhino

These days, physicists visualize the earth creating a gravitational field, which in turn acts on the moon. The gravitational field is the mediator Newton longed for, to get around the absurd concept of action at a distance. And we are literally swimming in a sea of electromagnetic fields.

Dear reader, these are not mere words. The crucial, and meaningful statement, is that the field[7] as a physical entity is entirely real. As real as a rhino, according to the Indian American physicist Anupam Garg.[8] And so on the back cover of his textbook on electromagnetism, I blurbed that quantum fields are as real as quantum rhinos.

Cartesian coordinates

René Descartes, watching a fly while lying in bed, taught us that we could locate a point in the 3-dimensional space we were born into by 3 numbers (x, y, z), which we will write for short as $\vec{x}$. Consider the electric field we just talked about. At any instant in time, call it t, and at any point $\vec{x}$ in space, the electric field is specified by 3 numbers, (E_x, E_y, E_z), namely, the component of the electric force pointing in the x-direction, the component pointing in the y-direction, and the component pointing in the z-direction, respectively. In other words, $E_x(t, x, y, z)$ is a function of 4 variables: t and (x, y, z). It varies in time and in space. Similarly for E_y and E_z. Again, we could write all this for short as $\vec{E}(t, \vec{x})$. Similarly, physicists write the magnetic field as $\vec{B}(t, \vec{x})$.

Incidentally, Maxwell, working before some clever fellows thought of putting little arrows on top of vector quantities such as the $\vec{E}(t, \vec{x})$ and of using subscripts to distinguish the different components of $\vec{E}$, actually wrote out all six components $(E_x, E_y, E_z, B_x, B_y, B_z)$ of the electromagnetic field and all four spacetime coordinates (t, x, y, z). Thus, his treatise is almost impossible for contemporary physicists to read. That he managed to see the electromagnetic wave though this morass is almost a miracle.

In theoretical physics, a good notation is often said to be half the battle.[9]

A notational confusion

At this point, I must mention a notational confusion that has confounded and brought grief to generations of beginning students of quantum field theory.

In Newtonian physics, the motion of material objects is first abstracted to the motion of point particles. The position of the point particle being studied is denoted by (x, y, z), as Descartes taught us. The three numbers (x, y, z) change with time: that is what it means to say that the particle is moving around. The goal of Newtonian mechanics is to determine the three functions $x(t), y(t)$, and $z(t)$, each a function of the time t. These three functions, written more compactly as $\vec{x}(t)$, are the dynamical variables of Newtonian mechanics.

In contrast, the electric field is denoted by $\vec{E}(t, x, y, z)$. As you can see, (x, y, z) here label an arbitrary location in space. You specify (x, y, z) and $\vec{E}(t, x, y, z)$ tells you what the electric field is at that location in space at time t. Clearly, when discussing fields, we regard $\vec{E}$ as our dynamical variable, not (x, y, z).

These two conceptually distinct uses of the letters (x, y, z) do not pose a problem in introductory physics courses, but obviously would wreak havoc when we are fooling around with both particles and fields. In that case, it would be mandatory to specify the position of the particle by something other than (x, y, z). One standard choice is (q_x, q_y, q_z), or even better, (q_1, q_2, q_3) packaged as $\vec{q}$, or more precisely, $\vec{q}(t)$. In fact, I have already used this "more advanced" notation ($\vec{q}$ instead of $\vec{x}$ for the position of a particle) when I discussed the Heisenberg uncertainty principle back in the prologue.

You actually know what a field is, you just don't know that you know

In the popular imagination, the word "field" conveys a certain mysterious, perhaps even mystic, air. But physicists actually use the word rather broadly and loosely, even with abandon. Essentially, almost any physical quantity that varies in space and time, namely $(t, \vec{x})$, may be called a field.

For instance, suppose you are studying the temperature $T(t, \vec{x})$ of the earth's atmosphere, or the air's flow velocity $\vec{v}(t, \vec{x})$. The former is a scalar* field, the latter a vector field. Evidently, their dynamics (that is, behavior) can be described by classical physics, and hence these are known as classical fields. In everyday circumstances, the electromagnetic field is also a classical field.

Sound furnishes another everyday example of a classical field, being a density wave in air. Humans can hear sound waves with frequencies between 20 Hz and 20,000 Hz.[†] Using the Greek letter ρ ("rho"), let us denote by $\rho(t, \vec{x})$ the deviation of the actual density of air from the quiescent density (that is, the density in the absence of sound). In other words, in the denser regions, the density fluctuation ρ is positive, while in the less dense regions, ρ is negative. In the absence of sound, $\rho = 0$. For ease of writing, I will henceforth often drop the extra words "deviation" and "fluctuation."

---

*This simply means that temperature is a number and does not have a direction in space.

[†]A hertz, denoted by Hz, is defined as 1 cycle per second

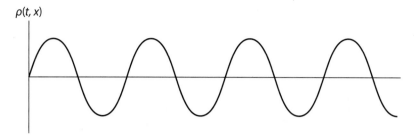

$\rho(t,x)$

Figure 3. A sound wave characterized by a single frequency or equivalently, a single wavelength. Such waves are called "monochromatic," evidently a term originating in the study of light waves.

Sound normally propagates in 3-dimensional space, but to keep the discussion as focused and as simple as possible, consider sound traveling down a long tube, so that space is effectively 1-dimensional, and we can write $\rho(t,x)$, with x measuring the distance along the (infinitely) long tube and t the time.

I plot in figure 3 the density fluctuation in a sound wave. I have intentionally not labeled the horizontal axis. You might have naturally interpreted the coordinate along the horizontal axis as x. Then the figure represents a snapshot of the density fluctuation at an instant in time. The figure is literally a picture of the density fluctuation in the tube, positive here, and negative there. An instant later, the figure would be different. It changes with time.

One figure could be interpreted in two different ways

Interestingly, you could have equally well taken the coordinate on the horizontal axis in figure 3 to be time t, so that the figure represents the density fluctuation at a fixed location in the tube. To the ear of an observer at that location, the density goes up and down, periodically denser and then less dense. The density changes, positive now, negative later, and then positive again yet later. Now the figure shows the entire history of the density fluctuation at one particular location.

That one figure could be interpreted in two different ways is an important point which we will exploit later, in chapter III.2, when we explain the origin of forces in quantum field theory. Keep that in mind!

Frequency and wave number

A wave is characterized by its frequency and by its wavelength (namely, the distance from crest to crest), in other words, by its variations in time and in

amplitude

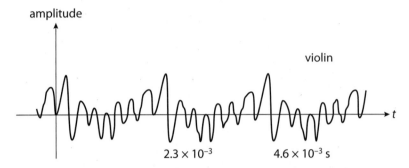

violin

2.3×10^{-3} 4.6×10^{-3} s

Figure 4. The note A, played on the violin, is formed by superposing a fundamental and a few harmonic waves. The pattern repeats itself after 2.3 milliseconds.
Modified from H. C. Ohanian and J. T. Markert, *Physics for Engineers and Scientists*, Norton, 2007.

space. Physicists denote[10] frequency by ω, but instead of wavelength, prefer to use the inverse of the wavelength, known as the wave number and written as k. The relationship between frequency and wave number, that is, the function $\omega(k)$, is characteristic of the wave.

If the only sound wave you could produce is that shown in figure 3, people would shun you as a rather monotonous person. That sound wave consists of one single frequency. Interesting sound waves are composed by superposing many different frequencies, as would be familiar to those readers who are musicians. See figure 4. Indeed, a chord consists of sound waves of several different frequencies that are in agreement, or accord, with each other.

Physicists sometimes call the sound wave shown in the figure a wave train. Of course, pleasing music cannot consist of a single note, but rather consists of a sequence of notes arranged cleverly to follow one upon another. Similarly, speech or song. It is almost miraculous, that the human vocal cord could exercise such fine muscular control, capable of rapidly producing one syllable after another.[11] No other life form on earth has mastered this "trick."

Beating between two waves with slightly different frequency and wave number

The wave train in figure 4 exhibits periodically a wave of higher amplitude than the others. You might have observed this same phenomenon at the seashore.[12] Waves come in sets, with a large wave followed by smaller waves and then larger waves, and then the cycle repeats, as shown in figure 5.

This pattern can be understood by picturing the interference of two waves with the same amplitude, but with slightly different frequency and wave number. Imagine a moment in time when the crest of one wave is matched with

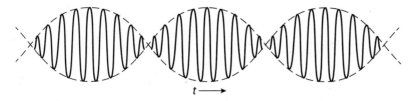

$t \longrightarrow$

Figure 5. Beating between two waves.
Redrawn from A. Zee, *Fly by Night Physics*, Princeton University Press, 2020.

the crest of the other wave, say, with slightly lower frequency. The two waves add, resulting in an exceptionally large amplitude; we have constructive interference.

Higher frequency means less time between crests. Thus, the next time the crest of the higher frequency wave arrives, the crest of the slightly lower frequency wave is not quite there yet. After each cycle, the lag grows a bit larger. Eventually, the two waves are totally out of phase, leading to destructive interference, thus explaining the pattern shown in the figure.

How long would that take? Well, after each cycle, the time lag is given by $1/\Delta\omega$, where, as explained in the prologue, $\Delta\omega$ denotes the difference between the two frequencies.[13] So, the time it takes for the phase lag to build up to π is given by $\pi/\Delta\omega$.

One key result to remember from this: The time between two big waves equals twice the time period worked out above, and thus $\Delta T \simeq 2\pi/\Delta\omega$.

Incidentally, waves on the beach commonly originate from storms at sea. By counting the number of waves between two large ones (as shown in figure 5) you could actually deduce how wide an area over which the storm occurred if you knew from the weather report where the storm was.[14]

Fourier and the frequency spectrum

We just illustrated the musical phenomena of beating by adding two waves with the same amplitude but different frequencies* We are certainly allowed to add waves with different amplitudes. For that matter, we could add a third wave with yet a different amplitude and frequency. Then, how about a fourth? Carrying this to its logical conclusion, Joseph Fourier showed in 1822 that, by adding many (possibly infinitely many) waves with different frequencies, we can construct a sound wave with any shape we like. Fourier, son of a tailor, was orphaned at a young age. Due to his lowly birth, he was excluded

*Note that frequency and wave number (or wavelength) are not independent variables. The relation $\omega(k)$ is determined by the physics relevant to the wave; for instance, for sound, it depends on how compressible air is.

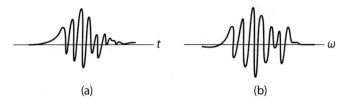

(a)   (b)

Figure 6. (a) A wave packet consisting of the superposition of waves with many frequencies. (b) The same wave packet as in (a) analyzed into its frequency components.

from being an officer in the artillery corps, and instead was assigned to teach mathematics.[15]

Thus, a sound wave may be characterized by its frequency spectrum. The sound wave shown in figure 3 consists of a single frequency. The wave train in figure 4 is composed by adding several waves with different frequencies that are multiples of each other.

Consider a pulse of sound (known as a wave packet in physics since it is composed of many waves), each with a definite frequency. At a given location, the density profile of air as a function of time t may look like what is shown in figure 6a. Before the pulse arrives, the density fluctuation ρ of air is zero by definition. Then it rapidly oscillates, varying between positive and negative values.

A pulse, such as that in figure 6a, may be analyzed into its frequency components, as shown in figure 6b. This important idea, universally used in the physical sciences and in engineering, is known as Fourier analysis.[16] Switching back and forth between figures 6a and 6b is known as a Fourier transform. Since in quantum physics, particles are revealed to be waves (much more on this later), the Fourier transform is an essential mathematical tool in quantum mechanics and in quantum field theory.

Or, as before, you could regard the figure as a snapshot of the density fluctuation as a function of space x. Then you would be talking about wavelength, or better yet, wave number k, instead of frequency. Simply change the labels on the horizontal axis in figures 6a and 6b to x and k, respectively. To summarize, Fourier transform allows us to hop back and forth between (t, x) and (ω, k).

The result we obtained earlier can now be extended. Let ΔT denote the duration of the sound pulse and $\Delta \omega$ the spread in its component waves in frequency. Then

$$\Delta T \Delta \omega \simeq 2\pi$$

Does this remind you of anything you have seen before? Yes: the uncertainty principle.[17]

And yes, the "other" uncertainty principle, relating the uncertainties in position and momentum, is also a manifestation of the Fourier transform.

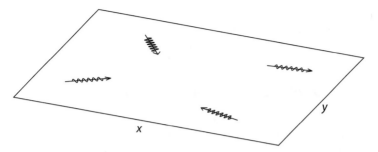

Figure 7. Wave packets in a 2-dimensional universe moving about and colliding with each other, possibly merging or producing other wave packets.

Wave packets could collide and scatter off each other

We could readily extend this discussion of wave packets, in 1-dimensional space to 2-dimensional space. Consider an elastic membrane, such as the surface of the drum.* By banging appropriately, we could create vibrational waves on the membrane, even construct wave packets.

Any point on this elastic membrane could be identified by two numbers, namely, (x, y), as Descartes taught us. Denote the deviation of the membrane from its equilibrium position, that is, its position in the absence of a wave, by $\varphi(t, x, y)$. Again, we could characterize the wave by its frequency ω and wave number. The discussion proceeds just as before, except that the wave number k has to be generalized to $\vec{k}$, a wave number vector or wave vector for short, with the vector pointing in the direction of propagation of the wave. (By the way, you see why the wave vector is a more useful concept than wavelength: it tells us about the direction of the wave also.)

Once again, we can form wave packets zinging around on the elastic membrane. The new concept that comes in when we move from 1-dimensional space to 2-dimensional space is "direction." Picture wave packets moving around in this two-dimensional universe in different directions, scurrying here and there, and occasionally even colliding with each other (figure 7).

By now, it takes no effort to move up to 3-dimensional space. Picture space filled with an elastic medium, perhaps a jello-like substance in which we could

---

*A word of caution and clarification here. At the mention of a drum, a "normal person" thinks of an everyday drum, surrounded by air, so that the vibration of the drum surface produces sound. A theoretical physicist, in contrast, immediately thinks of a drum of infinite extent, surrounded by nothing, existing by itself as a 2-dimensional universe. I merely said "drum" to help you fix in mind what I meant by "elastic membrane." We are thinking about the rippling waves on the surface of the drum, not the sound wave in the air enveloping the drum.

set up density waves. Simply write $\varphi(t, x, y, z)$, or more compactly, $\varphi(t, \vec{x})$ to denote density fluctuations. Once again, we could have wave packets moving around, but now in the 3-dimensional space we live in.

The power of free association

What do you think these wave packets remind physicists of? If you say elementary particles such as electrons, you may have what it take to be a theoretical physicist. The power of free association! These wave packets can move around in space, collide with each other, scatter and move off in different directions. They walk and talk like particles. Keep that thought in mind! We will come back to it.

Notes

[1] A lay reader to whom I sent the manuscript said that he was kept sleepless, not so much by this, but by the gravitational force between two bodies blowing up to infinity as they approach each other. In fact, many great physicists shared his worry. However, in classical physics, bodies have finite sizes, and a point particle is merely a convenient idealization. In quantum mechanics, this problem is obviated by quantum fluctuations. However, it is in some sense the origin of a notorious difficulty in quantum field theory involving the somewhat obsolete concept of "renormalization," a difficulty that has long been overcome, in spite of what you might have read elsewhere. Some voices on the web are decades behind the times.

[2] For a brief biography, see *Fearful*, pages 58–62. Rising from poverty, Faraday managed to find a job (without which we almost certainly would have never heard of him) working for the famed chemist Humphrey Davy. "Sir Humphrey's wife found Faraday physically awkward, and even irritating. He was small and stocky—not more than five foot four—with a large head that always seems slightly too big for his body. He spoke all his life with a flat London accent and had difficulty pronouncing his 'r's, so that as he himself said, he was always destined to introduce himself as Michael Fawaday." See R. Holmes, *The Age of Wonder*, page 352, Pantheon Books, 2008.

[3] In fact, I already snuck the word "field" past you in the prologue.

[4] A. Einstein, *Out of My Later Years*, Philosophical Library, 2015.

[5] Those of you who were bottle fed may be excused.

[6] Perhaps they were bottle fed.

[7] Read about how the earth's magnetic field preserved a memory of when the Babylonians torched Jerusalem in 586 BCE. https://www .timesofisrael.com/burnt-remains-of-586-bce-de struction-of-jerusalem-help-map-physics-holy -grail/.

[8] A. Garg, *Classical Electromagnetism in a Nutshell*, Princeton University Press, 2012.

[9] I could hardly believe it, but it is true. Einstein in his 1905 paper on special relativity still wrote out all 4 components of x explicitly.

[10] The frequency f (and its reciprocal, the period T) in everyday usage differs from ω by a factor of 2π: $\omega = 2\pi f = 2\pi/T$. We are certainly not going to quibble about such details here. Similarly, the wave number is defined by $k = 2\pi/\lambda$, with λ the wavelength.

[11] For a detailed analysis of an American politician's voice, see https://www.youtube.com /watch?v=waeXBCUkuL8.

[12] See *FbN*, chapter VII.2.

[13] For readers who remember some high school trigonometry, start with the identity

$\cos A' + \cos A = 2 \cos \frac{A'+A}{2} \cos \frac{A'-A}{2}$. Set $A = \omega(k)t - kx$ and $A' = \omega(k')t - k'x$, with $k' = k + \Delta k$ and $\omega(k') = \omega(k) + \frac{\Delta \omega}{\Delta k} \Delta k$. Thus

$$\cos(\omega(k')t - k'x) + \cos(\omega(k)t - kx)$$
$$\simeq 2 \cos(\omega(k)t - kx) \cos \left(\frac{1}{2} \left[x - \frac{\Delta \omega}{\Delta k} t \right] \Delta k \right)$$

We obtain a rapidly varying wave $\cos(\omega(k)t - kx)$ with large wave number k, with a correspondingly short wavelength, modulated by a slowly varying envelope

$$\cos \left(\frac{1}{2} \left[x - \frac{\Delta \omega}{\Delta k} t \right] \Delta k \right)$$

with small wave number Δk, and hence a long wavelength, as shown in figure 5.

[14]The point is that of the two interfering waves, one comes from the edge of the storm closer to us, the other from the edge farther from us. Noting that they arrive at the same time and knowing how the speed of ocean waves depends on frequency, we could estimate the difference between their frequencies. See *FbN*, page 280.

[15]Let me ask you: in the 21st century, how many remember Fourier and how many remember the commander of France's artillery corps at the end of the 18th century? Incidentally, Fourier was among the first to show that the earth would be much colder than it actually is given its distance from the sun and that the atmosphere, acting as a greenhouse, is crucial.

[16]For more, see https://en.wikipedia.org/wiki/Fouriertransform.

[17]In quantum physics, the energy of a particle is related to its de Broglie frequency by $E = \hbar \omega$ (see chapter I.5) with $\hbar$ being Planck's constant. Thus, multiplying the equation in the text by $\hbar$, we obtain $\Delta T \Delta E \sim \hbar$.

Time unified with space

In previewing our quest, I already mentioned Einstein's 1905 theory of special relativity, which, in total defiance of common sense, unified space and time into spacetime. Before heading toward the promised land of quantum field theory, we need to explore a bit this fabled region that we must cross. Ladies and Gents, I give you special relativity, in three ways! But first, some commonsense relativity.

Galilean, or commonsense, relativity

I remind you (see chapter I.2) that Descartes taught physicists to locate "events" in time and space by using coordinates (t, x, y, z), the when and where of happenings in our universe. But another observer is free to use a different set of coordinates (t', x', y', z'). How one set of coordinates is related to another is known as "relativity" to physicists.

One misconception is that relativity started with Einstein, but in fact, Galileo was the first, to quantify what we might call commonsense or everyday relativity (see figure 1).

Commonsense relativity states that the temporal and spatial coordinates of the two observers gliding by each other are related by the Galilean transformation:

$$t' = t, \quad x' = x + ut, \quad y' = y, \quad z' = z$$

In particular, the point assigned coordinates $x = 0$, $y = 0$, $z = 0$ by one observer would be assigned coordinates $x' = ut$, $y' = 0$, $z' = 0$ by another. The equation $x' = ut$ asserts that the two coordinate frames are moving with velocity u relative to each other along the x-direction.

To be a bit more concrete, go back to the duo from the prologue, with Ms. Unprime riding on a train smoothly gliding through a station, and Mr. Prime,

Figure 1. Galileo observing a butterfly flying normally and the smoke rising vertically from a candle on a smoothly moving ship.
Reproduced from A. Zee, *Group Theory in a Nutshell for Physicists*, Princeton University Press, 2016.

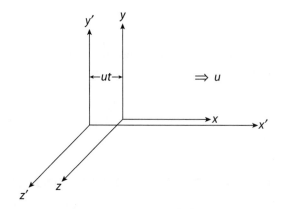

Figure 2. Two coordinate frames moving with velocity u relative to each other along the x-direction.
Redrawn from A. Zee, *Group Theory in a Nutshell for Physicists*, Princeton University Press, 2016.

the station master, standing on the platform. The point underlying relativity is that, while Mr. Prime could say that Ms. Unprime is moving, Ms. Unprime could equally well say that Mr. Prime is moving in the opposite direction.

Of the four equations displayed, the first, $t' = t$, states that time is universal: when one second has passed for Ms. Unprime, one second has also passed for Mr. Prime. Pure common sense.

The second equation defines the velocity u. The point designated by Ms. Unprime as $x = 0$ is seen by Mr. Prime as moving according to $x' = 0 + ut = ut = ut'$. See figure 2.

The third and fourth equations say that the two coordinates perpendicular to the direction of motion are not affected. This follows from a foundational principle of physics, that only relative motion could be defined, not absolute motion.[1]

Relativity simply states that the laws of physics cannot distinguish between coordinate frames moving smoothly relative to each other. On a jet moving along with no turbulence in sight, you could pour yourself a drink just as if you were at rest[2] at home!

The everyday rule for adding velocities, which we invoked in the prologue, is easily derived from the Galilean transformation. (Notational alert: Physicists use the Greek letter delta Δ to mean different things. In the prologue, it means "uncertainty;" here, "the change in." Even with the Greek alphabet included, there are only so many letters.) Suppose that in the time interval $\Delta t' = \Delta t$, Ms. Unprime sees an object moving through Δx and hence velocity $v = \frac{\Delta x}{\Delta t}$, Mr. Prime would see it moving with velocity

$$v' = \frac{\Delta x'}{\Delta t'} = \frac{\Delta (x + ut)}{\Delta t} = \frac{\Delta x + u\Delta t}{\Delta t} = \frac{\Delta x}{\Delta t} + u = v + u$$

Incidentally, in my experience, American physics students might be more familiar with moving sidewalks in airports than smoothly moving trains. In that case, Mr. Prime would be the guy in the souvenir shop watching people go by, and Ms. Unprime would be standing on the moving belt. (Another traveler walking on the belt would be the object moving with velocity v relative to Ms. Unprime. Mr. Prime sees this traveler moving with velocity $v + u$, as indicated by the calculation we just did.)

Does the universe have a speed limit?

I promised you three ways to special relativity, which I will now discuss in turn.

First way: Should we sit in on some physics crazed sophomores arguing in a beer soaked late night bull session, or listen with an air of feigned reverence to some chaired philosophers from America's most elite universities at a symposium to discuss deep truth? Your choice.[3] The topic: Is there a speed limit in the universe?

One philosopher intones, "There cannot possibly be a speed limit."

Proof by contradiction. Suppose nothing could go faster than the speed c. Suddenly, we see a spaceship zoom by with speed almost equal to c. Consider an observer moving by with speed u in the opposite direction. To this observer, the spaceship is receding in his rearview mirror with speed $c + u$. But this violates the assumed speed limit. "No speed limit! Quod erat demonstrandum," the philosopher crows.

Another philosopher, stunned by this argument, gravely nods in agreement, but could not resist muttering: "But what would Aristotle and Kant say?" Appeal to authority, yeah! A third philosopher muses. "With a speed limit, the notion of simultaneity becomes suspect. Communication between distant points necessarily takes time. We would have to conclude that absolute time does not exist, which is manifestly absurd! What kind of universe would that be if you can't tell what time it is?"

The three philosophers concluded by pure thought that physicists would never find a speed limit, but physics is not philosophy. Proof is by experiments. Well, physicists did find that nothing[4] could go faster than light. Hence, no universal time.

Dueling thinkers: the fall of simultaneity

Time is in the eyes of the beholder. The notion of simultaneity crashes and burns.

The second way: To see why simultaneity fails, let's watch[5] Professor Vicious and Dr. Nasty.[6] They have been at each other's throats for decades. Theoretical physicists are forever fighting over "who did what when." They are constantly bickering, telling each other (as the joke goes), "Nyah, nyah, what you did is trivial and wrong, and I did it first!"

Of course, the fight for credit goes on in every field, but in theoretical physics it is almost a way of life, since ideas are by nature ethereal. And the stakes are high: the victor gets to go to Stockholm, while the loser is consigned to the dustbin of history, a history largely written by the victor with the help of an army of idolaters and science journalists.

We are finally going to settle matters between Vicious and Nasty once and for all. The two of them are seated at the two ends of a long hall, Vicious at $x = 0$ and Nasty at $x = L$.

We now tell Vicious and Nasty to solve the basic mystery of why the material world comes in three copies.[7] As soon as they figure it out, they are to push a button in front of them. When the button is pushed, a pulse of light is flashed to the middle of the room where, at $x = L/2$, our experimental colleague, an electronics wiz, has set up a screen. When the screen detects the arrival of a light pulse, all kinds of bells and whistles are rigged to go off. In particular, if, and only if, two light pulses arrive at the screen at precisely the same instant, a huge imperial Chinese gong will be bonged.

"Fair is fair, any and all priority claims will be settled," we told Vicious and Nasty. "Now go to work and explain why quarks and leptons come in three sets." The dueling duo immediately assume the Rodinesque pose of the deep thinker and lock themselves in a "think to the death."

Meanwhile, you are sitting on a smooth train, moving relative to the dueling thinkers. Denote the time and space coordinates in your rest frame by t' and x'. In the Newtonian universe, time is absolute, and so we have $t' = t$. In your frame, you are sitting at $x' = 0$, but Vicious and Nasty are moving by according to $x' = ut'$ and $x' = L + ut'$ respectively. Of course, in the duelists' frame, with

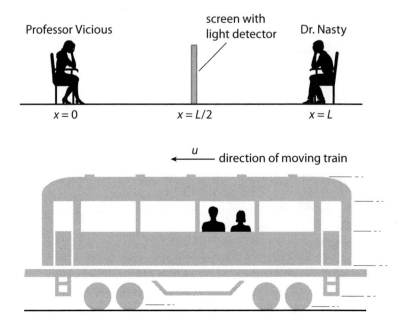

Figure 3. Professor Vicious and Dr. Nasty locked in a "think to the death."
Modified from A. Zee, *Group Theory in a Nutshell for Physicists*, Princeton University Press, 2016.

time and space denoted by t and x, you are the one who appears to be moving, gliding by at $x = -ut$. See figure 3.

Some time passes, and all of a sudden we hear a loud bong of the gong. "The best possible outcome, you solved the problem simultaneously!" we exclaim joyously with much relief. "You guys are equally smart and you could go to Stockholm together!"

The arrangement is fool proved electronically. We won't have either of them gloating, "I did it first!" Peace shall reign on earth.

But guess what? A Swede is sitting next to you. He too heard the gong. That's the whole point of the gong: you either heard it or you didn't. It's all admissible in a court of law. Now, not only is the Swede on the Nobel Committee, but he also happens to be an intelligent Swede. He reasons as follows.

Professor Vicious is gliding by as described by $x' = ut'$. When Professor Vicious pushed the button, she sent forth a multitude of photons surging toward the screen at the speed of light c. But the screen was also moving forward, away from the surging photons. Of course, light moves at the maximum allowed speed in the universe, and it soon catches up with the screen. The opposite is true for Dr. Nasty. The screen is moving toward the photons he sent forth. Thus, to reach the screen, Nasty's photons have less distance to cover than Vicious's photons. (Draw some photons moving toward the screen. It will clarify what you just read.)

Hence, reasons the intelligent Swede, for the two bunches of photons to reach the screen at the same time and so cause the gong to bong, the photons sent out by Vicious must have gotten going earlier. Thus, Vicious solved the problem first. With malicious glee, the Swede solemnly intones, "After Professor Vicious is awarded the Nobel Prize, she will kindly help us stuff Dr. Nasty into the dustbin of history!"

As Vicious enjoys her fleeting immortality, we bemoan or toast, as our taste might be, the fall of simultaneity. Nasty, trying to climb out of the dustbin, insists that he and Vicious had been sitting still, thinking hard, and it was the Swede that was moving. Since the gong had bonged, Nasty is absolutely sure that he and Vicious hit their buttons at the same instant and is entitled to half the prize, while the Swede is equally sure that Vicious hit her button before Nasty hit his.

The very notion of simultaneity depends on the observer!

Meanwhile, another Swede, also on the Committee, also intelligent, is moving by on another train described in the duelists' frame by $x = ut$. You can fill in the rest. He solemnly announced Nasty's destiny in Stockholm and Vicious's fate in the dustbin. Do you see why?

Young Einstein has bent the stately flow of time out of shape. Albert himself thought up this Gedanken experiment—I have merely added a few dramatic details—showing that the constancy of the speed of light necessarily has to alter our commonsense notion of simultaneity.

In Maxwell's electromagnetic wave, a varying electric field generates a magnetic field, and a varying magnetic field generates an electric field, with the cycle repeating indefinitely, moving the wave along. The rate at which a varying electromagnetic field generates a varying electromagnetic field has nothing to do with observers.

In theoretical physics we say, "Mind boggler in, mind boggler out!" We feed the mind-boggling fact that the speed of light does not depend on the observer into the wondrous machinery of logic and out pops another mind-boggling fact, namely, that simultaneity is, alas, no more.

The patent clerk uses a high tech clock to discover a secret about spacetime

Next, the promised third way: I sketch for you how Einstein deduced special relativity. Remarkably, of all the developments in theoretical physics since Newton, this requires the least amount of mathematical knowledge, only a tiny bit of high school algebra. Let's follow Einstein and consider a clock consisting of two mirrors separated by distance L , between which a light beam

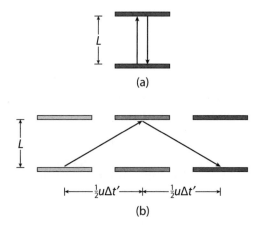

Figure 4. Einstein's clock (a) in its rest frame and (b) in a moving frame.
Reproduced from A. Zee, *Einstein Gravity in a Nutshell*, Princeton University Press,
2013.

bounces back and forth. See figure 4. Einstein was, after all, a patent examiner
living in a time of technological innovations[8] of all sorts, including ever-better
chronometers.[9]

Ms. Unprime, sitting on a smoothly moving train, has one of these high-
tech clocks with her.[10] For each tick-tock, three events occur: $A =$ light leaves
the lower mirror, $B =$ light bounces off the top mirror, and $C =$ light arrives
back at the lower mirror.

Let us write down the separation between events A and C in space and time.
Denote these separations in space and time by Δx, Δy, Δz, Δt. (A reminder:
Δ means "difference" or "the change in.") Since the pulse of light gets back to
where it started, clearly $\Delta x = 0$, $\Delta y = 0$, $\Delta z = 0$, that is, no change in the spatial coordinates. By construction, $\Delta t = 2L/c$, namely the distance $2L$ traveled
by light divided by its speed c.

Mr. Prime watches the train with Ms. Unprime on it moving by with speed
u in the x' direction and sees a pulse of light bouncing up and down in the y'
direction. What is the separation between A and C as seen by Mr. Prime?

Let's figure that out in the coordinate system he uses. Since he sees the clock
moving along the x-axis, he notes that $\Delta y' = 0$, $\Delta z' = 0$, (that's what "moving
along the x-axis" means), But $\Delta x'$, unlike $\Delta x = 0$, is nonzero and given by
$\Delta x' = u \Delta t'$ (that's what "moving with speed u" means). In the duration $\Delta t'$,
the train has traveled the distance $\Delta x'$.

But how do we determine $\Delta x'$ and $\Delta t'$ separately?

Use the fabulously astonishing equation $c = c$!

The distance traveled by the light pulse equals $c \Delta t'$. But what is $\Delta t'$?
Ask Mr. Pythagoras for help! We have two right angled triangles back to
back, each with right sides (figure 4(b)) with length $\frac{1}{2} u \Delta t'$ and L, and so the
hypotenuse equals[11] $\sqrt{\left(\frac{1}{2} u \Delta t'\right)^2 + L^2}$. So, from tick to tock, light travels twice

this distance, and hence

$$c\Delta t' = 2\sqrt{\left(\frac{1}{2}u\Delta t'\right)^2 + L^2}$$

Anybody who got a passing grade in high school algebra could solve this equation to determine $\Delta t'$ (hint: square both sides). Instead, let us follow Einstein. Noting that $\Delta x' = u\Delta t'$, we write the right hand side of this equation as $2\sqrt{\left(\frac{1}{2}\Delta x'\right)^2 + L^2}$. Squaring both sides, we obtain $(c\Delta t')^2 = 4\left[\frac{1}{4}(\Delta x')^2 + L^2\right] = (\Delta x')^2 + 4L^2$. Hence,

$$(c\Delta t')^2 - (\Delta x')^2 = 4L^2$$

But, remembering that $\Delta t = 2L/c$, we also have $(c\Delta t)^2 - (\Delta x)^2 = (c\Delta t)^2 = 4L^2$ since $\Delta x = 0$. Thus, no need to solve for $\Delta t'$. We can already see that

$$(c\Delta t')^2 - (\Delta x')^2 = (c\Delta t)^2 - (\Delta x)^2$$

even though $\Delta t' \neq \Delta t$ and $\Delta x' \neq \Delta x$. Since $\Delta y' = \Delta y$ and $\Delta z' = \Delta z$, we could also write this as $(\Delta x')^2 + (\Delta y')^2 + (\Delta z')^2 - (c\Delta t')^2 = (\Delta x)^2 + (\Delta y)^2 + (\Delta z)^2 - (c\Delta t)^2$.

Since this equality does not depend on u, the relative velocity between Mr. Prime and Ms. Unprime, we could imagine yet another observer named Double Prime, moving relative to Ms. Unprime with some other velocity along the x-axis. By the same reasoning, $(\Delta x'')^2 + (\Delta y'')^2 - (\Delta z'')^2 + (c\Delta t'')^2 = (\Delta x)^2 + (\Delta y)^2 + (\Delta z)^2 - (c\Delta t)^2$.

Conclusion: Even though different observers in uniform motion relative to each other observe different values for Δx and Δt, they all see the same value for the combination $(\Delta x)^2 + (\Delta y)^2 + (\Delta z)^2 - (c\Delta t)^2$. This combination of Δx, Δy, Δz, and Δt is the same for all observers. We have discovered an invariant of relative motion, namely, a quantity that is the same for all observers in relative motion!

By this clever thought experiment, Einstein used the Pythagoras theorem for space to obtain a sort of generalized Pythagoras theorem for spacetime.

Distinction between a very good physicist and a great physicist! A very good physicist knows math (high school algebra in our case) and can solve equations (solve for $\Delta t'$ in our example) till the cows come home, but a great physicist listens to what the equations are telling him or her (that Nature likes Pythagoras theorem so much that She wants to generalize it!)

Lorentz transformation

He meant more than all the others I have met on life's journey.
Einstein speaking of Lorentz

Let us now find a transformation from the spacetime coordinates (t, x, y, z) of one observer to that (t', x', y', z') of another observer, such that $(\Delta x')^2 + (\Delta y')^2 + (\Delta z')^2 - (c\Delta t')^2 = (\Delta x)^2 + (\Delta y)^2 + (\Delta z)^2 - (c\Delta t)^2$.

Two trivial comments to start with. Since $\Delta y' = \Delta y$ and $\Delta z' = \Delta z$, we can forget about them and simply insist that $(\Delta x)^2 - (c\Delta t)^2$ remains unchanged under the transformation. Furthermore, in studying the separation between two events in space and time, we could take one of the events to occur at the origin, that is, at $t = 0$, $x = 0$, $y = 0$, $z = 0$. Thus, $\Delta x = x - 0 = x$, etc., and we could stop writing Δ and lessen clutter.

Here is an algebra homework problem for a bright high school student: Find the relations between (t', x') and (t, x) such that such that $x'^2 - (ct')^2 = x^2 - (ct)^2$.

Even a dull high school student could already see by eyeball that the Galilean transformation $t' = t$, $x' = x + ut$ given earlier ain't gonna cut it: $x'^2 - (ct')^2 = (x + ut)^2 - (ct)^2$, which is most certainly not equal to $x^2 - (ct)^2$. We could already see that t' cannot possibly be equal to t: universal time does not exist!

Meanwhile, the bright kid[12] turns in the answer:

$$ct' = \frac{ct + \frac{u}{c}x}{\sqrt{1 - \frac{u^2}{c^2}}} \quad \text{and} \quad x' = \frac{x + ut}{\sqrt{1 - \frac{u^2}{c^2}}}$$

(plus $y' = y$, $z' = z$, of course.) This is the celebrated Lorentz transformation.[13] You could verify[14] that this indeed satisfies $x'^2 - (ct')^2 = x^2 - (ct)^2$.

You certainly do not have to study this Lorentz transformation in detail; this is not a textbook. I merely ask you to note three points:

- In the domain of everyday experience, namely, when u is much much less than c, $\frac{u}{c}$ is approximately 0, so that $\sqrt{1 - \frac{u^2}{c^2}}$ is almost equal to 1. Hence, $ct' = ct$ and $x' = x + ut$. The Lorentz transformation reduces to the Galilean transformation given earlier, as it must.
- Since $\sqrt{1 - \frac{u^2}{c^2}}$ becomes imaginary for $u > c$, we have learned that a universal speed limit c exists. The train cannot go faster than the speed of light without all of our equations breaking down.
- Surprise, t and t' are definitely not equal! The commonsense fallacy was that we thought for sure that when one second passed for us, one second had passed for everybody else.[15]

Take home message: There is no universal clock in the universe ticking off the same universal time for everyone.

As I said earlier, Einstein's special relativity is the subject in physics that requires the least amount of mathematics to understand, nothing beyond high school algebra.

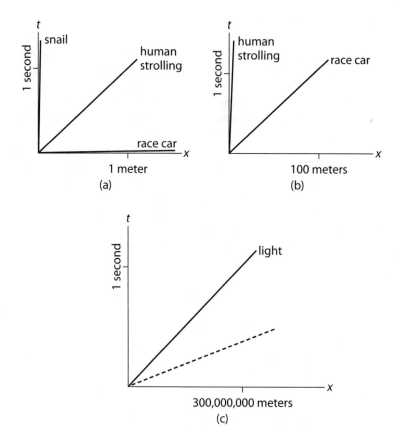

Figure 5. Three different spacetime diagrams, differing by the units on the horizontal axis.

Spacetime diagrams: using units with $c = 1$

One of my childhood memories was being told that in the time it takes me to say "tick tock," light would have traveled a distance equal to going around the earth seven times. For years, I tried to imagine how fast that would be. (Seven turns out to be about right: the earth's circumference is $\simeq$40,000 km, while the speed of light c is $\simeq$ 300,000 km/sec.)

Suppose we make a plot of where we are in our daily lives as time goes on. For simplicity, confine our movement to one dimension. With everyday units, meter and second, a plot might look like figure 5(a).

A human walking along at 1 m/sec corresponds to the 45° line. (In discussing special relativity, physicists plot time along the vertical axis, contrary to the everyday practice of putting time on the horizontal axis.) Note that

1 m/sec = 3,600 m/hour = 3.6 km/hour is a very leisurely stroll indeed. In this plot, an extremely fast car moving at 360 km/hour would be shown as an almost horizontal line at approximately $45°/100 = 0.45°$ from the x-axis.

Physicists call the line traced by a moving object in a spacetime diagram such as figure 5(a) a "worldline." Thus, if the worldline of a strolling human is the 45° line, then the worldline of the experimental car would be almost indistinguishable from the horizontal axis. At the other extreme, the worldline of a snail would be very close to the vertical axis. Note that the worldline of an object at rest, which is physics talk for "not moving," is just a vertical line.

Next, consider figure 5(b). For unit of distance, we now use 10^2 m. The worldline of the race car is now the 45° line, and the worldline of the strolling human is barely distinguishable from the vertical axis. (Incidentally, straight lines are drawn merely for simplicity; the worldline of an object with a varying speed would be curved.)

The expressions thus far in this chapter show that keeping c around merely adds to the clutter. Clearly, physicists living in the relativistic world would be wise to use, for the distance unit, the light second, that is, the distance light travels in one second, so that $c = 1$. Light would now be moving along the 45° line and the worldline of the race car is indistinguishable from the vertical axis, let alone lumbering enormities such as humans.

That nothing could move faster than light translates into the statement that no worldline could make an angle of less than 45° from the x-axis, as shown by the dotted line in (c).

Preview of an exciting development to come in part III. In the quantum world, worldlines of particles could have a slope of less than 45°! (In truth, yes, but not really. Stay tuned!)

Setting $c = 1$

"An inch of time is worth an inch of gold,
An inch of gold cannot buy an inch of time."
Chinese adage[16]

Clearly, to describe the relativistic world, it pays to measure space and time using the same unit, that is, to set $c = 1$. The Lorentz transformation displayed earlier in this chapter simplifies to the more eye-pleasing form

$$t' = \frac{t + ux}{\sqrt{1 - u^2}}, \qquad x' = \frac{x + ut}{\sqrt{1 - u^2}}$$

The square root implies that the relative velocity u between the two observers is limited by $u^2 \leq 1$.

Compare this with the Galilean or commonsense transformation given at the beginning of this chapter. The key difference is that t' now depends on both t and x, rather than being simply equal to t. Absolute time is dead. Long live relative time!

As the Chinese adage shows, human languages tend to measure time figuratively in terms of space. In English, we say that while our past is now behind us, the future is still ahead of us, as in a queue of people. For us humans, time is more difficult to visualize than space.

Notes

[1] To see this, have the two observers each build a fence out of sticks of a specified length, pointing in the y-direction, say, and such that the two fences are of the same height. Relativity states that the two observers are not able to tell who is moving and who is at rest. Thus, the top of the two fences must continue to coincide as they glide by each other; otherwise, if one fence is seen to be taller, the other shorter, the two observers could be distinguished from each other.

[2] Hardly at rest! You may not be aware of it, so smooth is the motion, but the entire galaxy you are in is hurtling toward a neighboring galaxy at high speed.

[3] The joke is that the two discussions are the same.

[4] You know the famous riddle due to R. Smullyan? One of my favorites. "What is greater than God, and if you eat it, you die?"

[5] This section is taken from *GNut*, p. 7.

[6] I love making up names for the characters recurring in my books. Readers have told me that their favorite is Confusio and his struggle to earn tenure. Not long after writing this chapter, I was astonished to read in the *New York Times* about a certain Professor Vile. A Google search shows that he actually exists.

[7] I am referring to the fact that quarks and leptons come in three families. See chapter V.4.

[8] According to the literary scholar Dame Gillian Beer, around 1865, when Lewis Carroll, an early practitioner of photography, wrote *Alice in Wonderland*, photography "froze or made portable a moment and a place." To me, that could have easily led to the concept of events in spacetime. Carroll was notoriously concerned with the notion of time, for example, with the white rabbit constantly consulting his pocket watch, an affectation and necessity when railways, with timetables and Einstein's trains, came into common use. To a physicist like myself, the two Alice books are full of allusions to concepts from physics: gravity, scale transformation, and mirror reflection, to name a few.

[9] P. Galison, *Einstein's Clocks and Poincaré's Maps: Empires of Time*, Norton, 2004.

[10] This story is adapted from chapter III.2 of *GNut*.

[11] Actually also known in several other ancient civilizations, Babylonian, Chinese, Egyptian, and so on.

[12] An even brighter kid might see that the best approach would be to recognize that we could write $x^2 - (ct)^2$ as $(x - ct)(x + ct)$. Then clearly, the desired transformation is simply to multiply $(x - ct)$ and divide $(x + ct)$ by the same arbitrary number. The Lorentz transformation derived in one line!

[13] As is often the case, the history is a bit convoluted. In 1887, when Einstein was 8 years old, the German physicist W. Voigt proposed an erroneous version of this transformation. Not knowing Voigt's work, Lorentz derived in 1895 the transformation in a better form than Voigt's, but still not quite in the form we now know. Then J. Larmor found the correct form in 1900. Not knowing Larmor's work, Lorentz reproduced it in 1904. In 1905, H. Poincaré, knowing only of Lorentz's work, developed the transformation further and named it the Lorentz transformation. As for Einstein, he only knew the 1895 version of the Lorentz transformation. The term "Lorentz transformation" is an example of the Matthew principle in theoretical physics: Whoever has will be given more . . . . Whoever does not have, even what he has will be taken from him (Matthew 13:12).

[14] Note, by elementary algebra,

$$x'^2 - (ct')^2$$

$$= \left((x + ut)^2 - \left(ct + \frac{u}{c}x \right)^2 \right) \bigg/ \left(1 - \frac{u^2}{c^2} \right)$$

$$= \left((x^2 + 2uxt + u^2t^2) - (c^2t^2 + 2uxt + \frac{u^2}{c^2}x^2) \right) /$$

$$\left(1 - \frac{u^2}{c^2}\right) = \left(1 - \frac{u^2}{c^2}\right)(x^2 - c^2 t^2)/\left(1 - \frac{u^2}{c^2}\right)$$
$$= x^2 - (ct)^2$$

[15]Thus, the commonsense addition of velocities we mentioned in the text is modified, on the application of the Lorentz transformation, to

$$v' = \frac{\Delta x'}{\Delta t'} = \frac{\Delta x + u\Delta t}{\Delta t + \frac{u}{c^2}\Delta x} = \frac{v + u}{1 + \frac{uv}{c^2}}$$

The third equality follows by dividing through with Δt. For u and v much less than c, the denominator is very close to 1, and this reduces to the commonsense $v' = v + u$. But for $v = c$, we see that this gives $v' = (c + u)/(1 + \frac{u}{c}) = c$ also. Thus, $c = c$ regardless of observer! This is the strange addition law for velocities alluded to in the prologue.

[16]Nicole Mones, *Night in Shanghai*, 2014.

The geometry of spacetime

Invariants

One fruitful and powerful line of thought in both physics and mathematics is to ask what is left unchanged, that is, invariant, under various specified transformations.

For the simplest example, consider two neighboring points P and Q in 2-dimensional space, aka the plane, with coordinates (x, y) and $(x + dx, y + dy)$, respectively. Pythagoras taught us that the distance ds between the two points is given by $ds^2 = dx^2 + dy^2$. (Incidentally, we have trivially changed notation from that used in chapter I.3: instead of Δx, Δy, we now write dx, dy, where d stands for difference. (See figure 1.) Some readers may know that in differential calculus, dx stands for an infinitesimally small quantity, essentially what Δx becomes when it is very small. We take P and Q to be infinitesimally close to each other.)

Another way of looking at the Pythagoras theorem is to look for expressions that are invariant under rotations. Readers who remember a bit of trigonometry from high school might know that if we rotate our coordinate system in the x-y plane through an angle θ, in the new coordinate system, the spatial coordinates (x', y') of a point are given by (see figure 2)

$$x' = \cos\theta\, x + \sin\theta\, y, \quad y' = \cos\theta\, y - \sin\theta\, x$$

In fact, you do not need to remember any trigonometry, since it suffices[1] to take the angle θ to be small, for which we have the simple relations $x' = x + \theta\, y + O(\theta^2)$, $y' = y - \theta\, x + O(\theta^2)$. The notation $O(\theta^2)$ indicates that the approximation is good to order θ^2, which for θ small is much smaller than θ (for example, for $\theta = 0.1$, $\theta^2 = 0.01$).

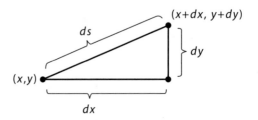

Figure 1. Two nearby points have Cartesian coordinates (x, y) and $(x+dx, y+dy)$, respectively. Pythagoras tells us how to determine the distance ds between the two points. In the text, dx and dy are described as very small, infinitesimal in fact. They are blown up here for clarity.
Reproduced from A. Zee, *On Gravity*, Princeton University Press, 2018.

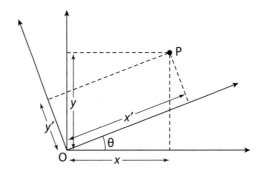

Figure 2. The point P is assigned different coordinates, (x, y) and (x', y'), in the two systems. Trigonometry tells us how (x, y) and (x', y') are related.
Reproduced from A. Zee, *Group Theory in a Nutshell for Physicists*, Princeton University Press, 2016.

Now let us ask what expressions are left invariant. Well,

$$x'^2 + y'^2 = x^2 + 2\theta xy + O(\theta^2) + y^2 - 2\theta xy + O(\theta^2) = x^2 + y^2 + O(\theta^2)$$

Precisely, Pythagoras's expression $x^2 + y^2$ remains the same in the new primed coordinate system. It is said to be a "rotational invariant."

Indeed, we could turn this around. By requiring the distance ds between the two neighboring points, given by $ds^2 = dx^2 + dy^2$, to be invariant, we could determine rotations. Indeed, this invariant defines Euclidean geometry. Everything we learned in geometry class in school (for example, the area of circles) follows from this invariant, as we will discuss in chapter V.5.

But this parallels Einstein's logic! By requiring $x^2 - t^2$ to be invariant, that is, by requiring $x'^2 - t'^2 = x^2 - t^2$ (we have now switched from everyday units to more sensible units with $c = 1$), he obtained, as was discussed in chapter I.3, the Lorentz transformation:

$$t' = \frac{t + ux}{\sqrt{1 - u^2}}, \qquad x' = \frac{x + ut}{\sqrt{1 - u^2}}$$

with u the relative velocity between the two observers and with $y' = y$, $z' = z$ suppressed. For small u, this reduces to $t' = t + ux + O(u^2)$, $x' = x + ut + O(u^2)$:

		for small angle or small velocity	
Euclid rotation		$x' = x + \theta y$, $y' = y - \theta x$	
Lorentz transformation		$t' = t + ux$, $x' = x + ut$	

A clear and striking parallel between rotation and Lorentz transformation! You should be able to see that the Lorentz transformation has the same, but not exactly the same, form as a rotation in the t-x plane through some "funny" kind of angle[2] determined by the relative velocity u between two observers, for example, Ms. Unprime and Mr. Prime.

It is straightforward to generalize Pythagoras's theorem to 3-dimensional space. Imagine a point R a distance dz directly above Q, that is, perpendicular to the paper. The distance ds between P and R with coordinates (x, y, z) and $(x + dx, y + dy, z + dz)$, respectively, is then given by[3] $ds^2 = dx^2 + dy^2 + dz^2$. Indeed, you could generalize this to D-dimensional space if you like. For instance, in 4-dimensional space, the Pythagorean theorem would state that $ds^2 = dx^2 + dy^2 + dz^2 + dw^2$. Since with x, y, z we have reached the end of the alphabet, we have to assign a letter to the 4th coordinate, and it might as well be w, but it could be any letter you like.

Pythagoras would have loved it

We learned in chapter I.3 that the combination $dx^2 + dy^2 + dz^2 - dt^2$ is the same for all observers in uniform motion relative to each other. It is an invariant of spacetime. Thus, the only candidate for the separation ds between two neighboring points in spacetime is ds as determined by

$$ds^2 = dx^2 + dy^2 + dz^2 - dt^2$$

This expression defines the geometry of spacetime. It differs from Euclid's geometry by a crucial sign distinguishing time from space. Depending on how you look at it, the difference is "huge" or "tiny."

If somebody asked you to generalize the geometry of space to include time, your first guess might be $ds^2 = dx^2 + dy^2 + dz^2 + dt^2$. But then how would time differ from space? You and I, he and she, and they too, all know that we could go east and west, north and south, and up and down as we please, but we cannot go back to when we were young. We must somehow distinguish time from space in our equations!

The solution arrived at by the greats of physics might make non-physicists laugh. Instead of adding dt^2, how about subtracting dt^2 instead? It seems so naive and childish, but it turns out to be right. Intriguingly, the distinction between time and space is the simplest conceivable, just a flip of sign. Nature actually works that way, amazing! Pythagoras would surely get a kick out of this.

Einstein, the exterminator of relativity

Now, a bit of sad history. Einstein did not use the term "theory of relativity" in his paper! The German physicist Alfred Bucherer (a total nobody almost nobody today has ever heard of), while criticizing Einstein's theory, was the first to use, in 1906, the name[4] "Einsteinian relativity theory."

I must now vent my pet peeve. Physics contains a number of unfortunate names, some due to historical confusion long since cleared up. Probably the worst name ever is relativity, as it has spawned a swarm of nonsensical statements, such as "Physicists have proved that truth is relative" and "There is no absolute truth; Einstein told us so," uttered with smug authority by numerous ignorant fools. Only a slight exaggeration: that guy Bucherer had inadvertently messed up the minds of more than a few "eminent" philosophy professors. The reader should realize by now that physicists, as exemplified by Einstein, say the opposite. Yes, they talk about observers in relative motion, and hence unavoidably use the word "relative," but what they search for are the invariants, that which the observers could agree on. Indeed, they require that the laws of physics be invariant and independent of observers. I, and some in my world, like to call Einstein the exterminator of "the relativity of truth."

Later in life, Einstein said that he should have used the name "invariant theory."

A most valiant piece of chalk

> With this most valiant piece of chalk I might project upon the blackboard four world-axes. . . . Then we obtain, as an image, so to speak, of the everlasting career of the substantial point, a curve in the world a *world-line*. . . . The whole universe is seen to resolve itself into similar world-lines, and I would fain anticipate myself by saying that in my opinion physical laws might find their most perfect expression as reciprocal relations between these world-lines."
> Hermann Minkowski[5]

Historically, it was Hermann Minkowski (from whom Einstein took a math course) who first proposed treating time as the fourth coordinate,[6] after

Einstein established special relativity. By the way, Einstein said that this business of a 4th dimension had never occurred to him.

Henceforth, space and time are unified into spacetime.[7] The vector $\vec{x} = (x, y, z)$ with its three components is unified with time t, to become the 4-vector $x^\mu = (t, x, y, z) = (t, \vec{x})$. (The terminology 4-vector is to distinguish it from an ordinary 3-vector.) Note that in going from 3-vector to 4-vector we have abandoned the arrow notation for the more powerful index notation. Here the index $\mu = 0, 1, 2, 3$ takes on 4 values, and t gets renamed x^0. Just for the record, and to make sure that you are following, the coordinates of spacetime are described by $x^\mu = (x^0, x^1, x^2, x^3) = (t, x, y, z) = (t, \vec{x})$.

Promoting 3-vectors to 4-vectors

You might be wondering how $E = mc^2$, surely the most famous[8] equation in all of physics, emerges out of all this.

I now give you a simple derivation. But first, let me review the long trek we undertook. We input the amazing equation $c = c$, stating that the speed of light does not depend on the observer. Ms. Unprime and Mr. Prime see light zip by with the same speed. Together with Vicious and Nasty, we watched simultaneity crash and burn. Looking at Einstein's clock, we conclude that Ms. Unprime and Mr. Prime must observe the same value for the combination $ds^2 = dx^2 + dy^2 + dz^2 - dt^2$. The requirement that this combination is invariant suffices to determine the Lorentz transformation, with no more than a few lines of high school algebra. Minkowski unified space and time into spacetime, and showed that the geometry of spacetime is only slightly more involved than the geometry of space.

The bottom line: The 3-vector $\vec{x}$ is partnered with t to form the 4-vector x^μ.

But as soon as that happens, then every concept in physics possessing a direction would also have to be promoted from a 3-vector to a 4-vector[9] in spacetime. This self-evident requirement rests on the very definition of dimension. That space is 3-dimensional forbids us from having a fundamental concept in physics defined in terms of a 2-dimensional vector. Similarly, that spacetime is 4-dimensional behooves us to talk about 4-vectors rather than 3-vectors.

An intriguing side remark: Up till now we have not had to invoke various fundamental physical concepts, such as force, momentum, and energy. We only have to input $c = c$.

Consider momentum, which in Newtonian physics is defined as mass times velocity: $\vec{p} = m\vec{v}$. The momentum of a moving object is proportional to how massive it is and how fast it is moving. The energy, or more precisely the kinetic energy, of a moving object is given by $E = \frac{1}{2}m\vec{v}^2$, namely, half of its mass times its velocity squared. These two concepts are so closely intertwined that when the momentum 3-vector $\vec{p}$ goes off to find another physical quantity to partner with to form a 4-vector, it naturally looks for the energy. Who else is there?

$E = mc^2$ pops out!

I'm almost ready to give you an exceedingly simple argument for $E = mc^2$. Basically, the argument is to show that the Newtonian kinetic energy refuses to combine with the Newtonian momentum vector to form a 4-vector.

Consider an object at rest on the train. For the sake of definiteness, let's refer to it as a ball of mass m. (Again, from now on, take the train to be moving in the x-direction, so that we could omit the arrow on $\vec{p}$, and denote by p the momentum in the x-direction.) To Ms. Unprime, the ball has neither energy nor momentum: $E = 0$, $p = 0$. But to Mr. Prime, the stationmaster, the ball is moving with velocity u and so has momentum $p' = mu$. Notice that everything is Newtonian thus far.

Let's see what Mr. Lorentz has to say about this. Since we are using the Newtonian formula for momentum, we are assuming that this is an everyday train, moving very slowly compared to the speed of light $c = 1$. With $u \ll 1$, we could set the $\sqrt{1 - u^2}$ to 1, so that the Lorentz transformation reduces to $t' = t + ux$, $x' = x + ut$. (Note that this is definitely not Galilean, which would have $t' = t$.)

For energy and momentum to transform in the same way as time and space, we should have, just by copying,

$$E' = E + up, \qquad p' = p + uE$$

Energy corresponds to time, momentum to space, duh. (With the understanding that this is correct for small u, we have suppressed quantities of order $O(u^2)$.)

Since Ms. Unprime observes $E = 0$, $p = 0$, this implies that Mr. Prime observes $E' = 0 + 0 = 0$, $p' = 0 + 0 = 0$.

But that is wrong wrong wrong. We just said that Mr. Prime sees $p' = mu$, not 0. The train is moving with velocity u, and thus the ball is also moving with velocity u and hence has momentum $p' = mu$. Mr. Newton himself told us so!

Note that everything is moving slowly compared to the speed of light, so there could be no "monkey business" nor any inane excuse.

Stare for a while at the Lorentz transformation written down just now for slowly moving trains and balls, and see if you could find a way out. You are given $p = 0$ but want $p' = mu \neq 0$. Well?

Looking at $p' = p + uE$, with $p = 0$, we have $p' = uE$. We want to have $p' = mu$. So what gives?

You see that the only possibility is for E to equal to m! In other words, for Ms. Unprime, the energy of the ball sitting at rest should be $E = m$, not $E = 0$.

An utterly trivial remark. While we have used sensible units so that $c = 1$, the hoi polloi might still insist on using the stick some French revolutionary used to measure distance with (or even some English king's foot). It is simple[10] to put c back in, and then $E = m$ becomes[11] $E = mc^2$.

Trumpets please! $E = mc^2$, ladies and gentlemen.

An object of mass m just sitting there has an energy mc^2, known to physicists as the rest energy.[12]

An invariant built out of energy and momentum

Alternatively, we could ask, given that energy and momentum transform the way they do, what is left invariant? Well, in parallel with a calculation we did earlier, we observe that

$$E'^2 - p'^2 = (E^2 + 2uEp + O(u^2)) - (p^2 + 2upE + O(u^2)) = E^2 - p^2$$

Thus, to the order we are working,[13] $E^2 - p^2$ as measured by different observers is the same and hence must correspond to a property intrinsic to the particle. But that could only depend on its mass m. Thus, we obtain $E^2 - p^2 = m^2$. A mass at rest (that is, with $p = 0$) has energy $E = m = mc^2$. (That last step is just for those of you who insist on not using units in which $c = 1$.)

Incidentally, in modern physics, the relation $E^2 - p^2 = m^2$ is often taken as the definition of mass. When a hitherto unknown particle is discovered, experimentalists measure its energy E and its momentum p and declare its mass to be the m given by this relation. Of course, this also provides a test of Einstein's theory; every time experimentalists see this particle, its energy and its momentum may be different, but its mass better be the same, that is, an invariant.

The Lord did not lead him around by the nose

> One more consequence of the paper on electrodynamics has also occurred to me. ... The argument is amusing and seductive; but for all I know the Lord might be laughing over it and leading me around by the nose.
> Albert Einstein writing to a friend in 1905

Incidentally, Einstein didn't have $E = mc^2$ in his paper proposing special relativity. This famous relation appeared a few months later in a brief note. As we all know, the Lord did not lead Einstein around by the nose.

Einstein's original 1905 derivation, in the glare of hindsight, was unnecessarily complicated. The derivation given here is much simplified. Later in 1946, in a lecture at the Technion in Haifa, Israel, he gave an elegant derivation which surprisingly, is omitted from most textbooks[14] and so is in danger of being forgotten.

Interestingly, while Lorentz found the transformation, he was not able to work out the ramifications for physics. It is tempting to hypothesize that this is because Einstein was young at the time while Lorentz was old.

Deeper and simpler

As physicists explore Nature at ever deeper levels, Nature appears to get ever simpler. The story of relativistic invariance exemplifies this remarkable, and striking, phenomenon. I may surprise the reader by saying that Einsteinian mechanics, once mastered, is intrinsically simpler than Newtonian mechanics. After working with Lorentz invariant equations, I find equations in Newtonian mechanics awkward and malformed. Space and time are not treated on the same footing, and neither are energy and momentum. The equations do not please my eyes, understandably so, since the Newtonian equations are only approximate to the Einsteinian equations. Why should Nature care whether the results of an approximation imposed by humans look pretty?

Similarly, recognizing the relativistic invariance of electromagnetism, fundamental physicists now write Maxwell's equations more compactly as one equation. When I was a student, I had to memorize Maxwell equations before every examination. Mmm, let's see, a magnetic field changing in time produces an electric field changing in space—or, is it changing in time? With relativistic invariance, a single equation describes an electromagnetic field changing in spacetime. I find this completely symmetrical equation as easy to remember as the shape of the circle. Intrinsically, advanced physics is simpler than elementary physics—a little secret not often revealed to the layman. Many people are stumped by high school or college physics because they are presented with misshapen phenomenological equations having little to do with Nature's intrinsic essence, with Her beauty.

Notes

[1] Because a rotation through an arbitrary angle could be built up by repeatedly rotating through small angles. This is the foundational idea of Lie groups. See *Group Nut*.

[2] If you squint your eyes a bit, and if we define a "hyperbolic angle ϕ" by $\cosh \phi \equiv 1/\sqrt{1-u^2}$, $\sinh \phi \equiv u/\sqrt{1-u^2}$, you would see that the Lorentz transformation has the form $t' = \cosh \phi \, t + \sinh \phi \, x$, $x' = \cosh \phi \, x + \sinh \phi \, t$. This is almost the same as the expressions for rotation except for a minus sign and for replacing cos and sin by cosh and sinh, which are known as hyperbolic cosine and sine. For the reader who wants more, please see, for example, *GNut*, page 170.

[3]This could be shown easily by considering the right triangle with PQ and QR forming two of its three sides.

[4]In German, Einsteinsche Relativitäts-theorie.

[5]In an address delivered at Cologne, Germany, 1908. Reprinted in *The Principle of Relativity: A Collection of Papers by A. Einstein, H. Lorentz, H. Weyl and H. Minkowski, with Notes by A. Sommerfeld,* Dover, 1952.

[6]He actually wrote $x^4 = ict$ with $i = \sqrt{-1}$ the imaginary unit, so that $+(dx^4)^2 = -(cdt)^2$. Then $ds^2 = d\vec{x}^2 + (dx^4)^2$. But this proves to be a clumsy notation when and if we ever want to go to even higher dimensions. For instance, for string theorists living in 10-dimensional space-time, it would be a lot easier to write x^μ with $\mu = 0, 1, 2, \ldots, 9$, and $x^0 = t$.

[7]Another quote from Minkowski: "Hence-forth space by itself, and time by itself, are doomed to fade away into mere shadows, and only a kind of union of the two will preserve an independent reality." In one of my books, I pro-posed the neologism "zaum," from the German words for time ("Zeit") and space ("Raum").

[8]From how often $E = mc^2$ pops up in pop-ular culture, I believe that it has left Newton's $F = ma$ far behind in the dust.

[9]Or something more involved. This caveat is added here for later use when we come to electromagnetism in chapter IV.4.

[10]The easiest way is to use dimensional anal-ysis: see *FbN* chapter I.1. The other is to use the exact form of the Lorentz transformation given earlier. Since $u \ll c$, we could throw away those pesky square roots and write, for example, $ct' = ct + \frac{u}{c}x$. From this point on, keep track of the factor of c carefully, and you will see $E = mc^2$ emerge. Hint: Watch out for dimensional consistency.

[11]For the aficionado, one way of seeing the factor of c^2 is to note that x goes with ct and p with E/c.

[12]Whether this energy could be extracted is outside of the purview of special relativity and requires knowledge of another area of physics, namely, nuclear physics.

[13]By sticking in factors of $\frac{1}{\sqrt{1-u^2}}$, you could extend this derivation readily to all orders in u.

[14]A notable exception is the textbook by Baierlein, *Newton to Einstein: The Trail of Light.* I am grateful to R. Baierlein for providing me the original reference: A. Einstein, *Technion Yearbook 5,* page 16, 1946. I gave my version of it in *GNut,* pages 232–233.

The rise and fall and rise of particles

From Democritus to the birth of quantum physics

Since we could readily cut, tear, or break everyday matter[1] into smaller pieces, it seemed natural, since the time of Democritus,* to suppose that matter is composed of tiny particles.

The mysterious exceptions were light, and to a lesser extent, sound. Nobody was able to cut light or sound into bits. Nevertheless, Newton boldly speculated that light may consist of invisible "corpuscles."

In everyday life, even the most vacant vacationers can feel, and understand, the difference between water and sand. Particles of sand have an independent existence, but a water droplet falling into the sea loses its identity forever.

How water waves combine is clear. We already talked about sound waves and light waves interfering in chapter I.2, but a quick review and a picture here. When the crest of one wave meets the crest of another wave, they add to a higher crest. Similarly with troughs combining into deeper troughs. When a crest meets a trough, they more or less cancel out. Physicists say that water waves, while passing through each other, interfere (figure 1).

That water largely consists of empty space in between the jostling molecules still boggles my mind. Knowing and feeling are entirely different.

Fast forward to the 19th century. Light waves were observed to interfere just like water waves. The experiments of Thomas Young (said to be "the last man to know everything") and others convinced physicists that light is a wave, culminating in Maxwell's theoretical demonstration that light consists of an electric field and a magnetic field oscillating together in a precisely

---

*Plato allegedly wanted to burn Democritus's books. Well, theoretical physicists are much more into Democritus than Plato, so take that, Plato!

Figure 1. Water waves interfering.
Reproduced from A. Zee, *On Gravity*, Princeton University Press, 2018.

choreographed dance. Newton's particles of light were swept away into the dustbins of history.

At the same time, the 19th century saw mounting evidence that air, an apparently continuous medium, actually consists of myriad molecules. Boltzmann shows that the measured behavior of gases could be calculated in terms of the collision of air molecules with each other and with the walls of the container. Sound is a density wave produced by the collective motion of air molecules

Meanwhile, Mendeleev's periodic table showed conclusively that the bewildering variety of known substances is constructed by combining a mere ninety-two kinds of atoms in different ways. I found this conclusion absolutely amazing even in hindsight! Two hydrogen atoms and one oxygen atom produce a remarkable fluid that makes life possible.

And then, light was finally recognized as "merely" a kind of electromagnetic wave.

Thus, 19th century physics proceeded on dual tracks, an untenable clash between the discrete and continuous. Even today many students of physics, not to mention crackpots, continue to be confounded by this seemingly irreconcilable duality.

Does the world consists of discrete particles or continuous fields?

Discovery of the electron and the rise of particles

In 1897, around thirty years after Maxwell's theory of electromagnetism and the triumph of the field, J. J. Thomson discovered the basic unit of charge, the electron.[2] Not since Newton's point particles and corpuscles of light has the notion of particle once again surged to center stage.

Nevertheless, physicists continued to favor the continuum, proposing a "pudding model" of matter in which the negatively charged electrons were said to be like raisins embedded in a positively charged pudding. But no. In 1911, Ernest Rutherford discovered that the positive charges of matter are concentrated in a tiny atomic nucleus, around which the electrons revolved, much like the planets orbiting around the sun. The atom of Democritus turns out to be almost entirely empty space, with a tiny hard core at the center surrounded by point-like electrons.

How is all this possible? Matter is made of discrete point particles, interacting via the apparently continuous electromagnetic and gravitational fields.

Soon enough, the nucleus was found to consist of even tinier particles known as protons and neutrons,[3] but we are getting ahead of ourselves.

A divine madness

The dramatic denouement came in 1900. While studying the interaction of matter with the electromagnetic field, Max Planck found that, in order to reconcile theory with experimental observation, he was obliged to shock the physics community with the revolutionary notion[4] that energy, which had always been regarded as a continuous quantity, could behave as if it came in tiny discrete packets (figure 2).

The physicist Abraham Pais remarked that Planck's "reasoning was mad, but his madness has that divine quality that only the greatest transitional figures can bring to science."[5]

Einstein then extended and put Planck's proposal on a firmer basis in 1905, his annus mirabilis,[6] by applying it to the photoelectric effect, which had long puzzled physicists. The photoelectric effect[7] was discovered accidentally in 1886 by Heinrich Hertz in his effort to detect the electromagnetic wave[8] suggested by Maxwell. When light is shone on a metallic surface, a photoelectric current is generated. We now know that the electromagnetic wave is ejecting electrons from the metal. According to classical physics, the current should increase with the intensity of the wave, since intensity measures the strength of the electric field in the wave and hence the electric force exerted on the electron. On the other hand, the current is expected not to depend on the frequency ω of the wave, which measures how fast the wave is oscillating.

But experiments showed precisely the opposite. The photoelectric current does not depend on the intensity of the incident light but increases with its frequency ω, once a certain critical threshold frequency ω^* is reached. For example, let's say that for a given metal, red light does not produce a photoelectric current. Making the red light more intense does not do anything. Instead, crank up the frequency ω of the light, that is, change its color. At

Figure 2. Max Planck in 1878.
From Wikimedia Commons.

some critical frequency ω^*, corresponding to, say, yellow light, a photoelectric current starts to flow. The current keeps increasing as we shift from yellow to blue and then violet.

Einstein explained all this by proposing that the electromagnetic wave, which everybody had agreed was continuous, in fact consists of a stampede of light quanta, later named photons. He postulated that in a wave with frequency ω and wave vector $\vec{k}$, each of the photons carries energy $E = \hbar\omega$ and momentum $\vec{p} = \hbar\vec{k}$, with $\hbar$ the Planck's constant already mentioned in the prologue. To eject an electron, an incoming photon has to be energetic enough to kick an electron over the energy barrier binding it to the metal. This explains the critical frequency ω^*, which varies from metal to metal, depending on how tightly that particular metal wants to hold on to its electrons.

Increasing the intensity corresponds to increasing the number of photons, but if each one of them is too weak to kick an electron out of the metal, having lots of them would not do any good. However, increasing the frequency ω of the light increases the energy of the photons, and the more energetic the kicks administered the electrons, the more likely they will be ejected.

Incidentally, contrary to what you might think, Einstein's Nobel Prize was not for his better known contributions, such as $E = mc^2$ and his curved space-time interpretation of gravity, but for the photoelectric effect. The other two epoch changing ideas were apparently too far out for the Swedish Academy. By the way, Einstein later did say that asserting that light consists of photons was "the only truly revolutionary thing I ever did." But from my reading and from my visit to the Einstein archive at the Hebrew University in Jerusalem, I am under the impression that with his enormous fame Einstein was fond of occasionally pulling the leg of his adoring interviewers and readership. I also interpret this puzzling remark as a gentle rebuke to the Swedes, which is of course not to deny that Einstein's explanation of the photoelectric effect was a monumental advance in the development of quantum mechanics.

Elementary particles

The reader is familiar with the electron, the proton, and so on. Starting with J. J. Thomson's discovery[9] of the electron in 1897, experimentalists have found a whole zoo of elementary particles. I will talk about these particles at length in chapter V.2, in particular about Gell-Mann's celebrated prediction of a particle he named Ω^- (Omega minus) and its subsequent experimental discovery in 1964. I include a famous photo (figure 3) at this early stage to emphasize the reality of particles. An electrically charged particle zipping through a bubble chamber* leaves a track of bubbles, which allow us to visualize what is going on. (For our purposes here, you do not have to know the names of these "elementary" particles, nor what they are all about. Suffice it to mention here that this photo helped establish the notion of quarks. See chapter V.2 for details.)

Teams of humans, mostly of the female persuasion as reflective of that era in U.S. history, were trained to examine thousands or more of these types of photos to pick out anything unusual. I have no doubt that I would be promptly fired. I would never have picked out this photo. (Presumably, these workers were told to look for "horizontal" tracks slashing across the photo, like the one

---

*Invented in 1952 by D. A. Glaser, awarded the Nobel prize in 1960. By the way, he refuted the story circulating among physics students that he was inspired by a bottle of beer.

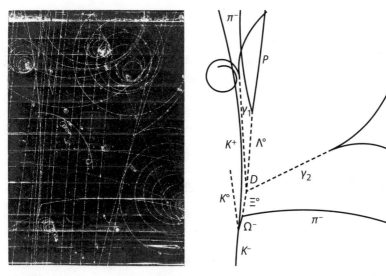

Figure 3. In this famous bubble chamber photo, as the right panel indicates, a K^- meson enters from the bottom left, collides with a proton in the superheated transparent liquid, and produces the hitherto unknown but fabled Ω^- together with a K^+ and a K^0 (which being electrically neutral leaves no track). The Ω^- did not get very far, before disintegrating into a pion π^- and a Ξ^0 (never mind what that is, but some readers at least could learn a seldom used Greek letter).
Image courtesy of Brookhaven National Laboratory.

left by the π^-.) Nowadays, computers are programmed for this job (and bubble chambers had long been supplanted by more sophisticated detecting devices).

Discrete, or continuous, or both?

Back to our narrative. The electromagnetic wave, thought to be continuous for so long, actually consists of a stampede of photons? Shocked? Yes, but soon the theoretical physics community digested this and settled down to the view that everything, even the electromagnetic wave, is ultimately made of discrete particles.

But then no, yet another shock! In 1924, Prince Louis de Broglie of France* proposed in his doctoral dissertation that the electron, evidently a discrete point particle, might actually be a wave, with frequency given by $\omega = E/\hbar$ and wave vector $\vec{k} = \vec{p}/\hbar$. The astute reader might have recognized that these relations are the same as those given above for the photon. But nobody had thought that discrete particles could also be described by waves!

*I once watched a play in which every time Schrödinger (in real life also a bon vivant) encounters de Broglie, he would address the latter as Dr. Broccoli, and de Broglie would correct Schrödinger in a haughty French accent, but to no avail.

De Broglie's thesis advisor, Paul Langevin,[10] did what befuddled senior professors are wont to do: defer to a higher authority. He sent the dissertation to Einstein, who recognized its importance immediately. On Einstein's say-so, Langevin gave de Broglie his degree.[11]

Recall that the wave number k, namely, the magnitude of $\vec{k}$, is defined to be the inverse of the wavelength λ. More energetic electrons have larger momenta, and hence smaller de Broglie wavelengths λ. Thus, energetic electrons have very small wavelengths and could pass as particles. In contrast, slowly moving electrons behave like waves,

The origin of the uncertainty principle

I can now give you a heuristic derivation of the uncertainty principle, as promised in the prologue. We may not consciously think about this, but in everyday life, when we locate an object by sight, we are arranging for a stream of photons, from the sun or some other light source, to fall upon it and for some of them to bounce back into our eyes. Since the wavelength of visible light is about a few times 10^{-5} cm, we are rarely aware of light being a wave, but a wave it is. Our eyes are not sensitive enough to detect a fuzziness of 10^{-5} cm, but that would be the best precision to which you could locate an object.

In the subnuclear world, how would you go about locating a proton? One possibility is to arrange for electrons to fall on the proton and for some of them to bounce back into a detector. According to de Broglie, the electron is also a wave, with wavelength λ inversely proportional to the electron's momentum p, that is, $\lambda \sim \hbar/p$. When you probe something with a wave, you clearly cannot locate its position with precision any better than the wavelength of the wave, the wavelength being the characteristic length scale of the waviness, so to speak. So, to locate the proton's position to within Δq, you would need to use electrons with wavelength λ at least as small as Δq. No problem, you say. "We will just use electrons with a larger momentum p and hence a smaller wavelength $\lambda \sim \hbar/p$." But when you bang highly energetic electrons off the proton, that pushes the proton around and causes its momentum to vary, that is, to become uncertain by an amount $\sim p$. Hence, the uncertainly in the momentum Δp and the uncertainty in the position Δq of the proton are related by $\Delta q \sim \hbar/\Delta p$.

To summarize, Heisenberg's uncertainty principle followed from de Broglie's relation, which the Prince postulated by analogy with what Einstein said about how the photon's momentum is related to the wavelength of the corresponding electromagnetic wave.

The birth of quantum mechanics

Now we reach a point of maximum confusion. What was thought to be continuous may actually be discrete and what was thought to be discrete may actually be continuous! How paradoxical and head splitting all this must have seemed to physicists at that time.

Finally, during 1925–1926 Heisenberg and Schrödinger set down quantum mechanics as we now know it. In the Schrödinger formulation, the electron is still envisaged as a particle whose position can, however, no longer be specified, but instead is governed by a probability amplitude with the property of the wave determined in the manner described by de Broglie. In this version of quantum mechanics, which is the version most often taught to physics students, the electron is still treated as a particle, while the electromagnetic field it interacts with is treated as a field. More on this in chapter IV.1.

Notes

[1] Etymologically related to "mother" (mater) and to "matrix."

[2] The electron was the first of many subatomic particles to be discovered.

[3] S. Weinberg, The Discovery of Subatomic Particles.

[4] I am simplifying a long, complicated story. See the books by B. Brown, *Planck*, and A. D. Stone, *Einstein and the Quantum*.

[5] See A. Pais, *Subtle Is the Lord*, page 371. For a fuller explanation of Planck's work, see *FbN*, chapter III.5. Most of my colleagues would agree with me that the crackpot letters we receive periodically are merely mad, with nothing divine about them.

[6] The miracle year during which he made several epoch changing discoveries, including special relativity.

[7] Unfortunately, I would have to be extremely brief in describing this important effect, but fortunately, it is treated in detail in every introductory book on quantum physics I know of. See for example, R. B. Leighton, *Principles of Modern Physics*, McGraw-Hill 1959, pages 67ff, with a helpful timetable of the important experimental measurements.

[8] See G, page 26.

[9] Nobel prize in 1906, ostensibly given for his study of the conduction of electricity in gases.

[10] He allegedly had an affair with Madame Curie, about which Einstein expressed an interesting contrarian opinion.

[11] A. Pais, *Subtle Is the Lord*, page 438.

Recap of part I

A point in space (the "where") is characterized by the Cartesian coordinates (x, y, z), which may be variously written as $x^i = (x^1, x^2, x^3)$ or $\vec{x}$. This is generalized to a point in spacetime (the "when" and "where") characterized by the coordinates (t, x, y, z), which may be variously written as $x^\mu = (x^0, x^1, x^2, x^3) = (t, x^i)$ or $(t, \vec{x})$.

Pythagoras told us that the distance ds between two infinitesimally separated points is given by $ds^2 = dx^2 + dy^2 + dz^2$, which may be regarded as defining the geometry of space. Minkowski generalized this to the distance ds between two infinitesimally separated points in spacetime, given by $ds^2 = dx^2 + dy^2 + dz^2 - dt^2$, thus defining the geometry of spacetime. Surprisingly, it differs from Euclid's geometry merely by a crucial sign distinguishing time from space.

Physicists call any variable that depends on (t, x, y, z) a "field."

The notion of field comes out of our discomfort with action at a distance. Look up at the moon, and we do not see a rope tying it to the earth. And yet the moon faithfully accompanies the earth as it moves around the sun. Faraday's notion of the electromagnetic field as an independent entity begets the "self-propagating" electromagnetic wave, whose speed is determined by how an electric field generates a magnetic field and vice versa, which manifestly does not depend on the motion of the observer. This led Einstein to modify our notions of space and time, and Minkowski to unify the two into spacetime. Henceforth, particles and fields have to abide by the geometry of spacetime.

The famous formula $E = mc^2$ just pops out!

Is matter ultimately discrete or continuous? From ancient times to now, the evidence pointed one way or the other in succession until it was settled by quantum field theory. How that was settled is part of the subject of this book.

The road to quantum field theory ▬

Preview of part II

Theoretical physicists have, centuries ago, formulated physics globally in terms of the action, instead of locally in terms of forces, as is taught in school. Instead of talking about a particle being acted on by a force at each instant in time, thus changing its velocity and its subsequent trajectory or path, we "allow" the particle to "choose" the path in space and time it would "like" to follow. Each of the possible paths a particle could follow is assigned a number, known as the action for that particular path. The particle follows the path which either minimizes or maximizes the action. Each particle in the physical world is striving to extremize* its action, or, more colloquially, constantly trying to find the best deal.

In chapter II.1 we explain in detail what these mysterious words mean. To get a quick overview, check out figure II.1.4 in that chapter.

If you find the action formulation rather bizarre, seemingly implausible, and barely credible, you are not alone. Upon first hearing that a particle could choose a path that it finds optimal, many physicists feel as you do. Yes, it does sound strange, but a bit of undergraduate level mathematical manipulation suffices to show that the action formulation reproduces the standard Newtonian view of the particle being accelerated by external forces. The action formulation and the Newtonian formulation are mathematically equivalent.

---

*"Extremize" is a generic word encompassing both "minimize" and "maximize."

Why then should physicists bother with the action formulation if it is equivalent to the more intuitive "everyday" Newtonian formulation? Indeed, many don't. They stick to the Newtonian formulation.

But the action formulation has two tremendous advantages. (1) It generalizes to fields immediately; we "merely" have to determine the action governing the field in question. (2) It allows us to transition from classical to quantum physics "easily." Thus, the action formulation could catapult us immediately to quantum field theory, the subject of this book.

All this and more will be explained in part II.

Getting the best deal: from least time to extremal action

Light in a hurry

Yes, light travels in a straight line, except when it doesn't. For you to read this book, eons of evolution have enabled your sainted mother to fashion a lens in your eyes, and to equip you with a brain and muscles to stretch and squeeze that lens just so, to compel the light rays coming from the black marks on the page to bend and focus on your retina. The cycle may well close. Reading this book might enhance your advantages, perhaps boosting your chance of being selected, which in turn increases your prospects for reproducing yourself.

The vanishing puddle of water

On a hot day, the highway beneath a distant oncoming car appears to be wet, but is in fact dry (figure 1). The speed of light in air depends on how hot, and hence how dense, the air is. A light ray leaving the hood H and headed downward encounters a layer of hot air near the road surface and bends upward. It ends up following path 2 to the observer's eye. The observer's brain, judging the direction from which the light ray comes, concludes that it came from H'. Another light ray goes directly from H to the eye, following path 1. This is repeated for light rays leaving every point on the car, causing a reflection of the car to be seen. The brain—what a marvelous organ—deduces[1] that the road must be wet.[2] Hence, the common mirage of the vanishing puddle of water on the highway up ahead.

Light saving time

Pierre de Fermat realized that this bending of light[3] could be explained by postulating that light is in a hurry. It wants to get to its destination as

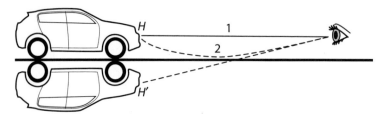

Figure 1. A common mirage.
Modified from A. Zee, *Fearful Symmetry: The Search for Beauty in Modern Physics,* Princeton University Press, 1986.

quickly as possible, given that the speed of light varies from medium to medium.

Perhaps I could best explain this by a parable, about a quest within a quest. As your tour group strides through the mythical "square of physics," steadily toward the northeast, you behold a shimmering lake, and there on a prominent rock in the middle of the lake a treasure beckons. A ring? No. A sword? No. A bar of gold? No. Something much more valuable: a shining suit of intellect you could download into your brain!

Everyone in your group is making a beeline for the treasure. Certainly you too! An action hero is as an action hero does: you immediately spring into action.

The others in your group, having heard Euclid say that the shortest path between two points is a straight line, are already proceeding in a straight line (starting from point F, where the group first espied the treasure, as shown in figure 2, going along the dotted line) toward the rock (at point G). That would be the path of least distance.

But no! You are smarter than these other numbskulls, and you have already calculated the path that would take you to the treasure in the least amount of time. Time counts more than space here: least time trumps least distance. You can run much faster than you can swim, just like any other human, for that matter. So you should spend more time running before plunging into the lake. A simple high school level calculation determines the best path to take (see the solid line in figure 2). Hurray, you beat the other guys and win a dazzling new intellect!

But you don't have to calculate to see that there is an optimal path. Clearly, only a cretin would follow the third path (the dashed line) shown in the figure.

And thus Fermat's least time principle for light explains why light bends* as it goes from one medium to another, for instance, from air through the

---

*Theoretical physics is famously a predictive science. It is not enough to explain why light bends, but the precise angle by which light bends has been shown to match Fermat's principle given the speed of light in various media, measured centuries later. At the time of Fermat, physicists could not even conceive of a way to measure these various speeds. Need I remind you that light covers a distance equal to the circumference of the earth in about 1/7 of a second?

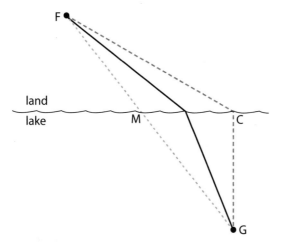

Figure 2. The bending of light told as a fable: a race is on to get from point F to point G. Modified from A. Zee, *Gravity in a Nutshell*, Princeton University Press, 2013.

aggregate of crystallins in your lens and then into the vitreous humor of your eyeball.

How to choose the winning path for a material particle traveling through spacetime

But this guiding principle for light led to an outstanding puzzle for theoretical physicists: How do you explain the trajectory of material particles? Particles, in contrast to light, do not travel at a fixed speed.

To explain to you the challenge facing theoretical physicists in the generation after Newton, I invite you to watch a race. Your task is not to crown, nor to choose, the winner. In fact, the runners all arrive at the finish line at precisely the same time. So who wins?

For races in everyday life, the competitors start at a specified location in space, all at the same time, but they cross the finish line one after another.[4] In other words, they start at the same point in spacetime but end up scattered all over the place in spacetime. In contrast, in our weird imaginary race, not only do the competitors start at the same point in spacetime, but they are required to end up at the same point in spacetime.

The finish line could be quite far from the start, and the runners don't even have to keep to the same route. They wander over dale and vale at will, but then all cross the finish line at the same instant: a photo finish every time. One runner starts out slow and then steadily accelerates, going faster and faster. Another dashes off as fast as she could at the pop of the gun, but then slows down. Yet another runs at a nice and steady pace. Some even stop to take a rest before continuing. And so on. In fact, an infinite number of runners are off and running, all reaching the finish line at the same instant.

Yet the race organizer can unerringly pick out one particular runner out of this multitude, and crowns him or her. Your challenge is to figure out how the winner is chosen. How?

Worldlines in spacetime

Before I answer that question, let me show you how physicists visualize the paths of particles as they traverse spacetime. Recall from chapter I.3 that Minkowski called these paths "worldlines." Again, I simplify by restricting space to be 1-dimensional.

Recall also from figure I.3.5 that the slower you move, the more vertical your worldline. A particle sitting still has a vertical worldline: it is moving in time, but not in space. Thus, in figure 3(a), the winner is runner 1, runner 2 comes in second, and runner 3 crosses the finish line at x_f last. For ease of drawing, I have made the runners run at a steady pace. In figure 3(b), runner 1 starts off fast, then slows down, runner 2 runs at a steady pace, and runner 3 starts off slow, then accelerates, dashing across the finish line.

If we measure the distance traveled in light units, for instance, light second, then light would move along the 45° line as shown in figure 3(c). In 1 second, light has moved a distance of 1 second. The statement that nothing can move faster than light means that everybody else's worldline has to have an angle of less than 45° from the vertical, such as worldlines 2 and 3. In these units, even the fastest jet has a worldline that's nearly vertical. Note that worldline 1 is strictly forbidden in classical physics: it has moved a lot in space but rather little in time.

To draw spacetime on a piece of paper, we have to take space to be 1-dimensional, as I noted. We could hardly represent the three dimensions of space plus one dimension of time on paper. The best we could do is shown in figure 3(d). Space is depicted as 2-dimensional, with axes x and y, and the worldlines of light leaving the origin lie on what is known as the light cone. Since material particles cannot move faster than light, their worldlines have to lie inside the cone.

Lagrange invented the Lagrangian

Thirty some years after Newton's death, after lots of confusion and priority fights among physicists, Leonhard Euler and his protégé Joseph Louis, Comte de Lagrange[5] finally figured out how the winner was chosen. They discovered that the race organizer calculates, for each runner at every instant, a number

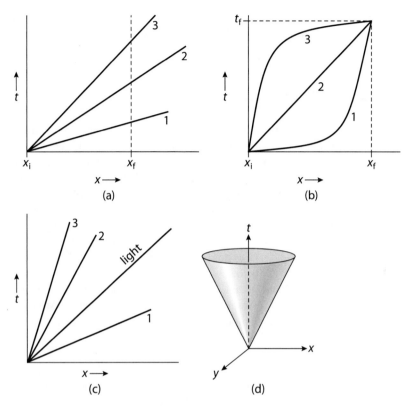

Figure 3. Paths in spacetime taken by the competitors (a) in an everyday race and (b) in the peculiar race incessantly staged in the universe. (c) The worldline of light when plotted with sensible rather than everyday units is a 45° line. The light cone is shown in (d).

known to physicists as the Lagrangian (naturally!), and then adds up* these numbers from start to finish. This sum is known as that runner's action. The runner who extremizes (meaning either minimizes or maximizes) his or her action wins.

An aside that will become important when we get to the quantum world: Fermat tells us that light minimizes travel time. It turns out that in some circumstances, material particles minimize the action, as we might have guessed, but in other circumstances, they maximize the action. Physicists have coined the word "extremize" to cover both "minimize" and "maximize." That the action principle is an extremal principle, rather than a simple minimal principle like Fermat's, remained a mystery till the advent of quantum physics.

---

*That is, integrates over time, for the cognoscenti.

This peculiar race is run over and over again, all over the universe, and yet that Race Organizer in the Sky is able to choose the winner every time without ever slipping up.

Choice of history

I will ask Humpty Dumpty to demonstrate the action principle. When Dumpty falls, he starts out at a leisurely pace, and then goes faster and faster.

A record of where Dumpty is and how fast he is falling at any instant in time is known to theoretical physicists as a history, that is, a path in spacetime. An infinite number of histories could be contemplated, but somehow only one history is actually realized. From everyday observation, Dumpty never starts falling fast and then slows down as if in fear of his imminent crack up.

What principle dictates Dumpty's choice of history? Indeed, this is the question at the heart of physics. How does anything choose its history? Fermat had answered this question for light.

At any instant during Dumpty's fall, he has both kinetic and potential energy. Call them K and V, respectively, for easy reference below. Allow me to remind you that, in Newtonian mechanics, the kinetic energy is simply the energy associated with the movement of the particle, while the potential energy is a kind of "stored" energy that is available for conversion into kinetic energy. For example, an object near the surface of the earth has potential energy because of the earth's gravitational pull. The higher the object is from the ground, the more potential energy it possesses.

Well, the Lagrangian, call it L, alluded to above, and which physicists struggled to find for over a century, is simply the kinetic energy minus the potential energy: $L = K - V$. Can't get any simpler than that!

Be sure to distinguish $L = K - V$, from the more familiar total energy $E = K + V$, namely, the sum of kinetic and potential energy. While L changes with time, E is conserved, that is, it does not change with time. As the object falls, its potential energy decreases, while its kinetic energy increases, keeping the sum of the two constant. In other words, potential energy is converted into kinetic energy. When we go downhill skiing, we pay the lift operator to provide us with lots of potential energy, which we then convert into kinetic energy as rapidly as we dare.

The action is then the result of adding up the Lagrangian from the start time to the end time.[6] In our example, these two times would be, respectively, the time when Dumpty leaves the security of the wall and the time when he spills his yolk on the ground.

The computation of the action is similar to that done by an accountant determining the total profit of a business for any given production strategy. He subtracts the total cost of production from the gross income on a weekly basis and then sums this quantity over the 52 weeks in the fiscal year. The

Lagrangian corresponds to the weekly profit, the action to the annual profit. The businessman naturally tries to maximize the total profit by following the most advantageous strategy or history.

An executive summary in a table:

business	gross receipts	cost of production	weekly profit	annual profit
physics	kinetic energy	potential energy	Lagrangian	action

Catch me if you can

Just like the businessman maximizing his profit, Dumpty chooses the history that would minimize his action. Since the action is equal to the kinetic energy minus the potential energy, $K - V$, summed over the duration of the fall, and since the potential energy increases with the distance from the ground, it clearly pays to spend more time high above the ground, so that a larger potential energy could be subtracted off. With the help of elementary mathematics, one can show that the best strategy for Dumpty is to accelerate at a constant rate.

In everyday life, that ancient Ming vase that you clumsily bumped into appears to hesitate for a moment or so, almost as if it is saying "Catch me if you can!", before gathering speed and crashing to the floor. That's Galileo's law of acceleration, of course. From the action point of view, we could understand what went on as the vase's attempt to minimize its action. The vase, by staying at high altitude for "as long as possible," maximizes its potential energy and thus lowers the action. But then it has to rush at the end to get to the floor in the allotted time, and thus pays the price of a larger kinetic energy.

The reader should not confuse extremization of the action with the everyday observation that all things desire to minimize energy, the principle underlying "water always flows downhill" and "a couch potato will stay on the couch." Throw a marble into a bowl. Come back later, and you would be astonished if it is not resting at the bottom of the bowl. The marble has minimized its total energy by setting its kinetic energy to zero and lowering its potential energy as much as possible. Similarly, a hanging elastic string affixed at its two ends assumes a familiar parabolic shape to minimize its total energy,* in this case the sum of its gravitational potential energy and elastic energy due to stretching.[7] (Occasionally, a student might wonder if this minimization of energy contradicts the conservation of energy. In fact, while the latter is absolute and sacred to physicists, the former is merely apparent because we

---

*It turns out that writing down, and then minimizing, the energy is significantly easier than figuring out the forces acting on each segment of the string. This example foreshadows the advantage of using the action principle rather than the differential equation of motion.

choose to ignore other forms of energy. By rattling in the bowl, the marble has generated sound and heat, both of which escaped into the environment.)

Physics is where the action is

The action is one of the most profound concepts in theoretical physics. So let us recap. For every possible path a particle could follow, we assign an action, traditionally written as S(path). The notation indicates that the action, namely, the real number S, depends on the path.

Let me introduce a mathematical term we will use later. In school, we learned about functions of numbers; for example, the path of a particle in 3-dimensional space is described by three functions of time $\vec{x}(t) = (x(t), y(t), z(t))$. If we input a number t, the function $x(t)$ outputs a number. For any time t we specify, these three functions tell us where the particle is at that instant. In other words, the three functions $x(t)$, $y(t)$, $z(t)$ determine the path.

Now go up one level of abstraction. In mathematics, a functional is a function of functions. Thus, the action S is a functional: it is a function of the three functions $\vec{x}(t)$. In other words, if we input $x(t)$, $y(t)$, $z(t)$, the action S outputs a number.

A story or fable might help. Suppose you are told to drive from San Francisco to New Orleans arriving exactly 56 hours and 12 minutes later, and to minimize the amount you pay for gasoline. You happen to know some towns in Nevada and in Kansas that sell a particularly cut rate fuel. You would have to determine your route taking into account the mileage per gallon you could get with the cheap fuel, the detour required, etc. Clearly, if you were told to minimize something else, such as wear and tear on the tires, you would choose a different route. And so on and so forth.

Different action would dictate a different path to follow. In real life, every person's utility function, as an economist might call it, may be enormously different.[8] See figure 4.

How do you find an extremum? Trial and error

The path actually chosen by the particle extremizes S(path). How do you find the extremum of a functional? Consider the simpler problem of finding the minimum of the function shown in figure 5. Easier to visualize a function than a functional!

Newton and Leibniz taught us to calculate the slope. The minimum (and also the maximum) is the place at which the slope is zero. In other words, an extremum occurs where the function is flat to leading order. Hikers (and

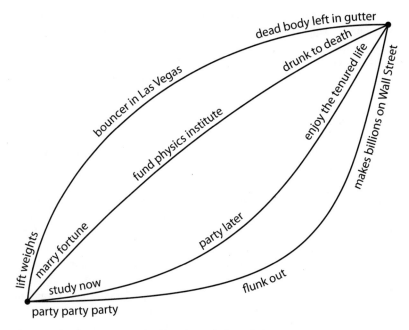

Figure 4. A physics student visualizes the path formulation.

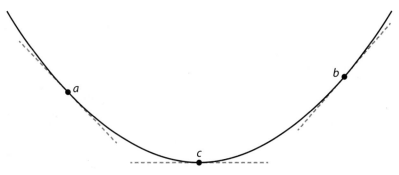

Figure 5. The slope is negative, positive, and zero at the points a, b, and c, respectively. The minimum occurs where slope changes from negative to positive, namely, at c.

anybody who can read a graph) know this of course without the benefit of calculus. If you are ascending or descending, you have not reached the valley floor. At the valley floor, your altitude stays more or less constant.

I already mentioned that theoretical physicists also refer to a path in space-time as a history. In our "fable," the history is the travelogue. You are in Liberal, Kansas, 41 hours and 27 minutes after leaving San Francisco; hmm, perhaps you would be better off in Beaver, Oklahoma? Follow this other path next time?

Similarly, to find the extremum of S(path), Euler and Lagrange taught us to vary[9] and wriggle the path or history a bit, to see if the action S(path) increases or decreases.

In everyday terms, compare with your neighbors. In physics, compare the path under study with the neighboring paths in spacetime.

Toy example

Let me illustrate the action principle with what theoretical physicists call a "toy example." Suppose you were told to solve the equation $x - 1 = 0$. Easy peasy, as my elementary school son would say, you move the -1 to the other side, flip sign, and obtain the answer $x = 1$. Roughly, this corresponds to Newton's formulation of mechanics. A more cumbersome approach would be to invite ourselves to contemplate the function $S(x) \equiv (x - 1)^2$ and look for its minimum. Since for any value of x other than 1, $S(x)$ is greater than 0, and since $S(1) = 0$, the minimum occurs when $x = 1$. (Indeed, the function $S(x)$ is a parabola of the form shown in figure 5.) The second approach, which corresponds to the Euler-Lagrange formulation, appears less straightforward, but in physics as in mathematics, things are not always as they meet the eye.

This almost laughably simple analogy captures a bit of the contrasting philosophy between the two formulations. More in the next two chapters.

Notes

[1] I wonder if babies and very young children would deduce this. Surely not before they associate reflected images with water.

[2] By the way, some readers may see that this example shows that light only cares about the local, not the global or absolute, minimum in time of transit. Path 2 is definitely longer than path 1, but it only needs to be better than its neighboring paths. The same phenomenon holds in human society. Most people only care that they are doing better than their neighbors, not compared to some insanely rich guy remote from their world.

[3] This phenomenon, known as refraction, could of course also be accounted for in Maxwell's theory of electromagnetism through a careful matching of the electromagnetic waves in the two media.

[4] See, for example, https://www.facebook.com/olympics/videos/10156329615364216. Never give up!

[5] Lagrange was a teenage prodigy, but fell into a deep depression in his old age. Fortunately for him, the daughter of Lemonnier, an astronomer friend of Lagrange's, managed to cheer him up. Almost 40 years younger than Lagrange, the young woman offered to marry him. Soon Lagrange was productive again. See *Mathematicians Are People, Too,* by L. Reimer and W. Reimer, page 88. "Grange" means "barn" in French. Strictly speaking, Lagrange was Italian by place of birth, but he was born before there was such a thing as Italy.

[6] For those readers who know some calculus, the action $S = \int_{t_i}^{t_f} dt \, L$ is the integral of the Lagrangian from some initial to some final time.

[7] Amusingly, the math needed to solve this problem is the same as that needed to solve for a falling object. See *GNut*. Compare figure II.1.1 on page 114 and figure II.3.1 on page 137. Change space to time and flip the string over!

[8]The analogy with physics fails because most people cannot predetermine the place and date of their demise. A distinguished French physicist once said to me, "You could only optimize your life locally, not globally." By the way, the word "locally" is used in everyday life only for space, but in physics circles, it is understood to refer to spacetime. By the way, this same French friend arranged for me to spend a year in Paris. Talk about local optimization!

[9]By the way, Euler and Lagrange named their work the "calculus of variation." For the reader who knows some calculus, it's not that big a leap from ordinary calculus to the calculus of variation. See *GNut,* chapter II.1.

Global versus local

Integral versus differential

For the motion of point particles, the action formulation is more compact and aesthetically more appealing than Newton's formulation. Yet the two formulations are entirely equivalent; we could mathematically derive one from the other. The outlook, or perhaps one could even say philosophy, however, is quite different in the two formulations. In the action formulation, one takes a global or integral view, comparing different ways by which a material particle could have gotten from here to there. In contrast, in Newton's formulation, the force at each instant of time determines the acceleration of the particle, which in turn determines the velocity and position at the next instant. There is no talk of there and then. Only the here and now, and force plays the deciding role.

Feynman was blown away, but the practical physicists don't care

Why is the action principle important, if it is entirely equivalent to Newton's differential formulation? It's not, if you are a practical guy. Indeed, many physicists do not even bother learning the action principle. Some physicists are incredibly (at least to me) practical minded.[1] However, with the advent of quantum field theory, the action principle has burst upon the stage. And that's part of the story of this book.

Richard Feynman recalled that when his physics teacher first told him about the action principle, he was blown away. I believe that many physicists, yours truly included, had the same, almost mystical, experience.[2]

Potential energy

The action equals the difference between kinetic energy and potential energy, integrated over time. I have already mentioned that potential energy is stored energy that could be released in the future, as in "That youngster has potential." You open a closet which you have not cleaned out for 20 years and a heavy box falls off the top shelf and smashes your toes. The work you performed 20 years ago lifting that box to the top shelf and leaving it there has been stored as potential energy. For 20 years, the conservation of energy guarantees that not a bit of the stored energy has been lost, and now, in an instant, faster than you could get out of the way, it is all converted into kinetic energy.

The variation of the potential energy from place to place gives rise to force. Perhaps even better: The spatial variation of the potential energy could be interpreted as a force. Historically, force came before potential energy, but mathematically, potential energy holds an advantage over force. The potential energy at a given point in space, denoted by $\phi(\vec{x})$, is a single number, but the force at a given point in 3-dimensional space consists of three numbers, one for each Cartesian direction: $F_x(\vec{x})$, $F_y(\vec{x})$, $F_z(\vec{x})$. In other words, you have to specify the magnitude of the force as well as the direction it acts in. Much simpler to keep track of one number and to follow how it varies in space, instead of three numbers. In math talk, a scalar function, such as $\phi(\vec{x})$, is simpler than a vector function, such as $\vec{F}(\vec{x})$.

Electromagnetic potentials

Essentially the same story holds in electrostatics. The potential energy of a charge q, located at $\vec{x}$, in an electric potential ϕ is given by the product q times $\phi(\vec{x})$. As you can see, physicists even used the same words as they moved on from mechanics to electromagnetism. Again, the spatial variation of ϕ gives rise to the electric force $\vec{E}$ acting on the charge. The charge q corresponds to mass m, while the electric potential ϕ is the direct analogy of the gravitational potential ϕ.

From everyday life, we are familiar with voltage, a kind of pressure exerted on electric charges: It pushes electrons through wires, driving them to do what your burning heart might desire. Voltage corresponds to the electric potential ϕ in physics.

Starring role in different formulations

In Newton's formulation of classical mechanics, force plays the starring role. We ask how the velocity of a particle exerted upon by a force changes from instant to instant. The concept of potential energy pops out of Newton's equation of motion. "Hey, look at the sum of these two terms, it does not change with time. Alright, let's call one term 'kinetic energy' and the other 'potential energy'." Potential energy is a derived concept, certainly convenient and helpful, but strictly speaking not necessary.

In contrast, in Euler's and Lagrange's action formulation, we cannot even write down the action without having the concept of potential energy from the start, action being the difference of kinetic and potential energies. Potential energy plays a starring role. No need to even mention force. It only pops up when we pick the extremal path. By extremizing the action, we find that the trajectory of the particle follows a path determined by Newton's equation of motion. "Hey, when we extremize the action, this equation comes out telling us that the acceleration is equal to something. Alright, let's call this something 'force'."

In Newton's formulation, force reigns primeval. In Lagrange's formulation, force is a derived concept, a descendant of the potential energy.

The story continues into electromagnetism. Just as the electric force $\vec{E}$ may be thought of as due to the spatial variation of electric potential ϕ, the magnetic force $\vec{B}$ acting on a charge may be thought of as due to the spatial variation of a "magnetic" potential $\vec{A}$. The alert reader might have noticed that the potential $\vec{A}$ is a vector (hence it is called the "vector potential" in textbooks), in contrast to the electric potential ϕ. (This interesting difference follows from how the magnetic force acts.)[3] This fact will play an important role in chapter IV.4.

Triumph in the quantum era and the drive toward unification

For decades after its discovery, the action principle appeared to the practical minded as a useless representation of known physics. But with the dawning of the quantum era, the action formulation emerged triumphant, lighting our way toward quantum field theory. But perhaps at this stage, you could already see a glimmer of how that would come to pass. With particles in the quantum world jitterbugging this way and that, the notion of force is not as clearcut as in classical physics. The picture of particles exploring different paths through life becomes suddenly more sensible.

Here I pause and emphasize an underappreciated[4] feature of the action principle. The history of physics may be viewed as a drive toward unification. Newton unified terrestrial and celestial physics. Maxwell unified electricity and magnetism. In our own time, we have witnessed the spectacular unification of electromagnetism and the weak interaction into an electroweak interaction. Furthermore, many believe that the electroweak interaction could also be unified with the strong interaction into a single grand unified interaction.[5]

Imagine that you are a bright young theoretical physicist in a galaxy far far away. You learn about the least time principle for light and the extremal action principle for matter. You become highly dissatisfied. Two principles on a fundamental level! You devote yourself to unifying these two disparate principles into a single principle. Finally, you succeed. Congratulations! You have discovered special relativity.

In our civilization, it did not happen that way. As is fairly well known, Einstein as a young man analyzed the apparent incompatibility of electromagnetism and Newtonian mechanics. By resolving this incompatibility through various thought experiments, he managed to discover special relativity.

But in our extragalactic fable, the unification could have happened at the level of the foundational principles. Least time and extremal action, the two look and sound completely different. Weird if you really think about it. Just to take one example, you might have learned in school that the kinetic energy of a material particle with mass m traveling at velocity v is given by $\frac{1}{2}mv^2$. The photon, to use the term anachronistically, was already known at the time of Einstein to be massless. So how does it even make sense to set the mass m of the material particle to zero? Its kinetic energy would go to zero. So, it is far from a straightforward problem to unify least time and extremal action.

In the glare of hindsight, theoretical physicists now know how to do this.[6] Highly satisfying.[7] Two foundational principles are now but one!

Notes

[1]As I said in chapter II.1, everybody's utility function differs tremendously.

[2]Fermat's least time principle has a strongly teleological flavor—that light, and particularly daylight, somehow knows how to save time— a flavor totally distasteful to the post rational palate. Things are teleological if they have a purpose, or at least act as if they have a purpose. That's a big no no in Western science. In contrast, at the time of Fermat, there was a lot of quasi-theological talk about Divine Providence and Harmonious Nature, so there was no question that light would be guided to follow the most prudent path. Indeed, to some, the least time and action principles provided comforting evidence of Divine guidance. A voice told each particle in the universe to follow the most advantageous path and history. Not surprisingly, the action principle has inspired a considerable amount of quasi-philosophical, quasi-theological writing, a body of writing which, while intriguing, proves to be sterile ultimately. Nowadays, physicists generally adopt the conservative, pragmatic position that the action principle is simply a more compact way to formulate physics, and that the quasi-theological interpretation suggested by it is neither admissible nor relevant.

[3]Some readers might know that the magnetic force acting on a charged particle depends on the velocity of the particle.

[4]By practical minded physicists, that is.

[5]See chapter V.4. Also, *Fearful*, chapter 14, and *QFT Nut*, chapters VII.5–7.

[6]See *GNut*, chapter III.5, especially page 212.

[7]"Philosophical" satisfaction matters to some, but certainly not all, theoretical physicists.

Enter the quantum

To say it better: the path integral

The history of physics is full of alternative formulations. Some thrive, some do not. We saw in chapter II.1 that classical mechanics could be formulated locally or globally. The two formulations are entirely equivalent.[1] Euler and Lagrange were still talking about Newtonian physics, but the way they said it had a profundity that resonated with 20th and 21st century physicists. I am reminded a bit of literature. Novelists and poets say what others have already said, but the good ones strive to say it better and to shed new light.

Quantum mechanics enjoys not one, not two, but three different[2] formulations.* The first, laid down in 1925 by Werner Heisenberg,† involved operators and matrices, mathematical concepts largely unfamiliar to physicists at that time. Less than a year later, Schrödinger proposed the equation, now named after him, governing how quantum probability changes with time. Then in 1932, Paul Dirac suggested, in a somewhat rudimentary form, yet a third formulation.[3] Dirac's idea appeared to be largely forgotten until 1941, when Feynman[4] developed and elaborated this formulation, which became known as the path integral[5] or sum over histories formulation. (In this book, I use the words "sum" and "integral" interchangeably: after all, an integral is essentially a fancy sum.)

While the three formulations are mathematically equivalent, the Schrödinger formulation is the easiest to grasp and the most convenient for practical

---

*Also often called "formalisms." But the word "formulation" is somewhat more accurate. The word "formalism" is sometimes used derogatorily.

†An often told story is that the 24-year-old Heisenberg, in order to escape a severe attack of hay fever, rented a room on Helgoland (Heligoland in German), a windswept (hence no flower and no pollen) 1.7 km^2 island, and there heeded Bohr's remark that thoughts of infinity most readily come while staring at the sea.

calculations. Therefore, it is typically taught to undergraduates first, and hence has become the best known. Indeed, when the young Heisenberg saw how simple Schrödinger's paper looked compared to his, he fretted that he wasn't going to land an academic position. Little did he know that he was destined to be on everybody's list of the ten greatest theoretical physicists of the 20th century.

In contrast to the Schrödinger formulation, the Dirac-Feynman formulation is almost unknown to the intelligentsia and is not even commonly taught to physics students. Feynman himself thought[6] in 1965 that it may be less deep than Heisenberg's operator approach. What irony! Now it looks like the path integral may ultimately turn out to be the deepest of the three formulations.

By the way, classical mechanics also enjoys three different formulations: Newtonian, Lagrangian, and Hamiltonian (which the reader may think of as a rewrite of the Lagrangian formulation). While they are mathematically equivalent, they are not morally equivalent, so to speak. For instance, it was the Hamiltonian[7] that opened the door to Schrödinger's equation and the quantum world.

At odds with common sense: not merely probabilistic, but probabilistic in a strange way

The reader is aware[8] that, in contrast to classical physics, quantum physics is probabilistic. But it is not merely probabilistic: It is probabilistic in a way completely at odds with common sense. As a simple example, the probability of obtaining a 1 on one throw of a die is $\frac{1}{6}$, and so the probability of obtaining either a 1 or a 2 on one throw of a die is clearly equal to $\frac{1}{6} + \frac{1}{6} = \frac{1}{3}$. We add probabilities in everyday life.

But in the quantum world, probability is determined by something called the "probability amplitude." To calculate the probability of either something happening or something else happening, the rules of quantum physics state that we add the probability amplitude for either of these events to occur, and then calculate the probability using the sum of these two probability amplitudes. In sharp contrast to everyday life, we add probability amplitudes, not probabilities:

everyday world	add probabilities
quantum world	add probability amplitudes

The framework for determining probabilities in terms of probability amplitudes is exactly the same in each of the three formulations that I mentioned. The difference between them lies in the way the probability amplitudes are calculated. For instance, in the Schrödinger formulation, the probability amplitude is also called a "wave function," which we obtain by solving the Schrödinger equation.

It is important to realize that probability arises in classical physics due to our ignorance. If we had known the initial orientation and momentum of the die precisely, and the position of the table, etc., we would be able in principle to predict how the die would land. In contrast, probability amplitude and probability are intrinsic to the quantum world.

A quick course on adding complex numbers

> People who wish to analyze Nature without using mathematics must settle for a reduced understanding.
> R. P. Feynman

A complex number may be thought of as an arrow in a plane, that is, as a 2-dimensional vector. Thus, a complex number, traditionally denoted by z, is characterized completely by the length of the arrow and by the angle θ the arrow makes with respect to some reference direction, usually taken to be the x-axis. The absolute value of z (denoted by $|z|$) is just the length of the arrow.

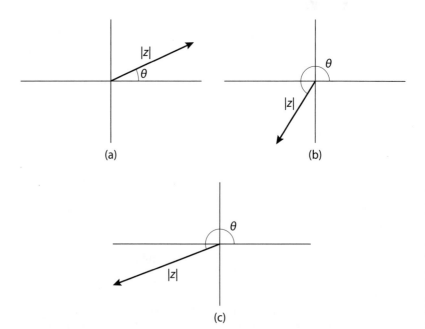

Figure 1. Three complex numbers, shown in (a), (b), and (c), each characterized by a length $|z|$ and an angle θ.

The angle θ is called the "phase." Just to be clear, a complex number with $\theta = 0$ is pointing east in the standard cartographic convention. In figure 1, three examples of complex numbers are shown. Note also that physicists and mathematicians measure the angle anticlockwise[9] starting from the x-axis.

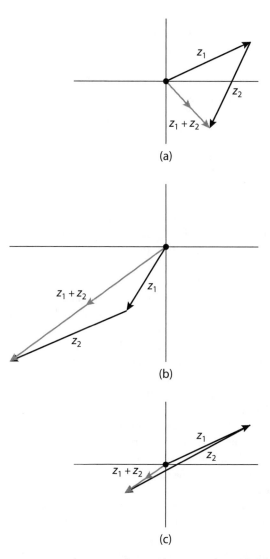

(a)

(b)

(c)

Figure 2. The sum $z_1 + z_2$ of two complex numbers z_1 and z_2. (a) Adding the two complex numbers shown in figure 1(a) and (b); (b) adding the two complex numbers shown in figure 1(b) and (c); and (c) adding the two complex numbers shown in figure 1(c) and (a). Note that in (c), the two complex numbers are almost the same length and point in more or less opposite directions. Hence their sum is rather short.

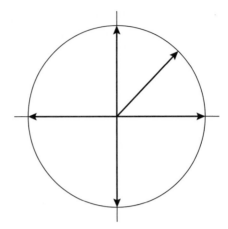

Figure 3. Complex numbers with length 1 sit on the circle with radius 1. The numbers $e^{i\theta}$ with $\theta = 0, \pi/4, \pi/2, \pi$, and $3\pi/2$ are indicated by arrows.

We add complex numbers in exactly the same way we add vectors or arrows. To add z_1 and z_2, think of them as arrows. Taking care to keep the direction of z_2 fixed (that is, keeping the arrow z_2 pointing in the same direction), move its "feathered end" to the "sharp tip" of z_1. Then the sum $z_1 + z_2$ is defined as the arrow going from the feathered end of z_1 to the sharp tip of z_2. Some examples are shown in figure 2.

The important point to grasp is simple. If z_1 and z_2 are more or less pointing in the same direction, then the length of $z_1 + z_2$ would more or less equal the sum of the length of z_1 and the length of z_2. In contrast, if z_1 and z_2 are more or less pointing in opposite directions, then the length of $z_1 + z_2$ would more or less equal the difference of the length of z_1 and the length of z_2, and hence could be quite small if these lengths are roughly the same.

To repeat, we add probability amplitudes, not probabilities, as indicated in the table given earlier.

A particularly simple class of complex numbers consists of complex numbers with unit length. These are characterized by a single number, the phase angle θ. A complex number with unit length and phase angle θ is denoted by $e^{i\theta}$ and known as a phase factor. For our purposes here, we don't even have to know what the symbol e and i represent.[10] I mention this mathematical notation just so that later we won't have to keep repeating the cumbersome phase "a complex number with unit length and phase angle equal to such and such."

Note that the angle θ is measured in radians, not some arbitrary Babylonian units called degrees.* See figure 3. For example, the arrow denoted by the complex number $e^{i\pi}$ points to 9 o'clock. In other words,[11] $e^{i\pi} = -1$.

---

*1 radian $= 360°/(2\pi)$. In other words, $90° = \pi/2$ radians.

In quantum physics, the probability amplitude of something happening is a complex number. The corresponding probability equals the absolute value squared of the probability amplitude, that is, the square of the length of that complex number.

In everyday life, there is no such thing as probability amplitude.

A quick review of classical physics

Recall that in classical physics, for each path a particle could possibly follow, we assign an action, traditionally written as $S(\text{path})$, indicating that the action, namely, the real number $S(\text{path})$, depends on the path. In other words, $S(\text{path})$ is a function of the path, duh. Recall also that a real number corresponds to a point on the straight line running from minus infinity to plus infinity.

Math summary. A complex number is an arrow in 2-dimensional space, namely a 2-dimensional vector, that could point in any direction. A real number has only two "directions," plus or minus. The addition of complex numbers is the same as the addition of 2-dimensional vectors (and readers who are into sailing would know that much of sailing[12] involves adding 2-dimensional vectors.)

Transition from classical physics to quantum physics

A tremendous advantage of the Dirac-Feynman formulation is that it renders the transition from classical physics to quantum physics particularly easy. You learned in the chapter II.1 that, in classical physics, of the infinite number of paths the particle could have followed getting from here to there, the path the particle actually follows is the path that maximizes or minimizes the action (that is, to put it more succinctly, extremizes the action):

classical physics	find the extremal path
quantum physics	add up the probability amplitudes of all possible paths

In the quantum world, a probability amplitude is assigned to each path. The probability of actually getting from here to there is then determined by the following rather bizarre procedure: You add up (more precisely, integrate[13] over) the amplitudes of all these paths.

To keep the discussion simple, introductory texts often talk about the two slit experiment. A beam of electrons is sent through two slits and then detected on a screen. A given electron goes through either one slit or the other. There are only two possible histories, and so the sum over amplitudes contains only two terms.

In the quantum world, unlike the classical world, all possibilities are possible. That's why it's called a possibility. For instance, in figure II.1.4, we could add another path: study quantum mechanics, get terribly frustrated, quit physics, play quarterback for a Super Bowl team, retire, and read popular physics books. Possible, but extremely improbable. The number of possible paths may not be countable, hence it is a path integral in general, and not just a sum.

A unit to measure the action with: $\hbar$

Recall from Dumpty's fall that his action equals his kinetic energy minus his potential energy integrated over time. So, action is basically some combination of energy times time. In classical physics, to get from here to there, the extremal path is followed. Yes, that includes you: The particles that compose you are all looking for the best possible deal.

A path has a large action if it involves lots of kinetic energy compared to potential energy, lasts a long time, or both. And the opposite is true for a path with a small action. Euler and Lagrange told us to compare the action of neighboring paths. But in classical physics, there is no intrinsic measure of action. In this respect, classical physics is not entirely satisfactory. Something is lacking.

Enter the quantum!

Recall from the prologue that the uncertainty in energy times the uncertainty in time equals Planck's constant, $\hbar$. Thus, $\hbar$ has the dimension of energy multiplied by time, just like the action. Mother Nature, in her generosity, gives us an intrinsic unit to measure action by.

What determines the probability amplitude for each path

Guess what determines the probability amplitude for each path!

Let's be clear that I'm not asking you to actually determine the probability amplitude. I am merely asking you which property characteristic of each path could possibly determine the probability amplitude.

Think about it for a minute and then tell me your guess. How about a hint? Try the what-else-can-it-be method. By the way, when all else fails, the what-else-can-it-be method often works surprisingly well in theoretical physics. Some[14] regard it as the "last resort of scoundrels."

Okay, time's up. Did you guess that the probability amplitude for each path is determined by the action assigned to that path? I did tell you to try the what-else-can-it-be method.[15]

Actually, guessing that the action determines the probability amplitude is fairly close to what Dirac said.[16] Indeed, the probability amplitude has to be determined by some quantity that plays an important role in classical physics and that is characteristic of each path. The action is the only quantity physicists know of that fills the bill.

A preternatural sense of the quantum physics

The action formulation turns out to be tailor made for the quantum world. Eerily, it almost seems that in 1760, Euler and Lagrange had a preternatural sense of the quantum physics to come.

So, we have learned that the probability amplitude for each path is determined by the action S(path) assigned to that path in classical physics. For the purpose of our discussion, you don't really have to know the precise formula relating the probability amplitude to the action; this is not a for-credit course on quantum mechanics. But for completeness, let me tell you anyway, in the form of a table.

probability amplitude for a path equals
the complex number with length $=1$ and angle $\theta = S(\text{path})/\hbar$

The probability amplitude is a complex number whose length is always 1 and whose angle is determined by classical physics: just the action divided by $\hbar$. Can't imagine it to be any simpler than that!

Ta dah, the equation representing the heart of quantum physics:

$$\text{probability amplitude of a given path} = e^{\frac{iS(\text{path})}{\hbar}}$$

Quantum physics is sort of like classical physics wrapped in a circle

So, whereas in classical physics, the action for each path is a real number, a point on an infinitely long line, in quantum physics, the probability amplitude for each path is a complex number, represented by a point on a circle of radius 1, or equivalently, by an arrow pointing from the center of the circle to that point.[17]

Executive summary: In classical physics, the action lives on a line. In quantum physics, the probability amplitude lives on a circle.

For those who prefer the Babylonian degree to the radian, the angle between the arrow and the x-axis is given by $\frac{360°}{2\pi} \times \frac{S(\text{path})}{\hbar} = 360° \times \frac{S(\text{path})}{h}$, with $h = 2\pi\hbar$. Just shuffling a factor of 2π around, utter triviality. To get at the profound truths, theoretical physicists are compelled to get rid of inessential and historical "nonsense." If you insist on following the Babylonians, you are merely making life complicated for yourself. By the way, I am astonished by how many beginning students get flustered by having to use radians instead of degrees.

To make sure everybody is following, let me give a few specific examples: for S(path) equals $\frac{1}{4}h$, the angle is $90°$; for S(path) equals $\frac{1}{3}h$, $120°$; for S(path) equals $\frac{3}{2}h$, $90°$; for S(path) equals $\frac{902}{3}h$, $240°$, and so on. Get that last one? It's important. Well, 902 divided by 3 gives 300 plus $\frac{2}{3}$. The 300 tells the arrow to spin around the circle 300 times, which effectively means "to do nothing." What matters is the fractional remainder $\frac{2}{3}$, which tells the arrow to spin through $\frac{2}{3} \times 360° = 240°$.)

Quantum versus classical physics at a glance: circle versus line

Whoa! That was a lot to absorb. Time for an executive summary.

	associated with each path		
classical physics	action	S(path)	a point on a line
quantum physics	probability amplitude	$e^{\frac{iS(\text{path})}{\hbar}}$	a point on a circle

In classical physics, each path is assigned an action, S(path), and the extremal path, namely, the one that extremizes the action, wins (that is, is chosen by Nature).

In quantum physics, each path is assigned a probability amplitude, $e^{\frac{iS(\text{path})}{\hbar}}$, and we are instructed to sum up all these probability amplitudes to determine the probability amplitude to get from the starting point to the ending point. Trumpets please! Watch quantum physics emerge:

$$\text{probability amplitude to get from here to there} = \sum_{\text{path}} e^{\frac{iS(\text{path})}{\hbar}}$$

(Here $\sum$ is the mathematical symbol for sum, and the entities the sum runs over are indicated beneath the symbol, paths in this case. Incidentally, the Greek letter capital sigma $\sum$ is supposed to represent a capital S for sum, while the symbol for integral $\int$ is allegedly a distorted capital S.)

I should emphasize that all of quantum physics is in principle contained in this statement, but to extract any specific result[18] may not be all that easy in practice. As was mentioned earlier, the three formulations, Schrödinger,

Heisenberg, and Dirac-Feynman, are equivalent. (How could they not be?) To derive the path integral as just stated starting from, say, the Heisenberg formulation is not that difficult, and is done in several textbooks.[19]

The magic and mystery of quantum physics is that paths with vastly different actions in classical physics may all have the same probability amplitude. As you know well, and as I already said, 360° is the same as 0°. Thus, for example, the paths with action equal to $-2h$, $3h$, $17h$, and 1 billion h for that matter, all have the same probability amplitude, characterized by an arrow pointing "east." We see that h really deserves its name "the quantum of action." It is literally the unit Mother Nature uses to measure action by, as I have already stressed.

It may seem paradoxical that different paths with vastly different actions in classical physics could have the same quantum probability amplitude, but that is how it is. Mysterious, eh?

The emergence of the classical world from the quantum world

We have seen how we could jump from classical physics to quantum physics, thanks in part to Euler and Lagrange and their action principle. But now an important question pops up. How does the classical world emerge from the underlying quantum world?

We inhabit a classical world with massive lumbering lumps like ourselves. Each time you turn a page in this book, or blink, or breathe a breath, not to mention any number of more strenuous activities, the action of your action is absolutely humongous compared to $\hbar$. Indeed, physicists discovered $\hbar$ only after they started exploring the microscopic world of atoms. In human made units, $\hbar$ is about 10^{-27} gram centimeter squared per second. Now imagine the typical mass, distance, and time in grams, centimeters, and seconds involved in each of your colossal actions. If you move a one-gram mass with a speed of one centimeter per second through a distance of one centimeter, your action amounts to a whopping $10^{27}\hbar$, or if you insist, 1,000,000,000,000,000,000,000,000,000 $\hbar$.

Picking out the actual path out of all the possibilities

Every path is surrounded by lots of neighboring paths, namely, paths that differ from it slightly. For a given path, call its neighboring paths simply its neighbors.

How does Mother Nature pick out the extremal path?

She has a very clever trick up her sleeve.

An extremal path is surrounded by paths whose actions differ from the extremal actions infinitesimally. In contrast, those paths that are not favored by the extremal action principle are surrounded by paths whose actions differ substantially.

As I have already belabored, the action S(path) for any action performed in the classical world is enormous compared to $\hbar$: S(path) typically equals a gazillion $\hbar$. Alternatively, theoretical physicists often think of $\hbar$ as a dial we can adjust. As we turn $\hbar$ down to zero, we should see the quantum world transition into our good old classical world. A hypothetical universe with $\hbar = 0$ is a classical universe. Either way, something should happen when the ratio $\frac{S(\text{path})}{\hbar}$ becomes huge.

Remember that to obtain the probability amplitude for the particle to get from here to there, we are supposed to add up the probability amplitudes of all the paths the particle could take. Perhaps you can start to see how Mother Nature picks out the extremal path actually realized in classical physics.

Little arrows adding up to a gigantic arrow

Consider a path that is not extremal. Its neighbors have their little arrows pointing every which way, almost completely randomly. The angle in degrees the little arrows make with the vertical is huge compared to 360, something of order 10^{27} or more. Remember, we don't care if you have gone around the circle 10^{27} times; it's the little bit that's left over that counts. The arrows are effectively pointing in every possible direction. All these arrows end up canceling each other. The total amplitude is zero. See figure 4(a).

In contrast, an extremal path is surrounded by neighbors whose little arrows are pretty much all pointing the same direction* as that of the extremal path. The little arrows add up to a gigantic arrow. See figure 4(b).

You see that whether the action is maximized or minimized for our extremal path does not matter, only that it and its neighbors are all pointing in the same direction. I consider this to be one of the great triumphs of quantum physics, the explanation of what would otherwise pose a puzzle in classical physics (namely, the puzzle of why the action is extremized, rather than minimized or maximized). Just go ask a classical physicist why the action is sometimes minimized and sometimes maximized; she would be stumped.

A little parable might help. Imagine that you were called in as an accountant to go over the annual account books of a small ma and pa company that closed after decades of operation. You flip through the several dozens of

---

*Incidentally, this is known as "stationary phase" to the mathematically knowledgeable, an approximation used in many areas of physics and engineering.

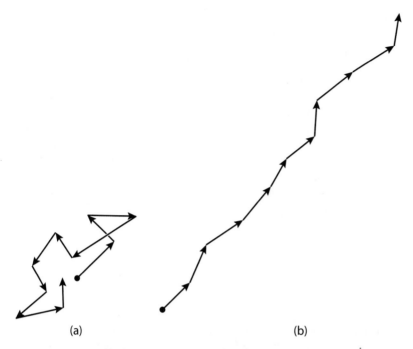

(a)   (b)

Figure 4. (a) Lots of little arrows pointing each and every way when summed up tend to cancel each other. (b) Lots of little arrows all pointing in the same direction add up to a big arrow. Keep in mind that in the path integral, the number of little arrows is effectively infinite.

books, each with thousands of entries with plus and minus signs all over the place, with plus for sales and minus for expenses. You were given only a couple of days. There is no way you could add up all these numbers. But after flipping through these books, you notice that each book contains almost an equal number of plus and minus signs, except for one book containing mostly pluses. Being a clever fellow, you realize that you only have to add up the entries in that one book, expecting all the plus and minus entries in the other books to cancel out.

That is more or less how Nature produces the world of the large and slow. Pick out the extremal path, and forget about the rest.

Recipe for quantum physics

There you have it, the fundamental postulate of quantum physics! A precise and accurate recipe for going from classical to quantum physics: Take the classical action S(path) associated with each path going from here to there, divide it by Planck's constant $\hbar$, multiply it by the imaginary unit i, and put all that in the exponential. Sum over all possible paths. This then gives the probability amplitude to get from here to there.

This recipe applies to both particles and fields. Trumpet blast, please. We are ready for quantum field theory.

It is perhaps worthwhile emphasizing what may be obvious to many readers, but is apparently still not clear to all who haunt the popular media. Semantically, the word "explain" is rather troublesome, and no doubt philosophers could obtain tenure by writing treatises about it. To me, "to explain" means "to relate the system under discussion to something you already know." You could explain to me how a refrigerator works: a gas is made to expand, et cetera. But nobody can explain quantum physics in the sense of relating it to something you already know, that is, classical physics. Physics can only tell you how the quantum world "works" by stating how the probability amplitude of any given process can be calculated, and that I have done here.

Notes

[1] This can be readily proved. A standard undergraduate level exercise shows that extremizing the action produces Newton's equation of motion.

[2] Some pedants even say nine, but to that I say phooey.

[3] P. A. M. Dirac, "The Lagrangian in quantum mechanics," *Physikalische Zeitschrift der Sowjetunion,* Band 3, Heft 1, 1933.

[4] Physicists sometimes wonder whether Feynman invented this formulation completely ignorant of Dirac's work. Historians of physics have now established that the answer is no. During a party at a Princeton tavern, a visiting physicist named Herbert Jehle told Feynman about Dirac's idea. According to the legend, the next day Feynman worked out the formulation in real time in front of the awed Jehle. See the 1986 article by S. Schweber in *Reviews of Modern Physics.* (Taken from page xv in my preface to Feynman's book, *QED: The Strange Theory of Light and Matter,* Princeton University Press, 2014.)

[5] The standard textbook for the path integral is Feynman and Hibbs, *Quantum Mechanics and Path Integrals.* I have never heard the legions of Feynman idolaters mention this book. To me, paraphrasing Einstein, the one comprehensible thing about the world of humans is that it is incomprehensible.

[6] See Feynman and Hibbs, page ix. I would like to think that it was merely Hibbs's opinion. Hibbs took notes during a course Feynman gave around 1950, and some 15 years elapsed before the notes were turned into a book. Allegedly, Feynman on various occasions lost interest in the project. Some physicists, paraphrasing what is sometimes said about the famous textbooks by Landau and Lifshitz, describe the book as containing "not a thought of Hibbs, and not a word of Feynman."

[7] More accurately, the Hamilton-Jacobi approach, which grew out of the Hamiltonian.

[8] Recall that I mentioned this in the prologue.

[9] The general public adopted the opposite convention, because sundials and clockmakers were mostly located in the northern hemisphere.

[10] For those readers who know about such things, e is Euler's number, and $i = \sqrt{-1}$ is the imaginary unit.

[11] Incidentally, when written in the form $e^{i\pi} + 1 = 0$, this identity, incorporating seven of the most important symbols used in mathematics, is often hailed as the most beautiful.

[12] Decades ago, I attended a sailing school in a Germanic country, and before we got anywhere near a boat, we were given a long winded and serious lecture about the addition of vectors, almost as in a parody of the local culture.

The instructor, sporting a haughty tone indicating that he didn't expect any of us vacationers to understand a word he said, drew some vectors on the blackboard and picked someone to come up and add them, fully expecting to humiliate the poor guy. For some reason, he picked me. When I did it quickly and correctly, the class burst into applause. No, I did not reveal that I was a theoretical physicist.

[13] You might wonder how the integral over continuous paths could be defined. The answer is that you do exactly what Newton and Leibniz would have done: You approximate the continuous paths by infinitesimal straight line segments, sum, and take the limit. See Feynman and Hibbs, figure 2–3, page 33.

[14] I don't. See *FbN*.

[15] Incidentally, if you guessed the length of the path in spacetime, you're not wrong. That is exactly the action assigned to a path in Einstein's theory of special relativity. But do notice that what I am describing here also applies to the nonrelativistic world.

[16] Dirac was inspired by some suggestive resemblance of the classical action to certain quantum quantities.

[17] Feynman likened the arrow to a hand on a clock (to be precise, a clock with only one hand). I heard that this analogy, unfortunately, led some readers to think that particles carry a little watch with them! One of my colleagues just told me, this very minute as I wrote this, that he for one was hopelessly confused.

[18] Some readers may find it instructive to watch Feynman derive the de Broglie relations for a point particle from its classical action using this path integral formulation. See Feynman and Hibbs, page 45. Or to derive the Schrödinger equation.

[19] For instance, the derivation given in *QFT Nut* (pages 10–12) takes up fewer than three pages!

Recap of part II

Starting from the everyday observation that light travels in a straight line, theoretical physicists have zoomed to an understanding of the quantum world as a sum of all possibilities.

First, we zip through space in the least amount of time. Then we move through spacetime extremizing our action. These two foundational principles, which superficially seem so different, later turned out to be one and the same. That they are the same amounts to special relativity.

Afterward, we watch the action ushering in the almighty quantum. The action in units of $\hbar$ determines the probability amplitude. Summing the probability amplitudes the way Dirac and Feynman showed us, we obtain quantum physics.

Becoming a quantum field theorist

Preview of part III

You are now ready to do quantum field theory! I outline for you the four steps you need to go through to evaluate the path integral for a field theory.

Your first discovery is that quantum fields can produce particles. For instance, the electromagnetic field when quantized can emit and absorb photons.

Your second discovery is that the quantum exchange of particles generates a force. This amazing secret was finally revealed to physicists after all these centuries. They finally understood the origin of the four fundamental forces that weave the universe together. You will also learn why two of these forces are long ranged and hence have been known to us since the moment of birth, and why the other two are short ranged and hence hidden from physicists for the longest time.

Your third discovery is that whether a force is attractive or repulsive can be explained simply by quantum field theory: This mysterious binary choice originates in the spin of the particle being exchanged.

The interplay between attraction and repulsion, between the long and the short, produces the universe as we know it.

How to become a quantum field theorist (almost) instantly

Four steps to doing quantum field theory

Now that you know how to set up the quantum mechanics of point particles in the path integral formulation, you are almost ready to become a quantum field theorist. Literally in an instant. Okay, that may be a bit of an exaggeration but not by much. All you have to do is to replace the point particle by a field.

Ironically, the Schrödinger formulation, although the easiest to teach and learn in an introductory quantum mechanics course, proves to be the most awkward and unnatural for quantum field theory.[1] In contrast, in the path integral formulation, you merely have to carry out four steps:

(1) Identify the relevant fields.
(2) Find the action governing these fields.
(3) Disturb the fields.
(4) Evaluate the path integral.

Interestingly, sometimes the steps are all difficult, sometimes they are all easy, and sometimes some steps are easier than the others. To explain, it may be best to actually perform these steps in a few simple cases.

A universe with only the electromagnetic field: as simple as possible

Consider a universe with only the electromagnetic field. We are not even including the electron, just the electromagnetic field by itself. Incidentally, in case you are wondering, this is how theorists working on fundamental physics approach problems. Keep things as simple as possible, but not any simpler!

In some sense, this is the simplest quantum field theory we could deal with. Thanks to Maxwell, we know the action for the electromagnetic field. So, just like that, we have already performed the first two steps! It's that easy. (Of course, Maxwell did most of the work for us.)

The mattress model

The third step in the list above, however, requires a bit of digression. Go back to the field $\varphi(t, \vec{x})$ describing an elastic membrane in chapter I.2. Recall that $\varphi(t, \vec{x})$ denotes the deviation of the membrane from its equilibrium position.

What is the action governing φ? Another professional trick: When theoretical physicists don't know what the action is, they often construct a model for which they know the action. In this case, a mattress! Just the classical mechanics of masses and springs, easy peasy.

The point is that physicists know how to write down the action of a square lattice of point masses connected by springs. You do, too! Simply recall the action for Humpty Dumpty: its action is equal to its kinetic energy minus its potential energy, integrated over time. So for the mattress, we add up the kinetic energy of the point masses and then subtract the potential energy stored in the springs.

We take the springs to be harmonic, a term borrowed from music, as you might have guessed. The force exerted by a harmonic spring is simply proportional to the amount it is stretched. Consequently, its vibration traces out a perfect sine curve of the type shown in figure I.2.3. In contrast, the force exerted by an anharmonic spring has a complicated dependence on the amount it is stretched.

The entirely reasonable attitude of theoretical physicists is that, when the spacing between the mass points in the mattress is much smaller than the length scale over which we are observing, we could not tell the difference anyway between a mattress and an elastic membrane (figure 1). We now know of course that an elastic membrane consists of a complicated mess of atoms interacting with each other, the details of which some material scientists could tell us about. But we don't care about such microscopic details—we are not into manufacturing and selling better membranes. We want to obtain an action of an elastic membrane, and in fact, the action S obtained in this way (pretending that the elastic membrane is actually a mattress with point masses tied very close together with harmonic springs) describes elastic membranes quite well.[2]

The field theory abstracted from a mattress whose springs are perfectly harmonic is known as a free field theory, and because of its enormous simplicity is usually presented at the beginning in quantum field theory textbooks.

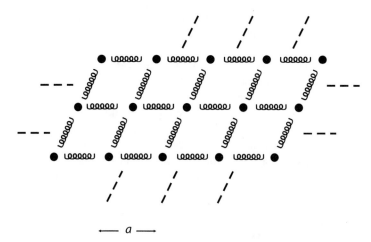

Figure 1. Masses connected by springs separated by distance a from each other. In the limit with a much smaller than the length scale of the phenomena we are interested in, this should be a good model for the elastic membrane.
Modified from A. Zee, *Quantum Field Theory in a Nutshell*, Princeton University Press, 2010.

Disturbing the field

I hope that you have not lost sight of why we detoured through the mattress. We had wanted to perform step (3) in the list near the start of this chapter and disturb the electromagnetic fields.

To do that, let us first learn how to disturb the φ field describing the elastic membrane. Well, we simply push and pull on the masses in the mattress.

The idea is simplicity itself: We want to disturb the membrane to find out how it responds and hence to learn something about it. Examples abound in everyday life: People in a mattress store push down on a mattress to see how it bounces back. Or, kick the tires on a used car before buying.

You might recall that the work done by a force pushing an object through a certain distance is given simply by the force times the distance. In any case, even if you don't recall, the statement that work equals force times distance sounds totally reasonable and may well be your first guess. Let us denote the external force acting at the position $\vec{x}$ at time t by* $J(t,\vec{x})$. (Note: external force, not the force exerted by the spring; we are the external agents pushing

---

*Some readers might be wondering why force is not represented by a vector here. The reason is simply that in this model, we are envisaging pushing down and pulling up on the mattress. (In real life, it would be difficult to pull up on a mattress, but never mind, this is just a theoretical model.) So, the magnitude of $J(t,\vec{x})$ corresponds to the strength of the force, and its sign distinguishes between up or down.

and pulling on the masses!) The amount by which the membrane is stretched from its equilibrium position is, by definition, $\varphi(t,\vec{x})$. Thus, the work done at that position is simply the product $J(t,\vec{x})\varphi(t,\vec{x})$, work that is stored as potential energy, which we have to sum, or integrate, over space $\int d^3x\, J(t,\vec{x})\varphi(t,\vec{x})$ to obtain the total potential energy.

Euler and Lagrange taught us to integrate this potential energy over time, namely, $\int dt \int d^3x\, J(t,\vec{x})\varphi(t,\vec{x}) = \int d^4x\, J(x)\varphi(x)$, and add it to the action S we obtained earlier. (Again, we use $x = (t,\vec{x})$ as shorthand for location in time and space, that is, in spacetime.) To repeat and to emphasize, $J(x)$ represents the external agents responsible for disturbing the membrane, namely, us. This extra term $\int d^4x\, J(x)\varphi(x)$ in the action has nothing to do with the springs, the masses, and so forth, in our silly microscopic model of the membrane.

It's a free country

The crucial point is that we are free to choose whatever function $J(t,\vec{x})$ we like. For example, by tapping the membrane around some point P with a frequency ω, we could send out a wave with frequency ω. Simply choose $J(t,\vec{x})$ to vanish for all $\vec{x}$ except for a small region around P and to vary in time with frequency ω. With more elaborate tapping, we could effectively construct any combinations of waves and thus form a wave packet as described in chapter I.2, and send it out propagating in a direction of our choice. In other words, by bouncing up and down on the mattress appropriately, we could set up waves and wave packets propagating in the mattress, or membrane, as depicted in figure I.2.7. The phrase "bouncing up and down" is represented by the function of space and time $J(x) = J(t,\vec{x})$, known as an external source, or "source" for short.

These wave packets running around reminded physicists of particles, as was mentioned in chapter I.2. We could imagine the source $J(t,\vec{x})$ as an abstract representation of an accelerator spewing out particles.

By reversing this process of emitting particles, we could also construct an appropriate $J(t,\vec{x})$ that would absorb these particles, known as a sink. Somewhat picturesquely, we speak of particles disappearing down a sink. In "real life," a sink would correspond to a detector located some distance away from the source or accelerator.

Instead of talking about sources and sinks, we use, for convenience, a single word "source" generically for both. By the way, I am describing the terminology invented by Julian Schwinger, who jokingly referred to this setup as sorcery.*

---

*So that his students at Harvard could inform their parents what they were learning. I personally think that Schwinger's sense of humor was more subtle than Feynman's, but chacun à son gout. Schwinger and Feynman are universally regarded as two of the greatest masters of quantum field theory ever. Much more about both of them in chapter IV.3.

Surely, most quantum field theorists are not into building accelerators, with humongous magnets, vacuum pumps, and whatnots, or into designing detectors. All this Schwingerian talk is an abstraction. The bottom line is that we should add a term $\int dt\, d^3x\, \varphi(t,\vec{x})J(t,\vec{x})$ to the action if we want to study how particles interact with each other in quantum field theory.

We spoke about mattresses and membranes, which span a 2-dimensional space, solely for the sake of simplicity. Now generalize to 3-dimensional space with less than the stroke of a pen. With merely a declaration! Just say that in $\varphi(x) = \varphi(t,\vec{x})$ and $J(x) = J(t,\vec{x})$, $\vec{x}$ is actually equal to (x,y,z).

Tickling the electromagnetic field

Now that we know how to disturb, or tickle if you prefer a more picturesque term, the scalar field φ, it's time to return to disturbing the electromagnetic field. Easy peasy! Simply add to Maxwell's action a term analogous to $J(x)\varphi(x)$ integrated over spacetime.

Unlike the hypothetical scalar field φ, the electromagnetic field is something we live with every day. How would you tickle the electromagnetic field? Exactly: You assemble a transmitter and drive current through it! The source. How do you construct a sink, down which a pulse of electromagnetic wave could disappear? Build a receiver with a fancy antenna!

Too many histories to sum over

Now we are ready for step (4) on our list. Evaluate the path integral.

Alas, easier said than done. Even in quantum mechanics, to obtain the probability amplitude to get from here to there, we have to evaluate a sum over all the paths a particle could follow, namely,

$$\text{probability amplitude to get from here to there} = \sum_{\text{path}} e^{\frac{iS(\text{path})}{\hbar}}$$

as explained in chapter II.3. But even Feynman could do this sum exactly for only a few cases.[3] (How few? Less than the fingers on one of your hands.) For quantum field theory, we have to sum over all the paths a field could follow, and it becomes even harder.

Let us be a bit more precise about what this means. Consider the scalar field $\varphi(t,\vec{x})$. Instead of $S(\text{path})$, we now use the the action of φ, namely, the action that we had just figured out by going to the mattress. When we advance to quantum field theory, path is naturally generalized to history. The starting position of the particle, "here," corresponds to the field at some initial time t_i, namely, $\varphi(t_i,\vec{x})$. The ending position of the particle, "there," corresponds to the field at some final time t_f, namely $\varphi(t_f,\vec{x})$. All histories, each corresponding to a possible $\varphi(t,\vec{x})$ with t denoting time in between t_i and t_f, are to be summed over. Hence, in quantum field theory, the path integral is sometimes called the "sum over histories."

So now we have

$$\text{probability amplitude to evolve from } \varphi(t_i, \vec{x}) \text{ to } \varphi(t_f, \vec{x}) = \sum_{\text{history}} e^{\frac{iS(\text{history})}{\hbar}}$$

As you can imagine, keeping track of an entire history of how a field evolves is orders of magnitude more difficult than keeping track of a path of a point particle. There are even fewer cases that can be evaluated, but happily, the two cases to be discussed in this chapter are among those, and that is of course why I discuss them.[4]

Theorist's freedom of choice

Our complete freedom in choosing the source function $J(t, \vec{x})$ is the reason we are able to squeeze out some important physics, in fact, I could say without exaggeration, some fundamental secrets of the universe.

For readers who know a bit of calculus (relax, I am not going to ask you to do an integral), or knew at some point but have now forgotten, the freedom I speak of is analogous to the following. Suppose you are given a function $f(s, u)$, just an ordinary function of two real variables, that is, with s and u two real numbers.* Suppose you are able to integrate $f(s, u)$ over s for any u. Once you obtain the answer, you are free to choose for u any value you want, for example, 7.69, π, $-\sqrt{3}$, any real number, and watch how the integral depends on u.

Same here. Since $S(\text{history})$ depends on both φ and J, after integrating over φ, we obtain a functional of $J(t, \vec{x})$. If we were able to integrate over all possible histories of φ for any J, we could then play around by choosing different $J(t, \vec{x})$ and watch how the quantum field responds. The key point is that our freedom is so much greater! Think of a real number (such as u in the analogy above) as a point on a line stretching from minus infinity to plus infinity. But in quantum field theory, we could choose $\vec{x}$ to be anywhere in spacetime! We could arrange for $J(t, \vec{x})$ to do something now in a region around here, and to do something later there. See figure 2. The phrases "here now" and "there later" describe different regions in spacetime.

The scalar field produces a particle!

So, let's play. Send out a wave packet to be detected somewhere else. Mathematically speaking, set $J(t, \vec{x}) = J_1(t, \vec{x}) + J_2(t, \vec{x})$, with $J_1(t, \vec{x})$ nonzero only in

---

*The choice of the two letters has absolutely no meaning; just whatever my fingers hit on the keyboard.

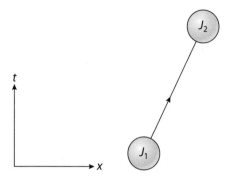

Figure 2. The source J_1 produces a wave packet subsequently absorbed by the sink J_2. Note that J_1 and J_2 are localized both in space and time.

region 1 in spacetime and $J_2(t, \vec{x})$ nonzero only in region 2 in spacetime. We see that $J_1(t, \vec{x})$ corresponds to an emitter and $J_2(t, \vec{x})$ corresponds to a detector. For a given wave vector $\vec{k}$, the probability P that the detector receives a signal exhibits a bump at a particular frequency ω, which depends on $\vec{k}$: it resonates. Again, physicists borrow from music. A detector tuned to a particular frequency "sounds again," exactly what you do with your radio.[5] Let's say it again. Normally, $P \simeq 0$ stays small, that is, the detector sits silent. As we tune it to some ω, which depends on $\vec{k}$, it suddenly becomes large, that is, it beeps, and everybody gets excited.

When we convert wave number $\vec{k}$ to momentum $\vec{p}$, and frequency ω to energy E according to de Broglie's recipe,[6] namely, momentum $\vec{p} = \hbar \vec{k}$ and energy $E = \hbar \omega$, we find that the resonance occurs at

$$E^2 = \vec{p}^2 + m^2$$

in units with c set to 1. But this is precisely what Einstein told us about how the energy E and momentum $\vec{p}$ of a particle with mass m are related, as was derived in chapter I.3.

In particular, for $\vec{p} = \vec{0}$, we have $E = m$, and with c put back in, this is just the familiar[7] $E = mc^2$, known even to people who do not know what c is!

Playing around with the field φ, we have produced and detected a massive particle, with mass m! In fact, this almost describes the way new particles are actually discovered in accelerators. The experimentalists detect a signal, and they measure the E and $\vec{p}$ of the hitherto unknown particle and use Einstein's relation given here to determine its mass m.

But what is m? Quantum field theory cannot tell you; m is just something we put into the action.[8] But of course, how could anyone expect us to calculate the mass of a hypothetical particle?

I must emphasize that this is an extremely important result for theoretical physics, that a field can produce a particle. Hence the exclamation point in the title of this section.

This resolves the long-standing tension between particles and fields. Theoretical physicists now understand that the particles we observe are all excitations in various quantum fields. In particular, as we will see presently, the photon is just an excitation in the electromagnetic field.[9]

The diagram in figure 2 gave us a foretaste of the Feynman diagrams to come. As you'll see, that is indeed the case. We have shown both the source J_1 and the sink J_2 to occupy a region in spacetime. The wave packets we are studying would then be spread out in space and in time. According to the discussion about Fourier transform in chapter I.2, they would have a spread in wave number $\vec{k}$ and in frequency ω, and hence in quantum field theory, a spread in momentum $\vec{p}$ and energy E. The Fourier transform will also be reflected in Feynman diagrams, as we will see in chapter IV.1.

Enter the photon

Emboldened, we now tackle the electromagnetic field. The complications I alluded to made the sum over histories somewhat[10] more intricate than what we just did for the scalar field φ, but still doable.

As mentioned earlier, the source now represents transmitters and receivers, like those built into your cell phone, for instance. We produce a packet of electromagnetic wave.* With the receiver set to the appropriate frequency and wavelength, we detect a particle, the photon! Again, applying de Broglie's recipe, we obtain the energy and momentum of the photon and deduce that it is massless.

Historically, that the photon is massless was anticipated by Planck and Einstein, and implied by Maxwell and de Broglie. (Maxwell knew the relationship between frequency and wave number of the electromagnetic wave he discovered, of course.)

By the way, the discussion here is also given in my textbook on quantum field theory QFT Nut, but there it is dressed in mathematical language. So, you are almost ready for it! Well, not quite. I am letting my enthusiasm carry me away. But in any case, I closed QFT Nut with an inspirational quote used by Feynman: "What one fool can do, another can." What better assurance could you have that, if you apply yourself, you could learn quantum field theory?[11]

---

*I am intentionally glossing over the source function for the electromagnetic field here, but will come back to it in chapter III.3.

Notes

[1] The Schrödinger formulation of quantum mechanics traffics in the wave function $\psi(\vec{x})$, where $\vec{x}$ is the position of the particle. With particle promoted to field, we have to deal with something like $\psi(\varphi(\vec{x}))$ (more properly written as $\psi(\varphi(\cdot))$ in more careful treatments), namely, a functional, defined as a function of a function, as was already mentioned in chapter II.1.

[2] Alternatively, we could guess the action by invoking various symmetries. Understand that we are not talking about real membranes and real mattresses!

[3] These are listed in Feynman and Hibbs, *Quantum Mechanics and Path Integrals*.

[4] For those readers who are able to evaluate integrals, I mention that both cases essentially involve a suitable generalization of the Gaussian integral in which the integrand is an exponential of a quadratic term plus a linear term, specifically, the integral $\int_{-\infty}^{\infty} d\varphi \, e^{-\varphi^2 + J\varphi}$, with φ and J real numbers. See *QFT Nut*, pages 523–524. In only two pages, you could learn how to do many integrals of this type.

[5] I am being a bit old-fashioned here.

[6] By the way, a particularly illuminating application of the path integral is the derivation of the frequency ω and wave number $\vec{k}$ of the de Broglie wave representing a particle with energy E and momentum $\vec{p}$. See Feynman and Hibbs, page 45.

[7] I'll let you in on a little secret! In my world, anybody who writes $E = mc^2$ would be laughed out of the room. In fact, the pros use an even more compact notation, as was mentioned in chapter I.4, introducing a 4-vector $p^\mu = (E, \vec{p})$ and defining $p^2 \equiv E^2 - \vec{p}^2$, and hence they write simply $p^2 = m^2$. Note that the arrow disappears.

[8] And in the mattress model, m depends on how stiff the springs are and things like that.

[9] We will also see in chapter IV.1 that the electron we know and love is an excitation in the electron field.

[10] An understatement! The woe and grief some graduate students go through are almost tragic to behold. They have to learn something called "gauge fixing." See chapter IV.4.

[11] I doubt that I am capable of inventing quantum field theory, as the greats did in the 1930s, but I can certainly learn it, and thus, as Feynman said, so can you.

Origin of forces: range and exchange

Fields beget particles and particles beget forces

From day one of physics, theoretical physicists have been talking about forces (as was mentioned in chapter I.1.) But the notion of force has also been among the most mysterious. Where do forces come from? Why is the universe filled with four of them? Thus, it was with considerable satisfaction that physicists finally understood that forces originate from the quantum exchange of particles. This chapter is devoted to what this concept means.

I pictured the atomic nucleus, back in chapter I.1, in terms of a contest between the strong and the electromagnetic interactions, the strong attraction between the nucleons (namely, the protons and neutrons) gluing them together versus the electric repulsion between the protons pushing them apart. A sporting contest of strength versus range: The strong interaction, while much stronger than the electromagnetic interaction, turns on only when the nucleons are literally on top of each other, while the electric repulsion is long ranged. I like to picture the nucleus as a boxing arena. One boxer, brawny and stocky, goes up against another, much weaker but with long arms (figure 1).

For those readers who must relate physics to human scales, the range of the strong interaction is about 10^{-13} cm. No wonder that, the moment we were born, we immediately knew about gravity and electromagnetism, the two long range interactions, while physicists had no inkling about the strong and weak interactions, the two short range interactions, until they started exploring the nucleus. The range of the weak interaction is shorter still, about a thousand times less than the range of the strong interaction.

Figure 1. A peculiar boxing match staged constantly in the universe.
Reproduced from A. Zee, *Fearful Symmetry: The Search for Beauty in Modern Physics*,
Princeton University Press, 1986.

A traveler in a strange land

> I desire … to be a traveler in a strange land [referring to
> physics]…[after publishing his seminal 1934 paper] I felt like a
> traveler who rests himself at a small tea shop at the top of a moun-
> tain slope. At that time I was not thinking about whether there were
> any more mountains ahead.
> —H. Yukawa, *The Traveler*

Let me start with a bit of history. The Japanese physicist Hideki Yukawa
(figure 2) proposed in 1934 that the quantum exchange of a hitherto unknown
particle, a meson, between nucleons could lead to the strong interaction. This
novel idea, that force is associated with a particle, was celebrated with a Nobel
prize in 1949, the first ever for a Japanese. In his autobiography, Yukawa
describes the "long days of suffering" he endured from 1932 to 1934 searching for a theory of the nuclear force. To calm himself, he tried sleeping in

Figure 2. Hideki Yukawa with family in 1949.
Photographer Unknown (Wikimedia Commons).

a different room every night. The crucial point came to him in a flash one night.

Incidentally, the name "meson" (from a Greek word meaning "middle," as in "mezzosoprano" and "Mesopotamia") has a convoluted history.[1] The experimentalists Carl Anderson and Seth Neddermeyer assigned the name "mesoton" to a particle that they had discovered, but Robert Millikan suggested changing it to "mesotron" to be consistent with the terms electron and neutron, but as Anderson remarked, not proton. The awful mesotron was later shortened to "meson" at the suggestion of the Indian physicist Homi Bhabha. According to George Gamow, some French physicists protested, fearing confusion with their word for house. Meson has the same sound as the Chinese and Japanese word for hallucination or illusion, which turned out to be appropriate because in the 1930s, Japanese physicists met regularly to discuss nuclear physics in what were known as illusion meetings. Later, the particle discovered by Anderson and Neddermeyer turned out not to have the properties predicted by Yukawa. To distinguish between them, the impostor was called the mu meson, and Yukawa's particle, the pi meson, since the Greek letter π looks like the Chinese character for mediator. Some time later, it was realized that the mu meson is not a meson at all; it is a cousin of the electron, and its name was shortened to muon. Similarly, the pi meson was shortened to pion, the name it proudly carries these days. Originally, Yukawa in his paper referred to the pion as the "U-particle."

Figure 3. A marriage broker bringing two families together.
Reproduced from A. Zee, *Fearful Symmetry: The Search for Beauty in Modern Physics*,
Princeton University Press, 1986.

Quantum exchange responsible for forces

To explain Yukawa's idea, I will use an analogy. Marriage brokers were common in older civilizations. We might picture a corpulent lady tirelessly traveling back and forth between two families, arranging for a union.[2] Effectively, the marriage broker produces an attractive force bringing the families together (figure 3).

One feature of this analogy is actually in accord with quantum physics. The distance over which the corpulent marriage broker could travel comfortably is limited by her weight. She can't bring together two families living three villages apart. Similarly, the distance over which a quantum particle could travel is limited by its mass, according to the uncertainty principle.[3] More on this below.

Force emerging out of quantum field theory

May I say the obvious? Professionals do not need analogies; they simply evaluate the path integral and see what they get. Fine, so let's get serious and actually calculate.

But wait, we have already done the calculation! At least we imagine that we did. We evaluated, in chapter III.1, the path integral for a harmonic scalar field $\varphi(x)$. Indeed, we took the answer, plugged in the source function $J(x)$ we

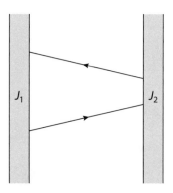

Figure 4. Two lumps, represented by the shaded regions. The lump J_1 emits a meson, an excitation in the φ field, indicated by the straight line, and absorbed by the lump J_2. Later, J_2 emits another meson, absorbed by J_1. This process of emission and absorption keeps on repeating itself, thus generating a force between the two lumps.

wanted, representing a source here and now, and a detector there and later. Then we witnessed $\varphi(x)$ producing a particle of mass m in front of our very eyes.[4]

All we have to do now is to plug in some other source function $J(x)$. This time, we want $J(t, \vec{x}) = J_1(\vec{x}) + J_2(\vec{x})$. Notice that $J_1(\vec{x})$ and $J_2(\vec{x})$ do not depend on time: They each represents a lump just sitting there, not only not moving, but also not vibrating or anything.

But due to the coupling $J(x)\varphi(x)$, the scalar field φ generates an attraction between the two lumps.

The two lumps, represented by $J_1(\vec{x})$ and $J_2(\vec{x})$, are not intrinsic to the quantum field theory. Rather, we put them in by hand, so to speak, and so they are called "external." As I already said, that they do not depend on time means that they do not move; they have no dynamics of their own. The general conception here is close to what Faraday had in mind when he introduced the electric field. An electric charge produces an electric field, which in turn acts on another charge. Charges do not act on each other directly, but through the electric field. Here the two lumps affect each other not directly, but through the scalar field φ.

Compare and contrast the two diagrams, figure III.1.2 in the preceding chapter, telling us about the particle produced by the field, and in figure 4 here, telling us about attraction between the two lumps.

Look at the meson lines in the two diagrams. In figure III.1.2, it is oriented in the time direction. In figure 4, they are oriented in some space direction. How a single path integral could inform us about the physics of two different situations depending on what we put in for the source $J(t, \vec{x})$ is of course a consequence of Einstein's unification of space and time into spacetime. In one case, the field is propagating in time, in the other, in space. More about this later.

At the risk of repeating, let me emphasize that we only need to evaluate the path integral once, but by choosing different source functions, corresponding to setting up different experiments or measurements, we could extract different physics generated by the scalar field φ.

Energy between two lumps

I emphasized to you in chapter III.1 that evaluating the path integral gives us the probability amplitude to go from an initial configuration to a final configuration. In quantum physics, the probability amplitude oscillates in time[5] with a frequency ω proportional to the energy E. (Remember that Prince de Broglie told us that $\omega = E/\hbar$.) Thus, the result of this calculation (which, as I emphasized, we already did in chapter III.1) gives us the energy E caused by the presence of the two lumps.

Our complete freedom in choosing the source function $J(t, \vec{x})$ is the secret to why we are able to squeeze out an important result, yielding up, I could say without exaggeration, some fundamental secrets of the universe. These will be revealed to you shortly. Yes, you! You who had paid for admission into the secret society by reading this book!

So, take the result from chapter III.1, plug in $J(t, \vec{x}) = J_1(\vec{x}) + J_2(\vec{x})$, and extract

$$E = -\mathcal{F}(J_1, J_2)$$

where $\mathcal{F}$ is a fancy functional* of the two functions $J_1(\vec{x})$ and $J_2(\vec{x})$. (Indeed, it is so fancy that I felt that I couldn't simply write plain F.) When you input the two functions $J_1(\vec{x})$ and $J_2(\vec{x})$, the functional tells you what E is equal to.

The key point is that we are free to choose $J_1(\vec{x})$ and $J_2(\vec{x})$, and hence free to vary the distance r between the two lumps and watch how the energy E varies.

A plot of how E varies with r is shown in figure 5. Physicists usually set E to 0 when the two nucleons are far apart. Thus, negative E indicates an attraction: The two nucleons could lower their energy by getting closer to each other. Sure enough, as suggested by the analogy, we find that E is negative, as written in the equation above, and that as we move the two lumps closer, E becomes even more negative, indicating an attraction between the two lumps.

As the two nucleons approach each other, when the separation r between them gets to be less than d, the range, the attraction kicks in and E abruptly goes negative.[6] (In reality, the transition is not so abrupt, but I wanted to emphasize the concept of range.)

---

*A word you learned in chapter II.1.

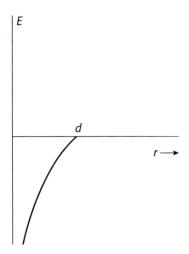

Figure 5. A highly schematic representation of the potential energy E between two nucleons as a function of the distance r between them.

Another commonly used analogy envisages two lumps lying on a mattress. This is perhaps, apt since the field φ was motivated by thinking about a mattress. In this analogy, a lump creates a depression around it, which might cause the other lump to roll nearer. The closer you move to each other, the more relaxed (that is, the lower your nervous energy) you become.

Furthermore, when the distance between the two lumps increases beyond a certain range, the attraction abruptly disappears. Two lumps lying far apart would hardly feel the presence of one another. It also makes sense that, since m has to do with how readily a wave packet would propagate on the mattress, it provides a measure of how stiff the mattress is.

I say "disappears" to emphasize the notion of a short range force. In reality, the force generated by Yukawa's meson decreases to zero exponentially fast rather than abruptly, but for practical purposes, the force appears to disappear abruptly. (This reminds me of a entire category of jokes circulating in the physics community that start with "A mathematician and a physicist ... " and end with the physicist saying "Yes, but for all practical purposes." Construct your own version.)

The scalar field φ has produced a short range attractive force! Both our mattress analogy and marriage broker analogy work.

Range and mass

The range of the strong force, as suggested by the marriage broker analogy, is inversely proportional to the mass m of the particle produced (as was discussed in chapter III.1) by the field. More precisely, the range d of a force produced by a field is inversely determined by the mass m of the particle associated with that field, thus,

$$d \sim \frac{\hbar}{mc}$$

We have included Planck's constant $\hbar$ and the speed of light c. This relationship is basic to physics and is used repeatedly these days.

Once again, this is yet another consequence of combining special relativity with quantum mechanics. We have moved into the northeast quadrant of the map of our quest. (Incidentally, one way you could tell where you are is to notice that this result contains both $\hbar$ and c, indicating that both quantum physics and special relativity are hard at work.)

Recall from chapter II.1, that with c imposing an absolute speed limit, your worldline is not allowed to slant more than 45° from the vertical. You cannot get from here to there instantaneously. You've got to propagate in time as well. It takes a long time to get to a far away place. (For the readers who have young children: "Are we there yet?") But in figure 4, the meson is propagating in a space direction, as noted earlier, and that is strictly forbidden by Einstein. So what gives?

Heisenberg's uncertainty principle comes to the rescue! Just as the embezzler in our prologue could get away violating energy conservation for a short time, the meson (that sneaky meson!) could propagate in space for a short distance without old man Einstein noticing. "Stop! You can't go faster than the speed of light!" Hence, the meson has a limited range.[7] The more massive the meson, the shorter would be its range, just as in our marriage broker analogy.

Historically, since the range of the strong interaction was known, Yukawa was able to predict the mass of the meson. A great triumph for theoretical physics when a meson with that mass was discovered! In sad contrast, these days people are far less inhibited and postulate the existence of unknown particles with abandon. In the vast majority of cases, they are not able to give a sharp prediction of their brainchild's mass, and when experimentalists fail to see the predicted particles, they simply shrug and say that their particles are too massive to have been produced in the specific experiment just performed.

Early in my career, during my first visit to Kyoto in 1980, the esteemed physicist Ziro Maki made an appointment for me to see the almost legendary Yukawa. The morning of the appointment, I put on a tie but was bitterly disappointed to be informed that Professor Yukawa was feeling ill. To my deep regret, he died a year later at the age of 74.

Dividing by 0 gives you ∞

Let us now apply our new found understanding to electromagnetism. As you probably know, the electric force between two charges decreases as the distance between them increases; more precisely, the force is proportional to the inverse of the square of the distance. Compared to the abrupt shut off of the strong force on the distance scale of the nucleus, this gradual decrease is extremely gentle. The electric and magnetic forces are said to have an infinitely long range.

But now we understand this difference between the strong force and the electric force in terms of the inverse relationship between mass and range. The photon is massless! Hence the infinitely long range of the electromagnetic force:

$$\infty \sim \frac{1}{0}$$

Didn't your teacher tell you not to divide by 0? (Ha! Quantum field theorists do it all the time. Everything in its proper context.)

Similarly for gravity. Newton told us that the gravitational force between two masses decreases as the distance between the them increases. Just as in the case of the electric force, it is proportional to the inverse of the square of the distance. Again, with this gradual decrease, gravity is also said to have an infinitely long range.

Now you understand why physicists are almost sure that the graviton, the analog of the photon associated with the gravitational field, is absolutely massless. The gravitational field can reach out across the vastness of interstellar space.

Virtual particle

A physics professor friend of mine, not a quantum field theorist, once told me that the concept most mystifying to him is that of a virtual particle. Good. You and I could now explain to him what quantum field theorists call a "virtual particle"! The particle produced and detected in figure III.1.2 in the preceding chapter is real. The particles emitted and absorbed in figure 4 are virtual. Einstein and Heisenberg would not allow them to exist for long. In contrast, the particle in figure III.1.2 actually exists between production, such as in an accelerator, and detection in some massive experimental set-up.

In quantum field theory, a particle whose wordline makes an angle of less than or equal to 45° with the time axis is said to propagate in a timelike direction; it could be real. In contrast, a particle whose wordline makes an angle of more than 45° with the time axis is virtual; it propagates in a spacelike direction.* The lumbering hulks that we are always propagate in a timelike direction.

---

*We will come across this distinction again in chapter V.1.

Note that a particle could be real or virtual, depending on the circum-stances. For example, the photons streaming into your eyeballs right now are real, produced by the sun or a light bulb. Real photons move at 45°, right between the time and the space axes, by definition. In contrast, the photons constantly being exchanged by atomic nuclei and the electrons orbiting them are virtual.

Understand clearly that usage of the words "real" and "virtual"[8] differs significantly between discourses by "normal everyday" people and by quan-tum field theorists. The effect produced by the virtual photons in your body (namely, electric attraction between unlike charges), directly holding the atoms in your body together and indirectly you together, is perfectly real. So, perhaps the origin of my friend's confusion is merely semantic. We are used to thinking of particles as "real." If you prefer, avoid the term "virtual particle" altogether and use "a quantum fluctuation in the field" in its stead.

Virtual particles[9] are not to be confused with the somewhat oxymoronic virtual reality so fashionable these days.

Notes

[1]This is taken from my book *Fearful,* page 335. I thank Satio Hayakawa and Laurie Brown for informing me of some of this history.

[2]For the modern Indian version, see https://www.youtube.com/watch?v=aZS2KbLAy5Y, recommended to me by a reader of the manu-script for this book.

[3]In one of George Gamow's popular physics books that I read as a student, he uses two dogs fighting over a juicy bone as an analogy. You can imagine that as soon as one dog grabs the bone, the other dog grabs it back. The two dogs are in effect brought together by the bone. That the bone is producing an attractive force made a deep impression on me. We could even imag-ine that the more massive the bone, the closer the two dogs have to get to each other. Perhaps we could interpret more massive as meatier. Be warned that these analogies are not exact, of course. In this connection, I think my marriage broker analogy is actually a bit more apt.

[4]Mathematical eyes, so to speak.

[5]You might wonder: "Where is time?" The answer is that in the Dirac-Feynman path inte-gral, we talk about the probability amplitude of a particle going from some initial position to some final position after time T has elapsed. Remember the "funny" race in chapter II.1? We

generalize from particle to the quantum field φ. What should be the initial state and the final state of φ? It's up to us to choose. The easiest choice is to take the field to be quiescent, that is, equal to zero, before and after the lumps are introduced.

The two lumps should sit there for a long time, for a time T much longer than the char-acteristic time scale of the quantum field φ (namely, $\hbar/mc^2$ with m the mass of the particle produced by φ). In other words, the time scale set by the uncertainty principle. How much the probability amplitude has oscillated after time T determines E, that is, the energy in the previously quiescent field φ caused by the intro-duction of the two lumps. All this sounds complicated, but it is not. I refer those readers interested in seeing how sausage is made to *QFT Nut,* page 28. Notice the 28; this is in a book with almost 600 pages. So, this is baby stuff! The students enrolled in my course have barely warmed up.

[6]As r goes to 0, $E(r)$ might bottom out rather than dropping to $-\infty$ as suggested by the figure, but that is not our concern here.

[7]Denote the range by d. Yukawa's result then reads $d \sim \hbar/mc$. The time needed for a me-son propagating at some fraction of the speed

of light to cover that distance is then $d/c \sim \hbar/mc^2$. Once again, we arrive at the energy-time uncertainty principle discussed in the prologue, namely, $\Delta E \Delta t \sim \hbar$ with $\Delta E \sim mc^2$.

[8]A digression into the rather twisted etymology of "virtual": Yes, it was related to "virtue," and then even earlier, to "virile." The meaning has shifted from manly, strong, to good, virtuous, and then to effective, to as good as actually doing it, and finally to not real. Quantum field theorists had used "virtual" in this sense for a long time before it somehow percolated into computer science around 1960. So, does that mean that a virtuous man is not real?

[9]Perhaps that notion is slowly seeping into our culture? On page 335 in David Mitchell's *Ghostwritten,* I read "I had to jettison matrix mechanics in favor of virtual numbers."

Attraction or repulsion: a mysterious but all important sign ▰▰▰▰▰▰▰▰▰

Dance of the universe: attraction or repulsion

Two humans could attract or repel, with neither rhyme nor reason. In the quantum world, in contrast, whether two particles attract or repel each other is governed by the iron laws of physics and mathematical logic.

The dance of the universe is finely choreographed with strong and weak forces, with attractive and repulsive forces, and with infinitely long range and extremely short range forces, as we had already discussed in chapter I.1. We saw in chapter III.2 that infinitely long range forces are mediated by massless particles, while short range forces are mediated by massive particles. The exchange of mesons between nucleons keeps the nucleus together.

But then a question may have naturally occurred to you. How are repulsive forces generated? To answer this question, I have to first remind you of a bunch of facts about electromagnetism.

The yin and yang of electromagnetism

First, electric charges can be positive or negative. Like charges repel, and unlike charges attract. A positive and a negative strive to get together: all that yin yang stuff.

As I mentioned already in chapters I.1 and I.2, electric charges generate an electric field, while moving charges—namely, electric currents—generate a magnetic field.* In general, it is convenient to talk about an electric charge

---

*Inside a common magnet, electrons are actually spinning like crazy, or moving around in some cases, to generate a magnetic field around the magnet, even though to our eyes, nothing seems to be moving.

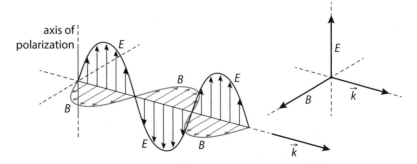

Figure 1. The right hand rule for a polarized electromagnetic wave.

density, denoted by $\rho(t, \vec{x})$, namely, the number of charges per unit volume at the point $\vec{x}$. Note that $\rho(t, \vec{x})$ could well depend on time t as the charges move about. Similarly, we talk about a current density $\vec{J}(t, \vec{x})$. Since the current is also characterized by the direction the electric charges are moving in, $\vec{J}$ is a vector.

The electric field $\vec{E}(t, \vec{x})$ and the magnetic field $\vec{B}(t, \vec{x})$ are both characterized by vectors also, as explained in chapter I.2. In contrast, Yukawa's scalar field $\varphi(t, \vec{x})$ does not point in a particular direction.[1] Perhaps you could even guess that this is the crucial difference that allows a repulsive force to be generated.

Polarization of light and photon spin

In an electromagnetic wave, $\vec{E}$ and $\vec{B}$ are perpendicular to each other and to the direction the wave propagates in, namely, the wave vector $\vec{k}$. Physics students are taught to point the index finger in the direction of $\vec{k}$, the middle finger in the direction of $\vec{E}$, and the thumb in the direction of $\vec{B}$, making all three perpendicular to each other. The electromagnetic wave is called "right polarized" or "left polarized," according to whether the student has to use his or her right hand or left hand, respectively. See figure 1.

I remark in passing that it is important to note that what we call "right polarized" or "left polarized" is just a matter of convention or mnemonic, just like clockwise versus anticlockwise. We expect interstellar beings to know that light has two possible polarizations, but which one is right and which one is left may well be totally meaningless to them. In my experience, surprisingly many laypersons, and even physics students, fail to distinguish laws of the universe from trivial conventions.

An everyday manifestation of this basic property of electromagnetic waves is wearing Polaroid sunglasses to cut down glare. Indeed, the trade name derives from the two polarizations of light. The lens is designed to allow only

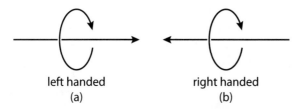

left handed
(a)

right handed
(b)

Figure 2. (a) A left-handed photon and (b) a right-handed photon. The circular arrow indicates the direction of spin, the straight arrow the direction of motion. You wrap your fingers of either your left or right hand around the circular arrow and your thumb would then point in the direction the photon is moving in.

one of the two polarizations through. By playing around with Polaroid sunglasses you could see that you can make the world less bright turning the lens this way and that.[2]

Moving on from classical physics to quantum physics, we see that the electromagnetic wave actually consists of a stampede of photons. Each photon is spinning, either clockwise or anticlockwise around its direction of motion. A given photon is said to have either right helicity or left helicity. See figure 2. The two polarizations of the electromagnetic wave in classical physics correspond to the two helicities of the photon in quantum physics.

Teams at a junior high school dance

So much for quantum physics; we now have to turn to special relativity.

To study the scalar field $\varphi(t, \vec{x})$, we have to couple it to its source $J(t, \vec{x})$ by adding to the action a term φJ. Similarly, we now have to couple the electromagnetic field to its source, namely, the charge density $\rho(t, \vec{x})$ and the current density $\vec{J}(t, \vec{x})$. (To avoid cluttering up the narration with too many unnecessary words, I will henceforth refer to current density simply as current.)

Now recall from chapter I.3 that Minkowski, with his "valiant piece of chalk" unified space and time into the 4-dimensional vector $x^{\mu} = (t, \vec{x}) = (t, x^{i})$ with the spacetime index $\mu = 0, 1, 2, 3$ and the space index $i = 1, 2, 3$. Furthermore, he showed that Pythagoras's formula, $ds^2 = dx^2 + dy^2 + dz^2$, for the distance ds between two neighboring points in space, should be generalized to $ds^2 = dx^2 + dy^2 + dz^2 - dt^2$ for the distance ds between two neighboring points in spacetime. Just as Pythagoras's formula defines Euclidean geometry, Minkowski's formula, with its infamous, almost mystical, minus sign distinguishing time from space, defines spacetime geometry.

So suddenly, every concept in physics possessing a direction has to be promoted from a 3-vector to a 4-vector in spacetime, or something even more involved known as a tensor. (Remember that back in chapter I.3, I said a caveat will be needed when we discuss electromagnetism? Well, the future has arrived. The electromagnetic field tensor F will be introduced shortly.)

Executive summary: Minkowski started a new fashion as soon as he partnered the 3-vector $\vec{x}$ with t to form the 4-vector $x^\mu = (t, \vec{x})$.

Let us picture a dance at a junior high school. Concepts defined by 3-vectors will be kicked out from this relativistic dance unless they can find someone to partner with to form 4-vectors. The electric current $\vec{J}(x)$ immediately latches onto the charge density $\rho(x)$ to form the so-called 4-current $J^\mu = (\rho, \vec{J})$. (Again, to lessen clutter, I am dropping the index μ on x^μ. I will also stop showing the dependence on x; it is understood that everybody varies according to where they are in spacetime.) In hindsight at least, the natural partner for the electric current $\vec{J}$ has to be the charge density ρ: $\vec{J}$ has to do with how many electrons are moving across a unit area in a unit time, while ρ has to do with how many electrons are sitting in a unit volume. With space and time unified, current and charge density also naturally unify.[3] And so it goes, people finding partners and forming teams like at a junior high school dance.

Partners in Minkowski's World

time	space
energy	momentum
charge density ρ	current $\vec{J}$
electric potential ϕ	vector potential $\vec{A}$
electric field $\vec{E}$	magnetic field $\vec{B}$

But look at that poor electric field $\vec{E}$ still milling around searching for someone to form a 4-vector with. And the magnetic field $\vec{B}$ is standing in a corner looking forlorn. You would think it completely natural for $\vec{E}$ and $\vec{B}$ to team up, particularly since we know that a moving magnetic field generates an electric field and vice versa. A match made in physics, if not in heaven! And thus $\vec{E}$ and $\vec{B}$ join hands to form the electromagnetic field $F = (\vec{E}, \vec{B})$.

To summarize, the source for the electromagnetic field is current $J^\mu = (\rho, \vec{J})$. Actually, most of us already know full well that the electromagnetic field is generated by electric currents and charges, so that is not a surprise. Ready to couple the electromagnetic field $F = (\vec{E}, \vec{B})$ to $J^\mu = (\rho, \vec{J})$?

A serious mismatch followed by a happy ending

But now we see a serious mismatch: the electromagnetic field F has $3 + 3 = 6$ components, while J^μ has only 4 components.[4]

One clue to the resolution of this mismatch[5] is that the potential energy of an electric charge is determined not by the electric field but by the electric potential ϕ, as we saw in chapter II.2. This suggests that the charge density ρ in J^μ is to be coupled to ϕ. So, add to the action a term like $\phi\rho$.

Good, the charge density ρ in $J^\mu = (\rho, \vec{J})$ is coupled to ϕ, but we still need J^μ to couple to a spacetime vector with 4-components. So now it is ϕ who has to find three partners to pair with $\vec{J}$.

To find these guys, our memory has to stretch back to chapter II.2. There I mentioned that, just as the electric field $\vec{E}$ is determined by ϕ, the magnetic field $\vec{B}$ is determined by the vector potential $\vec{A}$, a 3-vector. Just as $\vec{E}$ is determined by the variation of ϕ in space,[6] $\vec{B}$ is determined by the variation of $\vec{A}$ in space. "Hey $\vec{A}$, you there, you are the natural to dance with $\vec{J}$!"

A happy ending: ϕ is paired with $\vec{A}$ to form a 4-vector potential $A_\mu = (\phi, \vec{A})$ which $J^\mu = (\rho, \vec{J})$ can couple to: they each have 4 components.[7] The analog of the term φJ that we added in chapter III.1 to the action for the scalar field φ is a term of the form[8]

$$A_\mu J^\mu = A_0 J^0 + A_i J^i = A_0 J^0 + A_1 J^1 + A_2 J^2 + A_3 J^3 = \phi\rho - \vec{A} \cdot \vec{J}$$

Here I have introduced the Einstein repeated index summation convention (which some even consider as one of Einstein's greatest contribution to physics), according to which any repeated index is to be summed over. Thus, we sum the index μ over its range $(0, i)$ and the index i over its range $(1, 2, 3)$. Everything fits, as you would expect: The electric potential ϕ dances with the charge density ρ, and the vector potential $\vec{A}$ with the current density $\vec{J}$. So, this equation looks complicated at first sight and might frighten young children, but it is in fact a triviality. The first two equal signs just express the notation introduced by Einstein, while the third equal sign merely reminds us that A_0 and J^0 have other names, respectively, ϕ and ρ.

The story told here is in some sense the same story told in chapter II.2. The action for Newtonian mechanics proposed by Euler and Lagrange is given by the integral of the kinetic energy minus the potential energy over time. There is no room for Newton's force in the action. As explained earlier, in the action formulation, force pops out as a derived concept given by the variation of the potential energy in space. Similarly, there is no room in the electromagnetic action for $F = (\vec{E}, \vec{B})$ to be coupled to J^μ. The action is formulated in terms of A_μ, and the familiar electromagnetic field $\vec{E}$ and $\vec{B}$ pop out as a derived concept given by the variation of A_μ in spacetime.

I should mention that in Minkowski's world, the electric field $\vec{E}$ and the magnetic field $\vec{B}$ are not left out in the cold. They found each other in the relativistic dance to form what is known as the electromagnetic field tensor $F = (\vec{E}, \vec{B})$; it's just that they do not couple to the 4-current J^μ and hence are not part of the action. (Incidentally, the mathematical entity called "tensor" represents the simplest generalization of the concept of vector.[9])

Incidentally, we should not be surprised that the structure of electromagnetism is intimately tied to the structure of spacetime. After all, it was by studying electromagnetism and the invariant speed of light that Einstein discovered special relativity, which in turn enabled Minkowski to pick up his fabled piece of chalk.

The origin of repulsion

Phew! That was a long digression into electromagnetism. Let's remind ourselves where we were. Yes, trying to choreograph the dance of the universe. We wanted to understand the long range electric repulsion between two protons balancing the short range strong attraction in stars and in the atomic nucleus. And later, we want to understand how the universal attraction of gravity builds structure out of the primeval haze.

So, how can the photon produce a repulsion between the two protons while the meson produces an attraction between them?

But first, we need to look at what a repulsive potential between two lumps would look like.

As the result of some quantum field theory calculation, we have an attractive long range potential between two lumps, as shown in figure 3(a). The energy E is negative. As the distance r between the two lumps decreases, E becomes more negative, which makes the lumps want to get closer to each other. They attract each other. (This was essentially explained already in the preceding chapter, but in connection with the short range attraction between nucleons.)

For the potential in figure 3(b), the energy E is positive, and as the distance r between the two lumps decreases, E becomes more positive, which makes the lumps want to get away from each other.

The key observation: To go from attraction to repulsion, as in figure 3(b), we simply have to flip an overall sign. If we multiply the potential in figure 3(a) by a minus sign, it turns into the potential in figure 3(b).

Photons spin, mesons do not

So, to solve one of the mysteries of the universe, "all we have to do" is to find an overall minus sign. By now, with all this setup, you realize that the big difference between the strong force and the electric force must lie with the mediating particle.

The photon spins, while the meson does not. In the language of fields, this means that the meson is the excitation in a scalar field φ, while the photon is the excitation in a 4-vector field A_μ. Crudely speaking, to spin, you need to have a direction, and that calls for a vector field. A scalar field with "no direction home" won't cut it. The index μ on A_μ tells us that the field is a vector and hence produces a particle which spins.

I outlined a nifty 4-step program in chapter III.1 for you to become a quantum field theorist almost instantly. We started carrying it out for a

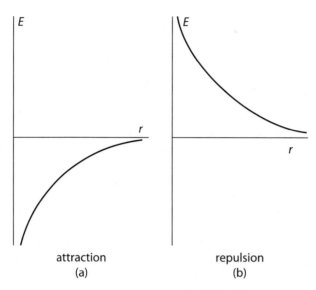

Figure 3. Attractive versus repulsive long range potentials: (a) attraction, (b) repulsion. Although we are not talking about gravitational potential here, some readers might find it helpful to think the potential depicted here as gravitational. Imagine a ball placed on the "slope." In (a), it would roll toward the origin. In (b), it would roll away from the origin.

universe containing only the electromagnetic field and some external lumps. Maxwell gave us the action, but then we got stuck on step 3, disturbing the field. We didn't know how to couple the lumps to the field. Instead, we went for the easy life and did the scalar field first. Well, that was a rather long detour, but now we finally know how to carry out step 3: Add a term of the form $A_\mu J^\mu$ to the action.

So, repeating the calculation Yukawa did for the meson but now for the photon, we obtain[10] $E = +\mathcal{F}(J_1, J_2)$, with a crucial flip of the overall sign from $-$ to $+$.

Where did that minus sign come from? Care to guess?

Yes, it came from the minus sign that distinguishes time from space in Minkowski's formula $ds^2 = dx^2 + dy^2 + dz^2 - dt^2$ for the geometry of spacetime.

I consider this as one of the greatest achievements of theoretical physics in the 20th century: understanding that like charges repel.

Executive summary: That A_μ carries an index while φ does not means that the photon spins while the meson does not. The spinning in spacetime drags in the minus sign Minkowski wrote in with his magic chalk. (He had to put in a minus sign in order to distinguish time from space!) That minus sign flips the energy from that shown in figure 3(a) to that shown in figure 3(b), thus changing an attraction to a repulsion. The meson produces an attraction, while the photon produces a repulsion.

Likes repel, unlikes attract

The folk dictum in many cultures suggest that likes repel while opposites attract. Regardless of whether it is invariably true in everyday life, the dictum holds in the subatomic world. Electrons repel, but an electron and a positron attract. Even more importantly for the evolution of the universe a few minutes after the Big Bang, the electrons and the protons found each other to form neutral hydrogen atoms, thus making the universe (to first approximation) transparent to photons, an important life stage for the baby universe.

But I digress. Now that we understand why like charges repel, we naturally want to know why unlike charges, one positive and one negative, attract. Again, good question. But this one is easy. If a positive charge is described by $J^\mu = (\rho, \vec{J})$, then a negative charge is described by $-J^\mu = (-\rho, -\vec{J})$: just flip the sign of both the charge density and current. Indeed, that is exactly what is meant by positive and negative charges.

So, we simply change J_2 to $-J_2$ in $\mathcal{F}$. The overall sign flips, thus giving us a negative energy between two unlike charges, which becomes more negative as they approach each other, as shown in figure 3(a). Unlike charges attract. That was by far the easiest thing I learned in graduate school.

Attraction or repulsion depends on spin

Analogies are just picturesque fables to help you understand the underlying physics, and are not meant as substitutes for real calculation, of course, as I have already emphasized. But still they might help some readers, or at least give us something to focus our minds on, especially since we have been talking somewhat abstractly for a while.

Imagine ourselves in the north country in the dead of winter. In the distance, we see an icy pond with two black objects on it. There appears to be a mysterious repulsive force pushing the two apart. As we get closer, we see that the black objects are actually two boys, and they are throwing a ball between them back and forth. We could also suppose that the boys are each standing on a plastic tray to minimize friction.[11] The exchange of the ball leads to a repulsive force between them.

Unlike the marriage broker analogy (see chapter III.2), this ball exchange has produced a repulsion, so this analogy actually applies more to the photon rather than the meson. However, the relation between mass of the particle exchanged and the range of the force produced still works, provided that we assume that the boys can throw the ball with the same speed regardless of how massive the ball is. Assume that balls with different masses are made of the same material. Then the more massive ball is, the larger its surface area, and so it encounters more air resistance, cutting down the distance the ball can travel.

To produce an attractive force, we have to cook up a slightly different analogy. On some other occasion, we approach the frozen pond and see that a mysterious attractive force between the two dark objects are bringing them closer together. As we get closer, we realize the two boys are throwing a boomerang to each other, with their backs facing each other. Now, when a boy throws a boomerang, the recoil pushes the boy toward the other boy, and when the other boy catches the boomerang, the momentum of the incoming boomerang pushes this boy, the catcher, toward the other, the thrower.

As I have said, take these analogies with a grain of salt. Or leave them. In reality, the meson produces an attraction while the spinning photon produces a repulsion. Our analogies lead to the reverse: the ball produces a repulsion while the spinning boomerang produces an attraction. But at least they have the feature that whether the force is attractive or repulsive has something to do with the spin of the object being thrown.

You fall because two negatives make a positive

But what about gravity? While the photon has one unit of spin, the graviton has two units of spin. (In quantum physics, spin is quantized.) Why so? Anticipating a bit, we will see in chapter V.5 that the field corresponding to the graviton has the form $h_{\mu\nu}(x)$. (I will explain why gravity needs to have two indices, but for now, let us simply note that the electromagnetic potential $A_\mu(x)$ carries only one index.) Heuristically speaking, the two spacetime indices on $h_{\mu\nu}$ compared to the one spacetime index on A_μ implies that the graviton spins twice as much as the photon.

Again, we can do the calculation Yukawa did, but with the particle being exchanged between the two external sources carrying two units of spin. More generally, if the particle being exchanged carries S units of spin, the energy due to the presence of the two external sources comes out to be[12]

$$E = -(-1)^S \mathcal{F}(J_1, J_2)$$

Quantum field theory produces the pleasing result that the force (between like objects) is attractive or repulsive according to whether the spin of the mediating particle is even or odd.

I can even try to give you a flavor of how this funny even-odd dependence on spin comes about. In the 3-dimensional space we live in, we think of spin as a vector whirling around. Picture that boomerang the boys were throwing around earlier in this chapter. That's a 3-vector of course, akin to $\vec{x}$. Now that we are in spacetime, this 3-vector $\vec{x}$ has been promoted to a 4-vector, akin to $(t, \vec{x})$. As it "whirls around," it mixes space and time, just as a 3-vector whirling around mixes x, y, and z. That churning stirs up the famous minus sign that Minkowski proposed to distinguish time from space.[13]

That was a spin 1 particle with a vector whirling around. A spin 2 particle is like something with two vectors whirling around. Twice the whirling, twice the fun. Thus, Minkowski's (-1) got stirred up twice. And now the magic that school kids learn, negative times negative equals positive,[14] $(-1)^2 = +1$, comes into play. Gravity is attractive.[15] Amazing how Nature works at the fundamental level!

This merits another trumpet blast for relativistic quantum field theory. Nowhere in nonrelativistic quantum mechanics can you find Minkowski's minus sign! In physics courses before relativistic quantum field theory, the professor can just tell us that charges repel, masses attract. If we ask why, the professor can only stare back, at a loss for words. The more patient professors might mumble something about quantum field theory, but the impatient ones just shout,[16] "That is a fact, OK?"

Physics students don't realize this, but within the confines of a specific course, many questions have no answer. I will come back to this issue in chapter VI.3 on intellectual completeness.

The four forces

To me, this almost mystic explanation, that the attractive or repulsive character of the fundamental forces originates in the distinction between time and space, came as an astounding revelation.[17]

Let me summarize what we know about the four fundamental forces in a table.

interaction	mediating particle	range	between like objects	strength
strong	meson	short	attraction	scary strong
electromagnetic	photon	long	repulsion	not that strong
gravity	graviton	long	attraction	pathetically feeble
weak	weak boson	extremely short	repulsion	weak is his name

I already mentioned the weak force in the prologue, but I will postpone a more detailed discussion until chapter V.3. The weak force was the last to be discovered, because its range is even shorter than that of the strong force. We will see in chapter V.4 that three of these four interactions could be unified.

It is the subtle interplay of this intricate web of forces, attractive and repulsive; long and short range; strong, weak, and feeble; which leads to the physical universe as we know it.[18]

Notes

[1]A legitimate confusion is that the displacement of the elastic membrane that gave rise to the scalar field φ also appears to have a direction, up and down. But remember that we were talking about a 2-dimensional space (namely, the surface of the membrane), in which the wave packets move around and do their thing. The direction perpendicular to the membrane has no meaning in this 2-dimensional universe. The scalar field φ could take on positive or negative values, but it cannot point. Incidentally, we already touched on this point in a footnote in chapter III.1.

[2]This is best done with two pairs of Polaroid sunglasses, so that two lenses could be superimposed on each other.

[3]If we denote length and time generically by L and T, respectively, then volume is L^3, and area multiplied by time is $L^2 T$. Speed is distance divided by time, and so c, the speed of light, has dimensions of L/T. Unifying space and time by setting $c = 1$ means that L and T are interchangeable.

[4]By the way, the mathematics behind these curious numbers, 1, 4, 6, 10, ... is known as group theory. Scalars have 1 component, vectors 4, tensors 6, and so on. See, for instance, *Group Nut*. More in chapters V.2, V.3, and V.4.

[5]It certainly sounds plausible that two entities with different numbers of components cannot be coupled together. Consider two vectors $\vec{a} = (a_1, a_2, a_3)$ and $\vec{b} = (b_1, b_2, b_3)$ expressed in Cartesian coordinates. As some readers know and as Descartes could tell you, upon changing the basis vectors (that is, the three directions in his room as he lay in bed), the components $a_1, a_2, a_3, b_1, b_2, b_3$ would all change, but the scalar product $\vec{a} \cdot \vec{b} = (a_1 b_1 + a_2 b_2 + a_3 b_3)$ would not change. (The same point was made in chapter I.3, when we discussed Pythagoras's relation.) This quantity could only be constructed for two vectors with the same number of components. Since, for a given history, the action in classical physics and the probability

amplitude in quantum physics certainly should not change under a change of basis vectors, the quantity describing the coupling of the source to the field we want to put into the action also should not change. Thus, we expect the source and the field to have the same number of components.

[6]I am glossing over some technical details having to do with the gradient versus the curl. See, for example, *FbN*, page 391.

[7]I won't explain the difference between upper and lower indices here. See, for example, *G Nut*, pages 182–184.

[8]Some readers might be worried about the extra minus sign after the third equality symbol. No need to worry. It's the same minus sign Minkowski drew with his valiant chalk and has to do with the difference between upper and lower indices. See the previous endnote.

[9]Another historical curiosity: the word "tensor" comes from 18th- and 19th-century studies of stress and strain in solids. In my experience, many undergraduates become inexplicably tense when first exposed to tensors.

[10]Incidentally, while all this may sound mysterious to some readers, nowadays a bright undergraduate is capable of doing this type of calculations that Yukawa pioneered almost a century ago. I can assert this, because a couple of months before I wrote these words, two undergraduates reading *QFT Nut* under my supervision did precisely this.

[11]One of my reader-friends thought that the tray had some deep significance. Not at all! The force due to the ball being thrown back and forth is rather small, yet I want the effect of the force pushing the boys apart to be immediately apparent even from a far distance. Hence the tray.

[12]See page 32 of *QFT Nut*. As I have remarked elsewhere, the calculation can't be too difficult if it appears on page 32 of a book with almost 600 pages!

[13]More precisely, but still at a heuristic level, when the field A_μ propagates across spacetime,

the time component A_0 knows that it should do a different dance from the space components A_i. See *QFT Nut*.

[14]Perhaps you've heard this joke. A linguistics professor was lecturing that there does not exist a language in which two positives make a negative. From the back row, somebody muttered, "Yeah, right!"

[15]See chapter V.5.

[16]A neurologist friend of mine who read the manuscript scrawled "Welcome to medical school!" on it at this point.

[17]It was revealed to me in graduate school at Harvard by Julian Schwinger, whom you will meet in chapter IV.3.

[18]For instance, we already saw in chapter I.1 how this interplay is manifest in nuclear fission and fusion.

Recap of part III

Carrying out the four steps I outlined in chapter III.1, you could be doing quantum field theory in no time. Unfortunately, the relevant path integral could be evaluated exactly only in the simplest cases, known as free theories. Remarkably, and even surprisingly, these simple theories already reveal to us some deep secrets of the physical universe.

First, fields create and annihilate particles. Second, the exchange of these particles between external lumps generates a force between the lumps.

Finally, finally, after so many years, physics professors know what they are talking about when they talk about forces in introductory courses.

We can even understand why some forces attract, while others repel, why some have long range, while others drop off abruptly. Once we understand this subtle interplay, the intricate dance of the universe becomes more comprehensible.

Sorry, I may have promised too much in the opening of chapter III.1. To become a professional quantum field theorist, you still have to read a textbook. (But isn't that true for many endeavors?) Some quantum field theory textbooks are simple but not substantive, some substantive but not simple. The simplest and yet substantive book I know of is my very own *QFT Nut*.

A universe of fields

Preview of part IV

The fields we discussed in part III are free, "free" being a technical word meaning that they interact neither with themselves nor with other fields. They couple only to external sources, or more picturesquely, lumps, put in at our pleasure. Away from these sources, free fields propagate freely through spacetime.

The first step toward a richer set of quantum field theories was taken by Dirac, who promoted the electron to a field. Then the electromagnetic field could couple to the electron field, and not just to some external fixed lump. Alas, then physicists could no longer evaluate the corresponding path integral, but could only treat it perturbatively, that is, by appealing to the coupling being small. This was carried out by a new generation of brilliant physicists, including Schwinger and Feynman, and eventually led to the calculation of the magnetic moment of the electron to unprecedented accuracy, the first great triumph of quantum field theory mentioned in the prologue.

I then explain the notion of gauge theory. As we will see in part V, the four fundamental interactions in the universe are now understood to all be based on gauge theories.

Everybody is a field: Dirac set the electron free ▬▬▬▬▬▬▬▬▬▬▬▬▬▬

Unnatural: A quantum particle coupled to a classical field!

In the hydrogen atom, the speed of the lone electron[1] is less than 1% of the speed of light, about $c/137$. Thus, we are safely ensconced in the southeastern quadrant of the map in the prologue, as was already mentioned there. Quantum mechanics is needed, but not special relativity.

Terminology: Physicists refer to fast particles as relativistic, "fast" meaning moving at speeds comparable to the speed of light, and slow particles as nonrelativistic.

So, the electron in the hydrogen atom is, to a good approximation, nonrelativistic. The Schrödinger equation suffices for determining the quantum states that the electron could be in. (A clarification here might be helpful. When physicists refer to the electron in the hydrogen atom as nonrelativistic, they do not imply that the electron does not obey Einstein's special relativity. They simply mean that the electron is moving so slowly that special relativity is not needed. The "non" does not connote antagonism.)

But from time to time, the electron jumps from one quantum state to another, emitting or absorbing a photon. The photon is moving, by definition, at the speed of light and demands a relativistic treatment. Indeed, the photon is the apotheosis of relativistic, he who sets the speed limit of the universe.

This problem, of calculating the electromagnetic radiation emitted or absorbed by atoms, is usually reserved for the end of an undergraduate course on quantum mechanics.[2] Actually, the photon is not even treated as a photon at all, but merely as a manifestation of Maxwell's electromagnetic field.

In contrast, the electron is treated as a nonrelativistic quantum particle, coupled to the relativistic but entirely classical electromagnetic field.[3]

The typical treatment in quantum mechanics textbooks is:

electron	quantum	particle	nonrelativistic
photon	classical	field	relativistic

In academic slang, what I just described is known as a half-assed treatment. But students don't seem to mind.[4]

Theoretical physics is much more than a bunch of calculations

Theoretical physics is much more than a bunch of calculations to obtain results in agreement with experiments. Beginning students of physics have to be disabused of this common misconception in due time. Aesthetics and balance count, as any number of physicists, Einstein and Dirac in particular, have preached.[5]

It seems terribly skewed to treat the electron and the photon so differently. And so, even with the roaring success of this approach during the atomic age, many leading theoretical physicists became dissatisfied by 1930 or so.

Then along came Paul Dirac, known to Feynman and to us for his path integral and to one of his biographers as "the strangest man."[6] What does not disturb the typical undergrad in the slightest bothered Dirac deeply. He felt that the electron should be treated as a field* also, on the same footing is the electromagnetic field. A kind of equal rights amendment for physics!

The strangest man set the electron free

A mind forever / Voyaging through strange seas of thought, alone.
W. Wordsworth

What Wordsworth said about Newton applies perhaps even more aptly to Dirac.[8] Setting the electron free! It put into motion a monumental advance for theoretical physics. Dirac invented, in a flash of insight, the much celebrated Dirac equation to describe the electron. And yet to think that the Nobel Committee[9] in its infinite wisdom almost passed him over for the prize!

---

*For the sake of historical accuracy, I have taken some liberty here in favor of a livelier narrative, as was noted in the preface. (The notion of a quantum field was far from clear at the time, and for some time, Dirac persisted in regarding the ψ in his celebrated equation as a probability amplitude,[7] much as the Schrödinger wave function in nonrelativistic quantum mechanics was regarded. In modern language, this would be called the "matrix element" of the quantum field between the vacuum and the single electron state.)

Allow me to show you the Dirac equation, said by some to be most beautiful in theoretical physics:

$$(i\gamma^{\mu}\partial_{\mu} - m)\psi = 0$$

The Dirac equation in a bit more detail

Now that you're done admiring this fabled equation, let me explain the notation at least. The reader wanting just an overview of quantum field theory could safely skim over or skip the following. Recall that I introduced the index notation in chapter I.4, in particular for the 4 spacetime coordinates $x^{\mu} = (t, \vec{x})$. Here, $\partial_{\mu} = \frac{\partial}{\partial x^{\mu}}$ denotes the partial derivatives with respect to spacetime. For instance, $\partial_0 = \frac{\partial}{\partial t}$ indicates differentiation with respect to time. For deep reasons[10] having to do with relativistic invariance, Dirac was forced to set ψ equal to a 4-component mathematical object known as a spinor, and as a result had to introduce γ^{μ}, 4 matrices now known as Dirac gamma matrices. Recall also that I introduced Einstein's repeated index summation convention in chapter III.3. Thus, $\gamma^{\mu}\partial_{\mu} = \gamma^0\partial_0 + \gamma^i\partial_i$. In the junior high school dance visualization I gave there, the temporal derivative ∂_0 and the spatial derivatives ∂_i have to team up to form ∂_{μ}, which then has to connect[11] with the 4 Dirac matrices γ^{μ}.

A friend who read the manuscript for this book interjected at this point that I owe it to the high school or college student and other young-in-spirit readers I addressed in the preface to display the Dirac equation explicitly. I quote from what he wrote: "Show the Dirac equation as a 9th grade student would write it, namely as four coupled very simple linear equations involving t, x, y, z and ψ. This I think would offset some of the terror that some individuals might have on seeing the $\gamma^{\mu}\partial_{\mu}$." Heavens to Betsy! My friend knows some pretty advanced 9th grade students. But he is persuasive, and I will do what he says in a endnote.[12] Take a peek to see whether it "offset" or intensified your "terror," if you had experienced any.

The corresponding action is given by

$$S_{\text{Dirac}}(\psi) = \int d^4x \; \bar{\psi}(i\gamma^{\mu}\partial_{\mu} - m)\psi$$

(By the way, the notation d^4x is simply a reminder that we are integrating over 4-dimensional spacetime.) You could (sort of) see that given the action, it is easy to obtain the equation of motion, and vice versa.

Fast forward to a decade or so later. Young Feynman got tired of writing $\gamma^{\mu}\partial_{\mu}$, and invented a funny slash symbol $\partial\!\!\!/$ as a shorthand for this. Even a mind as great as Fermi was puzzled by this notation. Returning from the historic (and fabled) 1948 conference at which Schwinger and Feynman announced their conquest of quantum electrodynamics, Fermi could only remember that Feynman had slashes all over the place.[13]

The two greatest appearances of 2 in the history of physics

Henceforth, we have an electron field, traditionally denoted by the Greek letter ψ, and an electromagnetic field, denoted by A, continually interacting with each other. Physicists say that the two fields are coupled. In other words, the action of the electron field depends on the electromagnetic field, and vice versa. The technical meaning of coupling is not far off from everyday usage of the word.

The Dirac equation, beautiful though it may be, must reduce to the (ugly!) nonrelativistic Schrödinger's equation when the electron is moving slowly, and indeed it does. (If it didn't, it would have long ago disappeared down the dustbin of history.)

Incidentally, nothing irritates me more than seeing the word "overthrow" in pontifications about physics. Established theories in physics, unlike those in other areas of human endeavor, are not overthrown so much as extended and generalized. The Dirac equation did not overthrow Schrödinger's equation any more than Schrödinger or Einstein overthrew Newton. It is almost the opposite: that the Dirac equation has to reduce to Schrödinger's equation under the appropriate circumstances imposes a powerful constraint on what it could possibly be. In contrast, Newton's equation for gravity in no way no how reduces to Aristotle's "rocks want to go home" hogwash.

Now I get to touch base with the prologue, in which I mentioned that the g factor, the so-called gyromagnetic ratio, of a spinning charge measures how fast it precesses* in a magnetic field. The calculation of g for a classical object is by now an exercise given to undergraduates, and comes out to equal 1 in suitable units. I also mentioned that by the time Dirac came along, the g factor of the electron had been measured to equal 2, a surprise and a mystery for theoretical physics.

So, after writing down his equation in a flash of inspiration, Dirac's pressing task was to calculate g_e. Just couple in the electromagnetic field and proceed! But Dirac was so nervous that his equation would not produce the observed 2 that he put off calculating g_e till the next day. He later said that he was worried that Nature would blow it, missing the opportunity to deploy a beautiful equation! Well, he did get[14] 2 instead of the classical 1. (Of course. Otherwise, we won't be talking about his equation any more. The lesson: merely saying that a theory is beautiful doesn't cut it; Nature has to like it.)

---

*The physics is essentially the same as that of the spinning tops that we all played with as children!

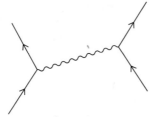

Figure 1. The Feynman diagram for two electrons scattering: an electron emits a photon, indicated by the wavy line, subsequently absorbed by the other electron. The electrons are no longer fixed lumps, as portrayed in figure III.2.4.

The title of this section is "the two greatest appearances of 2." But since the other famous factor of 2 in theoretical physics is not pertinent to quantum field theory as such, I have reluctantly put it into an endnote[15] in order not to interrupt the narrative.

Nothing is truly nailed down

Go back to the two sources in chapter III.2 interacting via the electromagnetic field, as shown in figure III.2.4. Remember that the two external sources are put in by hand and fixed in space, independent of time. They are nailed down.

But in physics, nothing is truly immobile. An ill fated car crashes into a wall. The wall is attached to the earth, which is so much more massive than the car that the earth's recoil is totally insignificant. But still,* the earth moves. Similarly, the sources, or lumps, in figure III.2.4 could represent two electrons, whose recoil is negligible if the mass of the electron far exceeds the energy of the photon.

So, replace the lumps by real life electrons with a will of their own, free to recoil as they emit or absorb a photon, as depicted in figure 1. Compare this with figure III.2.4. This photon exchange produces the interaction between the two electrons, as explained in chapter III.2.

Hopping back and forth between coordinate space and momentum space

We have been thinking of Feynman diagrams as little pictures showing how and where the particles interacted in spacetime, also called "coordinate space." (In particular, our convention is that time flows along the vertical axis.) Instead, we could, and more conveniently specify, the momentum and energy

---

*As a physics celebrity once muttered under his breath, so it is alleged.

carried by each particle. For example, in figure 1, we could label the two incoming electrons with momenta[†] p_1 and p_2, and the outgoing electrons with momenta p_3 and p_4. The Feynman diagram[‡] is then said to be drawn in (a mentally constructed) momentum "space."[16] Theoretical physicists are trained (thanks to Joseph Fourier, as was mentioned way back in chapter I.2) to hop back and forth (known as Fourier transform) easily between coordinate space (that is, the spacetime we live in) and momentum space. In general, calculations are easier in momentum space than in coordinate space.

A menagerie of fields

The electron is a field. So, then everybody[17] is a quantum field.

A quantum field, just like any continuous medium, be it water, air, elastic membrane, jello, whatever, can support waves. And the waves can form packets. In fact, you, by your talent for free association, already identified these wave packets as particles, back in chapter I.2.

At this point, you might object. Classical waves tend to spread out and disappear. But a particle like the electron persists forever, as far as we know. The secret behind the ability of theoretical physicists to think of wave packets as particles is due to the celebrated conservation laws of physics, one of which states that electric charge is conserved. The electron carries one negative unit of electric charge, and by its very name lives as the basic unit of electricity. Hence the wave packet representing the electron, which is in fact the electron, cannot simply disappear. If we want quantum field theory to describe fundamental particles such as the electron, we must, as a matter of first importance, build various conservation laws into the theory as cornerstones.

This is why, back in the prologue, when an electron pops out of the vacuum, it must be accompanied by a positron. The magician known as the vacuum could produce matter out of energy (Einstein taught her), but she can't produce an electric charge out of nothing. A negative charge must be accompanied by a positive charge. (Incidentally, I already alluded to this fact in the prologue.)

So, from the 1930s on, theoretical physicists talked about the electron field, the proton field, the neutron field, the meson fields, and of course also the photon field, the grandmommy of fields, the apple of Faraday's eyes.

We have earlier revealed that the electron field is the external source that the electromagnetic field A couples to. Similarly, we can now reveal that the

---

[†]Post Einstein, the word "momentum" is used generically in physics for energy and momentum.

[‡]I use the singular here, but yet another deep secret Mother Nature revealed to physicists is that there should be two diagrams. I will discuss this in part VI, but meanwhile, could you possibly guess why?

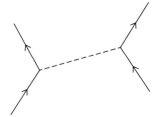

Figure 2. The Feynman diagram describing meson exchange between two protons. You might realize that this is essentially the same diagram as in figure 1. Conventionally, a meson is indicated by a dotted line, a photon by a wavy line (in honor of its origin in the electromagnetic wave, I suppose), but this convention is hardly universal—there is no government agency, fortunately, telling people how to draw Feynman diagrams. A word of encouragement: mastering Feynman diagrams may be a lot easier than you might think, as perhaps you could see from the examples given here.

proton (or the neutron) field is the external source Yukawa's meson field φ couples to. Yukawa's meson exchange as discussed in chapter III.2 can now be pictured as in figure 2.

By the early 1960s, every newly discovered particle had a field to its name. The situation clearly could not last, and so in 1964, Murray Gell-Mann proposed that all the particles that participate in the strong interaction are made of quarks. The action for the strong interaction is then written in terms of various quark fields, each of whose excitations is a quark. We now know that protons, neutrons, and mesons are made of quarks. Consequently, some fields, such as the proton field, have disappeared from the everyday vocabulary of most theoretical physicists.[18] (Into the dustbins of history?) Instead, we have quark fields.

As you would expect, the history of quantum field theory is inextricably intertwined with the history of particle physics. As physics moved from atomic physics through nuclear physics to particle physics, it became clear that the approach adequate for atomic physics, treating the electron as a particle and the photon as a field, would no longer be appropriate for particle physics. Almost by definition, in the subnuclear regime, particles governed by the quantum are moving at relativistic speeds more often than not. We will see this in more detail when we discuss the strong and the weak interactions in part V.

Feynman diagrams emerge rather naturally

When I was a student, I couldn't wait to learn about these diagrams that my elders were talking about. You can now see that Feynman diagrams (in figures 1 and 2) are just intuitive pictures depicting what is going on. They

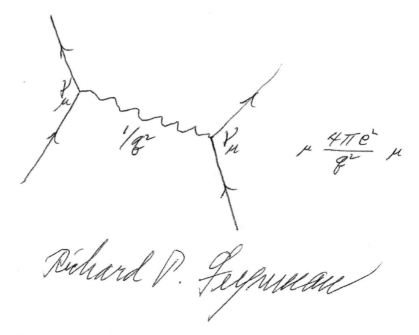

Figure 3. A Feynman diagram drawn and signed by Feynman. For some strange reason, for the amplitude (written to the right of diagram) Feynman chose to omit the symbol γ, not once, but twice. An amusing technical remark for any physicist who might be reading this: instead of the Heaviside-Lorentz convention now universally used, Feynman used the Gaussian convention, hence the factor 4π.

emerge rather naturally. With staircase wit,[19] you might even say that their introduction was more or less demanded by the theory, but hindsight is of course way too easy after 70 or so years.[20] Figure 3 shows one drawn and signed by the master himself.

Fields reign supreme

So, the perennial question that agitated 19th century physicists about whether particles or fields reign supreme has been resolved in favor of fields. Every known particle is now thought to be an excitation in the corresponding field.

In contrast with the half-assed treatment mentioned at the beginning of this chapter and taught to students, nowadays we place the electron and the photon on the same footing. Egalitarian treatment in quantum field theory textbooks:

electron	quantum	field	relativistic
photon	quantum	field	relativistic

Again, it may be worth emphasizing that relativistic encompasses nonrelativistic, and quantum encompasses classical.

Amusingly, the graviton, the particle associated with perhaps the oldest field known to theoretical physics, namely the gravitational field, has not yet been discovered. (When I say perhaps the oldest, I am interpreting Newton's letter to his friend Bentley rather generously. See chapter I.2.) More about the graviton in chapter V.5.

Discovery of a particle hardly means that experimentalists have captured one and put it in a bottle, or a cage, for that matter. Particles are often identified only indirectly through their effects, and this identification requires interpretation and a theoretical framework. Quarks and gluons* have been discovered in this sense.

Notes

[1]In fact, one of the first calculations students in an introductory course on quantum mechanics are taught to do is to determine this speed. See, for example, *FbN*, chapter I.3.

[2]For example, J. J. Sakurai and J. Napolitano, *Modern Quantum Mechanics,* pages 365ff.

[3]Let me quote Julian Schwinger here: "The evolutionary process by which relativistic field theory was escaping from the confines of its nonrelativistic heritage culminated in the complete reconstruction of the foundations of quantum dynamics." When Schwinger spoke fondly of the "daring escape" from our nonrelativistic heritage, that heritage is quantum mechanics as usually taught to undergraduates. Another unpleasant feature of this standard elementary treatment foisted on students is that time plays a privileged role, incompatible with Einstein's unification of space and time into spacetime.

[4]In physics courses, in contrast to humanities courses, students almost invariably accept whatever the professor says as the God given truth.

[5]And lumbering in their footsteps, me, too. The subtitle of my book *Fearful* is "the search for beauty in modern physics."

[6]G. Farmelo, *The Strangest Man.* See also https://www.wondersofphysics.com/2019/07/paul-dirac-stories.html?fbclid=IwAR37imhvI0

OLcEiXg51vfZ_CcwLqMWl-MQnYc_pGoupuwBD0oga5O5LK79w.

[7]Most readers can safely ignore this distinction. A probability amplitude is a complex number which may vary in space and time. In contrast, a quantum field is an operator in some formulations, and a variable to be integrated over in the path integral formalism. This confusion persists among many students, and even not a few professors.

[8]Note that the word "forever" was spelled as "for ever" in the original. The comma used by Wordsworth indicates, to me, that the truly greats, Newton, Einstein, Dirac, were destined to be alone, while those of us who did not end up "being alone" were most likely not qualified to join the true elite of theoretical physics.

[9]See M. Larsson and A. Balatsky, *Physics Today,* November 2019, page 46. Remember that Sweden is a small country.

[10]See for example, *QFT Nut,* chapter II.1, where all this is made clear.

[11]Incidentally, this is another way to understand why Dirac was compelled to turn the simple-Simon electron wave function of non-relativistic quantum mechanics into a 4-component spinor. Again, readers who want to know more are invited to look at *QFT Nut,* chapter II.1. Some readers might realize that this implies that the matrices γ^μ are 4 by 4.

*The fields that glued the quarks together into strongly interacting particles, the protons, neutrons, mesons, and the like. See chapter V.2.

[12]Write $\psi = \begin{pmatrix} u_1 \\ u_2 \\ v_1 \\ v_2 \end{pmatrix}$. This 4-component object

is known as a spinor. The Dirac field ψ, and each of its 4 components, vary in spacetime; thus, ψ is shorthand for $\psi(t, x, y, z)$, u_1 for $u_1(t, x, y, z)$, and so on. The Dirac equation describes how these 4 fields, u_1, u_2, v_1, v_2, vary in spacetime. It consists of 4 coupled partial differential equations as follows:

$$\frac{\partial u_1}{\partial t} + \frac{\partial v_2}{\partial x} - i\frac{\partial v_2}{\partial y} + \frac{\partial v_1}{\partial z} + imu_1 = 0$$

$$\frac{\partial u_2}{\partial t} + \frac{\partial v_1}{\partial x} + i\frac{\partial v_1}{\partial y} - \frac{\partial v_2}{\partial z} + imu_2 = 0$$

$$\frac{\partial v_1}{\partial t} + \frac{\partial u_2}{\partial x} - i\frac{\partial u_2}{\partial y} + \frac{\partial u_1}{\partial z} - imv_1 = 0$$

$$\frac{\partial v_2}{\partial t} + \frac{\partial u_1}{\partial x} + i\frac{\partial u_1}{\partial y} - \frac{\partial u_2}{\partial z} - imv_2 = 0$$

These 4 coupled linear partial differential equations describe an electron moseying along in spacetime minding its own business and not interacting with anybody, keeping extreme social distance.

Not that frightening, right? My friend is right. A 9th grader who knows what a partial derivative is would have no trouble with the Dirac equation!

One more remark for those who know what matrices are. As usual, you are free to choose a basis to write the matrices in. The equations as displayed here are written in the so-called Dirac basis.

[13]The full story is told in *QFT Nut*, page 105.

[14]See *QFT Nut,* chapter III.6, pages 194–195. The famous 2 pops up in equation (5). The calculation takes up exactly one page.

[15]The other famous factor of 2 in the history of physics: Newton, supposing that light consists of tiny "corpuscles" (as was mentioned in passing in chapter I.4), already theorized that light (such as that from a distant star) passing by a massive object (such as the sun) should be pulled by gravity toward the massive object. A German physicist named Georg Soldner, unaware of Newton's work, repeated the calculation, but his paper was swept away in the triumphant tide of Maxwell's electromagnetic waves. Much later, Einstein, fresh with his new theory of gravity but unaware of Newton's and Soldner's work, also calculated the bending of light by the sun, but erroneously obtained what we now call the "Newtonian value." Told that his predicted effect could only be detected during a solar eclipse, Einstein got together an expedition to the Crimea to observe the next scheduled eclipse. World War I broke out just then, and the Russians, quite reasonably, arrested this bunch of astronomers sneaking around with telescopes, claiming to measure the "curvature of spacetime" by "bending light." Obviously spying for Germany! Meanwhile, Einstein realized that he had made a mistake. Correcting his error, he found that his theory of gravity in fact predicts twice the Newtonian value. After the war, the English physicist Arthur Eddington organized two expeditions, one to Brazil and one to Africa. (See the Brazilian film "House of Sand.") Had World War I not happened and had Einstein not discovered this famous factor of 2, his theory of curved spacetime would have been a wet fizzle. Incidentally, this prediction is what made Einstein a worldwide celebrity: the public could hardly have been expected to care about the perihelion advance of Mercury (an almost incredibly tiny 43 seconds of arc per century!), and $E = mc^2$ had to wait until the atomic bomb to seep into the general consciousness. The final twist was that the Nazis, upon learning of Soldner's work, accused Einstein of stealing from a real German physicist. Little did they realize that the alleged thief, instead of stealing the Mona Lisa in the gallery, would have stolen the Mona Lisa in the gift shop.

[16]Perhaps reminiscent of how cyberspace is also a mental construct.

[17]I was told that it was Pascual Jordan, not Paul Dirac, who first proposed that every quantum particle is an excitation in a quantum field. I am not enough of a historian to verify this fact.

[18]But not the proton itself, of course.

[19]L'esprit d'escalier, Treppenwitz, firing the cannon after the cavalry had already charged you by.

[20]The history of Feynman diagrams has been ably chronicled by D. Kaiser in *Drawing Theories Apart: The Dispersion of Feynman Diagrams in Postwar Physics,* University of Chicago Press, 2005.

Theoretical physics, like music, starts with harmony but then tries to move on

Life is not easy

With quantum field theory formulated as a path integral, you might think that theoretical physicists have come to easy street. They could just sit back, evaluate the path integral, and understand the secrets held by quantum fields.

Alas, theoretical physicists are not able to evaluate the path integral, at least not those relevant to the real world. This inability to integrate would not surprise those readers who know some calculus.[1] Most integrals cannot be evaluated analytically. "Analytically" means evaluating the integral exactly[2,3] and by hand, in contrast to turning on your computer and evaluating it numerically and approximately for various choices of input numbers, that is, by merely crunching numbers, which one could always do. (See also chapter V.2.) Selection bias is at work here. Introductory textbooks on calculus focus, for obvious reasons, on those integrals which can be done, and may have given some abecedarians the erroneous impression that most, if not all, integrals may be evaluated analytically.[4]

Well, if so few ordinary integrals could be evaluated, you could imagine how much harder it would be to evaluate the path integrals defining quantum field theory.

To explain the difficulty and to introduce some terminology, we go all the way back to Newtonian mechanics and explain the role of harmony in physics.

Harmonic and anharmonic motion in classical and quantum mechanics

A stretched spring exerts a restoring force. If the restoring force F is proportional to the amount of stretch, call it q, the spring is said to be harmonic, another term borrowed from music. When we pull on, and then release, a mass attached to a harmonic spring, it oscillates nicely in time at a frequency characteristic of the spring. In Newton's celebrated equation of motion $F = ma$, both sides scale linearly with q: they match. (This somewhat mathematical sounding phrase merely means that if we increase the oscillation amplitude by multiplying q by a real number s, called the scaling factor, that is, if we let $q \to sq$, both sides are multiplied by s. To see this, note the particle's acceleration a is the change per unit time of its velocity v, while v is the change per unit time of its position q, so that $v \to sv$ and thus $a \to sa$. The force $F \to sF$ by assumption, and thus both sides of $F = ma$ scale in the same way.) The scaling factor cancels out, so that the oscillation frequency does not depend on the oscillation amplitude.

This mechanical setup, known as a harmonic oscillator, is a standard example in elementary physics, and even beginning students can calculate its exact motion, described by a sinusoidal wave (if you permit me to sling some jargon around) as was exemplified by the sound wave shown in figure I.2.3.

But if the restoring force exerted by the spring is given by, say, $F = -(kq + hq^2 + gq^3)$ with k, h, g some constants characteristic of the spring, then horrors, F no longer scales nicely: $F \to -(skq + s^2hq^2 + s^3gq^3)$, some gobbledygook not simply related to F at all. Regardless of whether you are a beginning student or the world's greatest theoretical physicist, you would not be able to solve analytically the motion of a mass attached to such a dreadful spring, said to be anharmonic.[5]

The same situation persists into quantum mechanics. Students in an introductory course are introduced early on to how the simple harmonic oscillator can be solved exactly in quantum mechanics. (By the way, the word "simple" is also commonly attached in textbooks, not just popular books on physics.) In contrast, nobody has been able to solve the anharmonic quantum oscillator (although tomes have been written about it).

Incidentally, since the harmonic oscillator, both classical and quantum, is so easy to solve, it has been used as a model or stand-in for systems whose physics we do not know or do not care to know. The modeling of an elastic membrane, in chapter III.1, by a mattress of masses tied together by harmonic springs is a prototypical example, leading to a crude first understanding of field theory. A historically significant example is when Planck discovered quantum physics by studying electromagnetic radiation in a heated cavity. He knew that the cavity was made of atoms in equilibrium with the radiation, but he obviously did not, and could not, know how to calculate the behavior of these atoms. How could he, before the age of quantum mechanics, which he himself was about to inaugurate? So, he modeled the atoms as a bunch of classical harmonic oscillators.

Action scaling quadratically is key

I mentioned in chapter II.2 three formulations of quantum mechanics, the Schrödinger, the Heisenberg, and the Dirac-Feynman. Interestingly, the mathematical fact that enables the harmonic oscillator to be solved exactly is, at least superficially, different in these three formulations. Let us focus on the Dirac-Feynman formulation here, as the discussion would then carry over readily to quantum field theory.

In the Dirac-Feynman formulation, we are instructed to integrate over the probability amplitude $e^{\frac{iS}{\hbar}}$. Recall that the action S is given by the kinetic energy K minus the potential energy V integrated over time. Recall also, from elementary physics and from chapter II.2, that the potential energy is just the work done in stretching the spring, stored until released. The work done is equal to the force exerted times the distance over which the force acts. Thus, for the harmonic oscillator, the potential energy V is proportional to $q \times q = q^2$. Under the scaling $q \to sq$ mentioned above, $V \to s^2 V$. What about the kinetic energy K? Again, recall from elementary physics that the kinetic energy of a mass m moving at velocity v is given by $K = \frac{1}{2}mv^2$. Thus, $K \to s^2 K$.

Very nice! Both K and V scale quadratically with s, and thus the action also scales quadratically: $S \to s^2 S$. (Incidentally, that K and V scale the same way is equivalent to the fact that both sides of Newton's $F = ma$ for the harmonic oscillator scale the same way.)

Gaussian integral and free field theory

At this point, physics and mathematics touch hands, rather mysteriously I might even say. Thanks to a procession of mathematical luminaries of the first order, from Laplace to Gauss to Poisson, the so-called Gaussian integral (familiar to students of the theory of probability and statistics)

$$\int_{-\infty}^{+\infty} d\varphi \, e^{-\varphi^2 + J\varphi}$$

can be evaluated analytically.[6] For the benefit of some readers, I will elaborate in an endnote.[7] For others, it suffices to take this as a mathematical fact.

I have intentionally written the Gaussian integral using notation reminiscent of the discussion in chapter III.1. In particular, the term $J\varphi$ in the exponential corresponds to the coupling of the meson field to the external source. This term turns out to be easy to handle.[8] The key observation is that the term φ^2 in the exponential scales quadratically.

This remarkable fact implies that the path integral over the probability amplitude $e^{\frac{iS}{\hbar}}$ could also be evaluated if S scales quadratically in the fields. (In light of the preceding paragraph, the fact that the coupling to the external source scales linearly may be ignored.) A quantum field theory with this property is said to be "Gaussian," or more commonly, "free."

But wait a minute, you say. The famous (and it is famous in some circles) Gaussian integral is merely an integral over a single real number φ, while in the path integral, we have to integrate over fields which are themselves functions of spacetime. Yes, you're right, it is not immediately obvious, but Dirac and

Feynman* showed us that it can be done. Again, I will sketch how for some of you calculus fans in an endnote.[9]

To recap, that the Gaussian integral could be evaluated and that the path integral with an action scaling quadratically in the fields could also be evaluated are both remarkable statements. This "explains," in particular, why the quantum harmonic oscillator could be taught to students in an introductory course.

Away from Gauss means toil and trouble

As soon as the integral is not Gaussian, then physicists (and mathematicians) are stuck. Consider the integral (I am calling it Y merely because I want to refer to it later)

$$Y = \int_{-\infty}^{+\infty} d\varphi \, e^{-\varphi^2 - \lambda \varphi^4 + J\varphi}$$

with λ a constant. I have added a quartic term φ^4 in the exponential. Now the quantity $\varphi^2 + \lambda \varphi^4$, the analog of the action in quantum field theory, no longer scales quadratically. The integral, known as non-Gaussian, cannot be done analytically, and the tricks outlined in endnotes 8 and 9 fail to work.

The corresponding functional integral in quantum field theory would be orders of magnitude more difficult than this integral, since the variable φ is to be replaced by a field $\varphi(x)$ varying arbitrarily over spacetime.

As long as the action is quadratic in the fields, Gauss could still help us

"Hmm, back in chapter III.2 we figured out why various fundamental forces were attractive or repulsive, and thus got a long way toward understanding how the physical universe works," you say. "Why were we able to evaluate all those path integrals then?"

Good question! Consider the three forces we discussed.

First of all, the action $S_{\text{Maxwell}}(A)$ is indeed quadratic,[10] and hence represents a free field theory. One way of understanding this from everyday experience is that light beams pass freely through each other (notwithstanding movies with light sabers). Photons left to their own devices do not interact with each other.

Second, Einstein's action is, unfortunately, not quadratic (which causes all the trouble you may have heard about quantizing gravity), but it is quadratic

*Feynman even wrote a whole book teaching people how to do it.

to a fantastic degree of accuracy due to the extreme feebleness of gravity discussed in chapter I.1. More on this in chapter V.5.

Third, we were able to evaluate the path integral over Yukawa's meson field because we "cheated"! Or, rather, Yukawa "cheated"! Because we simply assumed that the elastic membrane is harmonic. Recall that the model for the membrane is a bunch of mass points connected by springs. We took the springs to be harmonic and hence ended up with a quadratic action for the meson field. Also, more on this in chapter V.2.

To summarize, as long as the action is quadratic in the fields, Gauss could still help us evaluate the relevant path integral.

So that's how quantum field theorists are able to explain, as was discussed in chapter III.3, one fundamental mystery of the universe: between like objects, exchange of a spin 0 particle produces an attraction, exchange of a spin 1 particle produces a repulsion, and exchange of a spin 2 particle produces an attraction.

The Dirac action

Of the three quantum field theory actions discussed in chapter III.2, one is naturally quadratic (Maxwell), one is to an extremely high accuracy quadratic (Einstein), and one is adequately assumed to be quadratic (Yukawa). What about the Dirac action for the electron?

Well, look at the Dirac action $S_{\text{Dirac}}(\psi) = \int d^4x \, \bar{\psi}(i\gamma^\mu \partial_\mu - m)\psi$ displayed in chapter IV.1. You can see with your very own eyes that upon the scaling $\psi \to s\psi$, $\bar{\psi} \to s\bar{\psi}$, the action scales like $S_{\text{Dirac}}(s\psi) = s^2 S_{\text{Dirac}}(\psi)$, that is, quadratically. Indeed, the electron by itself lives totally free, and its motion can be solved by beginning students who can write down the Dirac equation.[11]

To recap, both $S_{\text{Maxwell}}(A)$ and $S_{\text{Dirac}}(\psi)$ are quadratic, describing a free photon and a free electron, respectively.

Free until coupled

But what happens when we couple the photon and the electron?

Back in chapter III.1, we learned to "disturb" the scalar field φ by adding to the action a term of the form $\int d^4x \, J(x)\varphi(x)$. Similarly, we could disturb the electromagnetic field A_μ by coupling it to the current density J^μ, that is, by adding to the action a term of the form $\int d^4x \, J^\mu(x)A_\mu(x)$. The integral over x just indicates that we are free to disturb the electromagnetic field at any location x in spacetime. The difference is that when Minkowski unified space and time, A_μ became a 4-vector, as you learned in chapter III.3. Recall also Einstein's repeated index summation convention mentioned there.

The next crucial step was Dirac setting the electron free by replacing the nailed down source J^μ for the photon by electron fields, as we saw in chapter IV.1. Dirac wrote $J^\mu = \bar{\psi}\gamma^\mu\psi$. (Recall the four Dirac matrices γ^μ already displayed earlier in the Dirac action.) In effect, the external source J^μ has been animated by the electron field ψ.

Thus, Dirac replaces the term in the action shown above by

$$S_{\text{coupling}} = \int d^4x \, e \, \bar{\psi}(x)\gamma^\mu\psi(x)A_\mu(x)$$

We put in an experimentally measured number* $e \simeq 0.303$ to indicate how strongly the electron and the photon fields are coupled together.

I have to pause to teach you how to read this hieroglyphic. If you open a textbook on quantum field theory, you would see these kinds of symbols scrawled all over. Back in chapter III.1, we talked about the absorption and emission of a photon. In the corresponding technical terminology, the electromagnetic field A is said to be capable of annihilating and creating a photon. Fine. Similarly, you would think that the electron field ψ should be able to annihilate and to create an electron.

But that's wrong because of charge conservation!

Annihilating an electron increases[†] the electric charge of the universe by 1, while creating an electron decreases the electric charge of the universe by 1. But a single field cannot be capable of doing both! Due to charge conservation, the field can do one or the other.

Let's say by convention that the electron field ψ can annihilate an electron. Then it cannot create an electron: It must create a particle with charge opposite to that of an electron. In other words, it can only increase the electric charge of the universe by 1.

Voilà, ladies and gentlemen, the positron! Indeed, this is the formal version of Dirac's argument for the existence of the positron, namely, the antielectron, and of antimatter in general.[12] We will give a more physical argument in chapter V.1.

So the field ψ annihilates an electron and creates a positron. To create the electron, we have to include the field $\bar{\psi}$, known as the conjugate of ψ, doing the opposite of what ψ does, namely, it annihilates a positron and creates an electron.

Note that because the photon is not electrically charged, the electromagnetic field A is capable of both annihilating and creating a photon.

I offer you a table[13] to help you remember all that. Final exam next week!

---

*This e is not to be confused with Euler's number, also denoted by e and equal to 2.71828..., which forms the basis of the exponential function and which also appears frequently, for instance, in the path integral.

[†]Increases rather than decreases for the trivial reason that the electron is defined to have one unit of negative charge. This infelicitous sign choice, which has bedeviled physics and engineering students ever since, goes back to one of the founding fathers of the United States. Thanks but no thanks, Ben Franklin! See chapter V.4.

	annihilates	creates
ψ	electron	positron
$\bar{\psi}$	positron	electron
A	photon	photon

Building blocks of Feynman diagrams

Now you can read what $e\bar{\psi}(x)\gamma^{\mu}\psi(x)A_{\mu}(x)$ says. Read from right to left. At the location x in spacetime, the electromagnetic field A annihilates a photon, and the electron field ψ annihilates an electron. Then the conjugate field $\bar{\psi}$ creates an electron. Finally, the number e, known as the electromagnetic coupling strength, fixes the probability amplitude for this process to occur. See figure 1(a).

An electron absorbs a photon. At the point x, the photon disappears, said to be annihilated. But then the electron is also allegedly annihilated, but the annihilation is immediately followed, at the same point x in spacetime, by the creation of an electron. Theoretical physicists do not ask the "metaphysical" question of whether or not the electron emerging from x is the "same" as the electron arriving at x. This description may seem rather childish, even

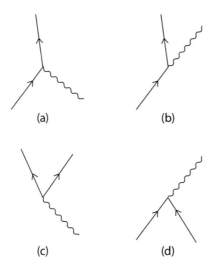

Figure 1. The coupling vertex of an electron to a photon can be read four ways: (a) the absorption of a photon by an electron, (b) the emission of a photon by an electron, (c) the production of an electron positron pair by a photon, (d) the annihilation of an electron positron pair into a photon. The wavy line represents the photon, the solid line the electron.
Redrawn from A. Zee, *Fly by Night Physics*, Princeton University Press, 2020.

odd, but as the fantastic agreement regarding the electron's magnetic moment mentioned in the prologue attests, that is how the theory works.

Similarly for the emission of a photon, as shown in figure 1(b). We see that this could be obtained from figure 1(a) by bending the photon line "to go forward in time." (This possibility of bending the lines in Feynman diagrams is a deep property of quantum field theory, known as crossing symmetry.)

Crossing could be applied to the electron as well. The resulting processes, shown in figure 1(c) and 1(d) describe the production of an electron positron pair by a photon and the annihilation of an electron positron pair into a photon, respectively.

You now can see that the Feynman diagram shown in figure IV.1.1 is "constructed" by putting together the building blocks, known in the jargon as the interaction vertices of quantum electrodynamics, shown in figure 1. So, join in the fun and build some Feynman diagrams!

Fun with Feynman diagrams

Put the interaction vertex in 1(a) and in 1(b) together to form the process in figure 2, showing an incoming electron absorbing a photon, continuing on for a while and then emitting a photon. In effect, a photon scatters an electron. This process, written as $\gamma + e^- \to \gamma + e^-$ and known as Compton scattering, provided the first hint of the existence of a quantum world.[14]

Next, rotate this Feynman diagram by 90° to obtain the one in figure 3(a). You see an electron and a positron coming together to produce two outgoing photons in a process called "pair annihilation." An electron positron pair annihilate each other, and poof! They turn into two photons. The science fiction literature is full of stories about a character encountering his anti-self and annihilating into a puff of "pure energy." Well, at the fundamental level, this is the process responsible.

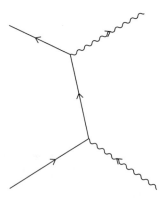

Figure 2. The Feynman diagram for the process $\gamma + e^- \to \gamma + e^-$, in which a photon scatters off an electron.

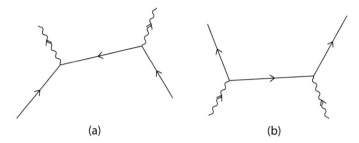

Figure 3. The Feynman diagram for pair annihilation (a) and for pair production (b).

What if we had rotated the Feynman diagram in figure 2 the other way and obtained the one in figure 3(b)? You can probably figure out what process this describes before reading on. Indeed, two photons collide to form an electron and a positron in a process called "pair production." For instance, a highly energetic photon going through an electromagnetic field can produce an electron positron pair, a process of astrophysical interest.

See how easy it is to draw some Feynman diagrams to describe various fundamental processes involving photons and electrons!

Living in our world (no other choice!), our experience of physical phenomena, other than our being rooted by gravity to the earth, originates essentially from zillions of photons interacting with zillions of electrons. It still boggles my mind, decades after all this stuff about quantum field theory, that, at the fundamental level, all of that fantastic richness of the world is determined by the interaction vertices shown in figure 1.

Quantum field theory just became almost impossibly difficult

So finally, behold the action describing how electrons and photons interact: $S_{\text{QED}}(\psi, A) = S_{\text{Maxwell}}(A) + S_{\text{Dirac}}(\psi) + S_{\text{coupling}}(\psi, A)$. The resulting quantum field theory is known as quantum electrodynamics, or QED for short.[15]

A huge difference! In our previous discussion, the action consisted of Maxwell's action describing the electromagnetic field A, plus the JA term. But now J^μ has been replaced by $\bar{\psi}\gamma^\mu\psi$. The electron field ψ has a life of its own, in contrast to poor little J, nailed down and put in by hand, unable to go anywhere. Mathematically, we now have to integrate not only over all possible histories of A, but also over all possible histories of ψ. Furthermore, the coupling between the electron and the photon, $\sim \bar{\psi}\psi A$, sure as the sky does not scale quadratically. Instead, $\sim \bar{\psi}\psi A \to s^3 \bar{\psi}\psi A$. We can see by eyeball that it contains three fields and scales cubically, that is, by s^3.

In the 1930s, theoretical physicists did not know how to evaluate the corresponding path integral systematically. Then World War II erupted, and physics had to wait until a new generation came on the scene. This story will be told in chapter IV.3. Meanwhile, I conclude this chapter by explaining why actions quadratic in fields are easy and by briefly mentioning another type of quantum field theory.

Why quadratic means freedom: just happily moving along

With your newfound knowledge of quantum fields, you can now understand why physicists could readily handle actions quadratic in fields. Look at, for example, the Dirac action $S_{\text{Dirac}}(\psi) = \int d^4x \; \bar{\psi}(i\gamma^\mu \partial_\mu - m)\psi$. Focus on the fields, and read from right to left, consulting the table given earlier in this chapter if you need to. The field ψ annihilates an electron and then $\bar{\psi}$ immediately creates an electron, at the same spacetime point no less. Surely a weird way of describing an electron happily moving along in spacetime. An electron disappears and appears, then disappears and appears, on and on.

You might think that this language is bizarre to the max, but hey, that's what I learned at a couple of fairly reputable universities. Doesn't it sound a tad like some crazed New Age babble? You constantly destroy your old self and then immediately create a new one as you move on in life.

Similarly, the Maxwell action S_{Maxwell} describes a photon disappearing and appearing, again and again.

So being "quadratic" means leading an uneventful life. The fun starts when we add the cubic coupling term

$$S_{\text{coupling}} = \int d^4x \; e \; \bar{\psi}\gamma^\mu \psi A_\mu$$

to the action. Then, for instance, an electron and a photon could disappear together, followed by a single electron reappearing, at the same spacetime point no less. In "everyday" language, the electron has absorbed a photon!

This reminds me of an academic joke. A man and a woman were seen to go into a hut. A while later, two men and a woman came out. The mathematician announced that the hut now contains a negative man. The biologist insisted that reproduction has occurred. The theoretical physicist thought that the observation provides an example of quantum field theory at work. Punchline: A little kid exclaimed, "A man was hiding in that hut!"

For future use when we discuss gauge theory in chapter IV.4, note that the actions $S_{\text{Dirac}}(\psi)$ and $S_{\text{coupling}}(\psi, A)$ can be combined neatly as

$$S_{\text{Dirac}}(\psi) + S_{\text{coupling}}(\psi, A) = \int d^4x \; \bar{\psi}\left(i\gamma^\mu(\partial_\mu - ieA_\mu) - m\right)\psi$$

The coupling of the electromagnetic field to the electron, on which so much of our existence depends, may be succinctly described as replacing the ordinary spacetime derivative ∂_μ by a fancier derivative defined by $D_\mu \equiv \partial_\mu - ieA_\mu$, a step above the kind of derivative you would encounter in a calculus course. When theoretical physicists say that electromagnetism has a "deep" geometrical origin, this is essentially what they mean.

When the going gets tough, the tough move to a simpler universe

In analogy with QED, instead of coupling the meson field φ to a nailed down external source J, we may try writing $\bar{P}P\varphi$, with P the proton field. When we take the source J apart, so to speak, we see that it is made of proton fields. This is essentially what was done in the early 1950s, but it was soon realized that the proton, being a participant in the strong interaction, has a complicated internal structure.

In some ways, theoretical physicists have an easy life compared to experimental physicists, not to mention engineers, doctors, and others dealing with the real world. Here is a standard strategy in theoretical physics. When the real world feels way too complicated, study a much simpler universe instead.

So why not replace the proton field P by another meson field more massive than φ, denoted by Φ, by writing $\Phi\Phi\varphi$ in the action? In fact, why not get rid of Φ and couple φ to itself and write φ^3 in the action, known as a self coupling? (This corresponds to anharmonicity in the elastic membrane that motivated the field φ in the first place.) Theoretical physicists are perfectly free to imagine a universe consisting of nothing else but a φ field coupled to itself. That would indeed be the simplest quantum field theory, which is not free, that we could imagine.[16]

Self coupling of the scalar field

Another simple quantum field theory enticing theorists consists of a scalar field $\varphi(x)$ coupled to itself via the term φ^4, known affectionately as the "phi 4 theory."[17] It is defined by a path integral of more or less the same form as the integral Y mentioned earlier and which nobody knows how to do. Since φ is capable of annihilating and creating a meson, this theory could describe the scattering of two mesons: of the four fields in φ^4, two of them annihilate two mesons, and the other two create two mesons, all at the same spacetime point. The net result is that of two mesons colliding and bouncing off each other, a phenomenon of great interest to physicists studying the strong interaction in the 1950s and 1960s. Again, see chapter V.2.

You might be astonished by the number of people who have devoted their lives to this theory and the tens of thousands of papers that have been written about it. But already this φ^4 theory is difficult enough, and the full solution is not yet known. However, people have deduced many of its properties, at least qualitatively.[18] Importantly for theoretical physics, the study of this φ^4 theory, with some extension and generalization,[19] has produced considerable insights into a variety of phenomena, of which I might mention, for the sake of completeness,[20] superconductivity, phase transition, critical phenomena, disordered growth, and the Higgs mechanism.

A pedantic triviality: Lagrangian versus Lagrangian density

I hate to interrupt this narrative for the sake of pedantic trivialities, but unhappily, the world does contain a (steadily diminishing, I dearly hope) band of pedants. Most readers can safely skim, or skip, this section. The action principle was first formulated for point particles and then generalized to fields. Recall that for a particle, its Lagrangian $L(t)$ equals its kinetic energy minus its potential energy, all evaluated at time t. The action $S = \int dt\, L(t)$ is defined to be the Lagrangian integrated over time. Since fields live everywhere in space, in a field theory, the Lagrangian $L(t) = \int d^3x\, \mathcal{L}(\vec{x}, t)$ is given by the integral of $\mathcal{L}(\vec{x}, t)$, known as the Lagrangian density, over space. The action $S = \int dt L(t) = \int dt \int d^3x\, \mathcal{L}(\vec{x}, t) = \int d^4x\, \mathcal{L}(\vec{x}, t)$ is then given by the integral of the Lagrangian density over spacetime. In a relativistic theory, we talk about spacetime, and so naturally about the Lagrangian density, not the Lagrangian.[21] Hence, in the quantum field theory community, the word "density" is almost always dropped, but not without fear of some pedant popping out of some cranny to shout that you guys are mixing up two distinct concepts. In this book, I will not bother too much about the distinction.

As an example, consider the Dirac action $S_{\text{Dirac}}(\psi) = \int d^4x\, \bar{\psi}(i\gamma^\mu \partial_\mu - m)\psi$. Most people in my community would mean by Lagrangian the expression $\bar{\psi}(i\gamma^\mu \partial_\mu - m)\psi$ rather than its integral over space.

Notes

[1] They would know that while differentiation is algorithmic, that is, by applying certain rules step by step one could readily differentiate any (reasonable) expression involving elementary functions. Integration, in sharp contrast, is not algorithmic. Until the early 1970s, when computers were able to perform symbolic manipulation, as distinct from mere number crunching, evaluating an integral was considered something of an art form, and often involved some cleverness.

[2] The reader into doing integrals for fun probably knows that whether an integral can be evaluated analytically is often far from immediately apparent. For instance, the integral $I = \int_0^{+\infty} dx\, e^{-\left(ax^2 + \frac{b}{x^2}\right)}$ can be evaluated analytically, while $J = \int_0^{+\infty} dx\, e^{-\left(ax + \frac{b}{x}\right)}$ cannot be. (Yes, it may be evaluated as a Bessel function of the second kind, but that's merely giving the integral a name, since one way of defining this function is by its integral representation.) By the way, in one of Feynman's books, he said that the integral I can be evaluated analytically without showing how. After reading that, I sat down and did it. This illustrates the point that, in many

areas of science and mathematics, hearing that something could be done is crucial. After I told a colleague what Feynman said, he was also able to figure out how to evaluate I, using a different method. (Surely some mathematician did it long ago.) We have no idea how Feynman did it, but we could make an educated guess.

[3]Computers are now able to evaluate certain classes of integrals analytically. See, for example, the Wikipedia articles on symbolic integration and on the Risch algorithm.

[4]For the example given in chapter III.1, analytic evaluation of the integral means to obtain the result as a function of J, as was emphasized there. In the example in endnote 2, an analytic evaluation means to obtain a result as a function of a and b that we could subsequently manipulate.

[5]An anharmonic violin string is not merely out of tune but vibrating uncontrollably.

[6]Over the centuries, mathematicians have devised at least 11 different ways for doing this integral. See K. Conrad, https://kconrad.math.uconn.edu/blurbs/analysis/gaussianintegral.pdf.

[7]Write $I = \int_{-\infty}^{+\infty} dx e^{-x^2}$ (here x is just a real number as in elementary calculus, not the spacetime coordinate. Following Poisson, we note that $I^2 = \int_{-\infty}^{+\infty} \int_{-\infty}^{+\infty} dx dy \; e^{-(x^2+y^2)} = \int_0^{2\pi} \int_0^\infty d\theta dr \; r e^{-r^2} = 2\pi \int_0^\infty du \; \frac{1}{2} e^{-u}$, thus reducing it to an elementary integral. In the second equality, we switched to polar coordinates, and in the third equality, we substituted variable $u = r^2$. See the next endnote for why doing this integral, simpler than the one in the text, suffices.

[8]We simply complete the square as in high school algebra. Write $\varphi^2 - J\varphi = (\varphi - \frac{1}{2}J)^2 - \frac{1}{4}J^2$. Thus, $\int_{-\infty}^{+\infty} d\varphi e^{-\varphi^2 + J\varphi} = e^{\frac{1}{4}J^2} \int_{-\infty}^{+\infty} d\varphi e^{-(\varphi - \frac{1}{2}J)^2} = e^{\frac{1}{4}J^2} \int_{-\infty}^{+\infty} d\zeta e^{-\zeta^2}$. In the last step, we shifted the integration variable to $\zeta = \varphi - \frac{1}{2}J$. The remaining integral over ζ can be evaluated as in the preceding endnote and is just a number independent of J. We see that it is crucial to have a quadratic action, which allows us to complete the square.

[9]In the exponential, we now have the integral over spacetime $\int d^4 x \; (-\varphi(x)^2 + J(x)\varphi(x))$ instead of $(-\varphi^2 + J\varphi)$. Follow the same step as in chapter III.1, replacing the membrane by a mattress, that is, by replacing the integral over the spacetime coordinate x by a discrete sum.

Label the mass points on the mattress, or more academically, the lattice, by j. The integral is replaced by $a^4 \Sigma_j (-\varphi_j^2 + J_j \varphi_j)$ with a the lattice spacing. But exponential of a sum is a product of exponentials! Hence $e^{a^4 \Sigma_j (-\varphi_j^2 + J_j \varphi_j)} = \Pi_j e^{a^4 (-\varphi_j^2 + J_j \varphi_j)}$. Thus, we have a giant (infinite, actually) product of ordinary integrals $\int d\varphi_j e^{a^4 (-\varphi_j^2 + J_j \varphi_j)}$, which we know how to evaluate, as explained in the two preceding endnotes.

[10]The reader familiar with Maxwell's equations may also note that they scale linearly upon scaling $\vec{E} \to s\vec{E}$ and $\vec{B} \to s\vec{B}$, provided that we also scale the charge and current densities, if there are any.

[11]If you are able to write it down (as we did in an endnote in chapter IV.1), you are able to solve it. In this sense, one could argue that the Dirac equation is the easiest equation in fundamental physics.

[12]The Dirac field is complex. The deeper reason has to do with the representations of the Lorentz group.

[13]This table is taken from *FbN*, chapter 10.1. For those readers interested in going beyond what is given here but are not quite ready to tackle a real textbook, such as *QFT Nut*, this provides an easier entry.

[14]As was pointed out by Einstein in connection with the photoelectric effect. See *G*, page 29.

[15]R. Feynman, *QED*, with a preface by A. Zee. 2014. As a result of this publication, I am amused that Google Scholar lists Feynman among my co-authors as a result of this publication.

[16]The excellent textbook on quantum field theory by Mark Srednicki in fact starts with this theory.

[17]For various reasons, this is preferable to the φ^3 theory.

[18]For instance, it is known that the φ^4 interaction produces a repulsion between the mesons. See, for example, *QFT Nut*, pages 192–193.

[19]For example, by making φ complex.

[20]See, for example, *QFT Nut*, chapters V.4 and VI.5.

[21]Absolutely nothing profound here: Lagrangian is to Lagrangian density as population is to population density.

Quantum electrodynamics, perturbation theory, and cultural taboos ▬▬▬▬

Once coupled, life becomes complicated

Once the electron field and the electromagnetic field are coupled together, the resulting path integral can no longer be evaluated. The action for quantum electrodynamics $S_{\text{QED}}(\psi, A)$ consists of two terms, $S_{\text{Maxwell}}(A)$ and $S_{\text{Dirac}}(\psi)$, both quadratic, plus the interaction term $S_{\text{coupling}}(\psi, A)$, which unfortunately, but by its very nature, is cubic in the fields. This cubic term renders the theory impossible to solve analytically, as discussed in chapter IV.2.

The way forward after World War II was driven by Richard Feynman, Julian Schwinger, and Shin'ichiro Tomonaga, who later shared a Nobel prize for their work on quantum electrodynamics. It's worth noting that Feynman and Schwinger were both 21 in 1939 (Schwinger was older by three months, if you want to know), while Tomonaga was 12 years older. Each had his own approach to quantum field theory. Since I have never met Tomonaga and since I was educated by Schwinger and Feynman, directly and indirectly, I naturally tend to describe the approaches I know best.

Undeniably, over the intervening decades, Feynman's diagrammatic method has become by far the most popular, because it is so easy to learn. The reader, however, should not get the impression, often propagated in popular media, that it is the only possible approach. How else could Tomonaga and Schwinger accomplish their great work?

During the war, Feynman worked on the atomic bomb, while Schwinger worked on radar. They both lamented later that the war had robbed them of potentially the most creative periods of their lives.

Do what we can do first

The idea for perturbation theory is entirely natural. When confronted by a problem you cannot solve, you first solve a simpler but closely related problem (were you so lucky!). You then hope that the solution of the problem you cannot do is close to the solution of the problem you can do. Then you attempt to calculate the difference between the two solutions as a series of small corrections.

For example, Newton worked out his theory of gravity assuming exactly circular orbits for the moon and the planets. In fact, the orbits are slightly elliptical, but this deviation from a circular orbit could be, and was, put in later.

I have often said in my books that Nature is unreasonably kind to theoretical physicists. It so happened that the first quantum field theory they encountered, namely, quantum electrodynamics, has a rather small coupling: $e^2 \simeq 0.1$ in fact. Were this much larger, Schwinger and Feynman would not have gotten very far.

Incidentally, you might recall from chapter IV.2 that the coupling strength[1] in $S_{coupling} \sim \int e\bar{\psi}\psi A$ is about $e \simeq 0.303$. Why e^2 instead of e? Because for the quantities we are interested in here (the electron's magnetic moment, for example), an emitted photon is soon absorbed, and so the coupling occurs twice. See below.

A whiff of perturbation theory

Let me give you a flavor of perturbation theory. Suppose you were told to divide 1 by 0.9 and you don't know how. But you exclaim, "I do know how to divide 1 by 1 though. The answer is 1! Since 0.9 is close to 1, the answer to the division problem I cannot do should be close to 1. I will try to calculate the correction later."

To calculate the correction, I have to invoke a teeny bit of high school algebra. Multiply $(1 - \varepsilon)$ by $(1 + \varepsilon)$:

$$(1 - \varepsilon)(1 + \varepsilon) = 1(1 + \varepsilon) - \varepsilon(1 + \varepsilon) = 1 + \varepsilon - \varepsilon - \varepsilon^2 = 1 - \varepsilon^2$$

Note that this result is exact.

The Greek letter epsilon, ε, is traditionally used to denote a small number. The key is that a small number squared is an even smaller number, a fact which we already made use of in chapter I.4. For instance, $(1/10)^2 = 1/100$, which is much smaller than $1/10$. Thus, if ε is small, we can safely drop ε^2 on the right side of this equation, and say that $(1 - \varepsilon)(1 + \varepsilon) \simeq 1$ (as always, using the symbol $\simeq$ for "approximately equal to"). We conclude that multiplying $(1 - \varepsilon)$ by $(1 + \varepsilon)$ gives 1 approximately. In other words, 1 divided by $(1 - \varepsilon)$ is approximately $(1 + \varepsilon)$. We have $1/(1 - \varepsilon) \simeq (1 + \varepsilon)$.

So, $1/0.9 = 1/(1 - 0.1) \simeq (1 + 0.1) = 1.1$. You announce triumphantly, "The first order correction to the answer I gave you before, namely 1, is 0.1!"

This hypothetical you could then go on and calculate the second order correction, third order correction, and so on, depending on your energy and longevity. Indeed, mathematicians had long ago worked out the infinite series to be $1/(1-\varepsilon)=(1+\varepsilon+\varepsilon^2+\varepsilon^3+\cdots)$.

The second order correction is just $\varepsilon^2=0.1^2=0.01$, the third order $\varepsilon^3=0.001$, and so on. In this simple example, you could of course, and so could we all, work out the exact answer, namely, $1/0.9=10/9=1.1111\cdots$, but the point is that, were we unable to find the exact answer, we could still come pretty close to it.[2]

Some readers no doubt know that many other useful infinite series are known, for example, $\sqrt{1-\varepsilon}\simeq1-\frac{1}{2}\varepsilon+\cdots$. (I mention this particular example, because it is crucial[3] for Einstein's special relativity and because as a kid I learned to take the square root of numbers, such as 24, by this method.[4])

Feynman diagrams

> Yes. I was seeing something in space and time. There were quantities associated with points in space and time, and I would see electrons going along, scattered at this point, then it goes over here, scatters at this point, so I'd make little pictures. . . . That's what those things were. Emits a photon, the photon goes over here —. . . . And I did think consciously: Wouldn't it be funny if this really turns out to be useful, and the *Physical Review* would be all full of these funny-looking pictures? It would look very amusing."
> R. P. Feynman, talking about how he invented his diagrams

So, similarly, expand the path integral for QED as an infinite series in e^2. Exploit the smallness of e^2 and just calculate the first few terms. For quantum field theory, unlike the simple division problem discussed above, we have the sum over all possible histories followed by the electron and the photon fields, so as you can imagine, the number of terms proliferates explosively. (For an example, see figure 3 later in this chapter.) Feynman proposed drawing diagrams to keep track of the various terms.

Indeed, you have already seen how Feynman diagrams emerge naturally. For any given process, there are an infinite number of Feynman diagrams, corresponding to the infinite number of possibilities. To illustrate, consider the two electrons exchanging a photon in figure IV.1.1. After doing that, one of the electrons could again emit a photon, just feels like it, which is then absorbed by the other electron, in a process known as two photon exchange. And the two electrons could exchange yet another photon. And yet another. Perhaps not surprisingly, theoretical physicists call the infinite set of diagrams in figure 1 "ladder diagrams."

Furthermore, the two electrons exchanging photons could well emit some photons which take off on their own. Since the photon is massless, it is readily emitted, at the slightest provocation. You are in all likelihood exploiting this basic fact[5] of the universe right this instant in order to read this book.

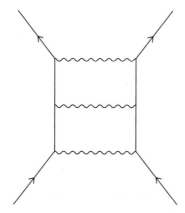

Figure 1. A ladder diagram for electron-electron scattering.

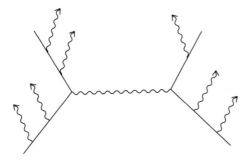

Figure 2. Two electrons scattering off each other, emitting photons like crazy in the process.

Again, in hindsight, perturbation theory seems straightforward. Far from it! Because there are so many possibilities that the fields could follow, the calculations often end up giving infinity as an answer. Indeed, it took Schwinger and Feynman, two theoretical physicists of prodigious abilities, to even calculate a few terms. Were it that easy, the generation of Heisenberg and Dirac would have done it before the war.[6]

Polarizing the vacuum

Look at the Feynman diagrams in figures 1 and 2. You might wonder that while the two electrons are busily exchanging and emitting, why can't the photon be engaged in some shenanigans of its own? Indeed it can. By staring hard, you might be able to see what the photon can do. Yes? If so, you have the mind of a theoretical physicist!

The photon could produce an electron positron pair, and then the electron and the positron could come back together and annihilate each other,

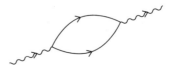

Figure 3. Vacuum polarization: A photon moving along creates an electron and a positron, which subsequently annihilate each other, turning themselves back into a photon. This is one example of the incessant fluctuations in the quantum world described in the prologue.

producing a photon. See figure 3. The net result is that a photon moving along in spacetime could metamorphoze into an electron positron pair for a short time, determined by the uncertainty principle and made possible by the confluence of special relativity and quantum mechanics, as described in the prologue.

The phenomenon I just described is known as vacuum polarization.[7] By creating an electron and a positron, the photon is effectively polarizing the vacuum, splitting the vacuum into a negatively charged region and a positively charged region. Perhaps this is reminiscent of splitting the people into the far left and the far right. A photon merrily moving along could do this again and again.

Incidentally, sometimes students in my quantum field theory course ask: How can the electron and the positron could curve back and meet each other again? Don't they zip off in straight lines, almost at the speed of light? The answer: remember that they are actually de Broglie waves created by the electron field. Everyday manifestation of waves: sound could go around corners.

In developing quantum electrodynamics, calculating the vacuum polarization correctly posed a major challenge for quantum field theorists. Unhappily, some complications[8] peculiar to electromagnetism are involved in guaranteeing that the photon stays massless. One way of looking at that mysterious quantity called mass is that it affects how a particle with mass would propagate, that is, move along, in spacetime. This was suggested by the discussion in chapter III.2. The difficulty is to maintain something called gauge invariance.* If you mess up, the photon could become massive, a disaster for physics.

Schwinger was the first to overcome the difficulty. In fact, Feynman was unable to do it. Schwinger liked to mention that Feynman in his paper simply and unjustifiably neglected vacuum polarization.

Fantastic agreement between theory and experiment

Finally, finally, we arrive at the most fantastic agreement between theory and experiment in physics, mentioned in the prologue. For your convenience, let me

---

*To be discussed further in chapter IV.4.

Figure 4. Julian Schwinger's tombstone

repeat here the calculated value for half the gyromagnetic ratio of the electron: $g_e/2 = 1.001\,159\,652\,181\,643(764)$.

In the previous chapter, I recounted the famous story of Dirac being too nervous to calculate this quantity after he wrote down the Dirac equation in 1928. Before Dirac, the theoretical prediction was $g_e/2 = 1/2$, off by a factor of 2. Dirac's prediction was $g_e/2 = 1$, namely the first digit in the present day theoretical value given above.

After the war, a new generation came into theoretical physics. The race was on between the two prodigies, Schwinger and Feynman, to calculate the deviation from Dirac's prediction of 1. In 1948, Julian Schwinger succeeded in calculating the first correction, of order e^2, obtaining* $g_e/2 = (1 + \frac{\alpha}{2\pi}) \simeq 1.001\,161\,4$. This result[9] is engraved on the great physicist's tombstone (figure 4).[10]

Feynman duly congratulated Schwinger for determining the first order correction to Dirac's calculation. The second order correction was calculated in 1957 by Charlie Sommerfeld in his Harvard PhD thesis supervised

---

*Here $\alpha \equiv e^2/4\pi \simeq 1/137$, defined by the coupling e between the electron and the photon, is known for historical reason as the fine structure constant.

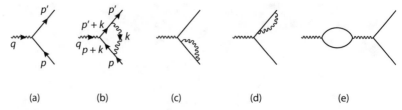

Figure 5. This is taken straight from my quantum field theory textbook. Of the five diagrams shown, (a) gave Dirac his triumph, the factor of 2 mentioned in the prologue; (b) gave Schwinger his triumph, the $\alpha/2\pi$ carved on his tombstone; while (c), (d), and (e) do not contribute to the anomalous magnetic moment. Incidentally, in (e), the photon is polarizing the vacuum.
Reproduced from A. Zee, *Quantum Field Theory in a Nutshell*, Princeton University Press, 2010.

by Schwinger. Both experimental measurements and theoretical calculations continue, with ever higher precision. More on this in an addendum to chapter V.3.

Two great triumphs: Dirac's and Schwinger's

> Feynman had found his vision in a paper of Dirac that gave a . . . setting for action, the natural invariant starting point of a relativistic theory. I found my vision in the same place.
> J. Schwinger[11]

To calculate the magnetic moment of the electron, you calculate the diagram shown in figure 5(b). (Of course, a fairly deep understanding is needed to deduce that only this diagram contributes.) The solid lines represent the electron, the wavy lines the photon. In 1948, it took a Schwinger to get through the calculation, but this many decades later, it has become a standard exercise in quantum field theory textbooks.[11]

In fact, I have now told you enough about Feynman diagrams that you could describe the corresponding physical process. Note that the lines are labeled by momentum, so these are momentum space diagrams. In figure 5(a), Dirac's triumph, an electron with momentum p absorbs a photon with momentum q and goes off with momentum $p' = p + q$. Try to determine what figure 5(b), Schwinger's triumph, describes before reading on.

OK, time's up. In figure 5(b), an electron with momentum p emits a photon with momentum $-k$ (which is the same as absorbing a photon with momentum k) and goes off with momentum $p + k$. It then encounters the incoming photon with momentum q, absorbs it, and goes off with momentum $p + q + k = p' + k$. On the way, the electron absorbs the photon with momentum $-k$ that it had emitted earlier, and ends up with momentum $p' = p + q$. See, it is very pictorial and intuitive.

In a bit of self-evident terminology, the diagrams in figure 1(a) is called a tree diagram, while those in figure 1(b–e) are called "loop diagrams." Speaking loosely, we could say that Schwinger and Feynman succeeded in tackling loop diagrams, while the generation before them, including such brilliant minds such as Dirac, could only calculate tree diagrams.[12]

Schwinger had developed his own approach, which I will not describe here. Considerable confusion resulted, but ultimately Freeman Dyson showed that the two rival approaches, Schwinger's and Feynman's, amount to the same thing.

Feynman rules

Feynman developed a set of rules for calculating the quantum amplitude associated with these diagrams. For instance, if you see a line representing a photon, you should write something called a "photon propagator;" if you see a line representing an electron, write down something called an "electron propagator;" and so on. If you see a point at which a photon line joins an electron line, called an "interaction vertex," you should write the coupling e. And so on. After you write all this stuff down, following these Feynman rules, you multiply it all together. For a tree diagram, you're done. For a loop diagram, you have to integrate over the momentum, with the help of a set of standard tricks Feynman invented.[13] The process became totally mechanical. You just memorize some rules,[14] literally like a cookbook recipe, and you can call yourself a quantum field theorist.

As Schwinger said, Feynman brought quantum field theory to the masses (figure 6).

One of my postdocs used to disparage people he did not like as "professors of Feynman diagrams." Sad but unfortunately true, many could learn the rules without much understanding, go through the motion, obtain an academic position, and teach the next generation to do the same.

Two prodigies of theoretical physics

Naturally, for many years at Harvard, it was implicit that Feynman diagrams were not needed and were even frowned upon. I attended Schwinger's course every year for four years in a row and I never saw a single Feynman diagram. I understood, without anybody having to tell me that only weaklings would need to use them, like mental crutches. But since I was not Schwinger's dissertation student, I was free to calculate as many Feynman diagrams as I pleased, and I did. I heard, however, that Schwinger's students would hide somewhere, calculate using Feynman diagram, and then translate the calculation back into

Schwinger:
"Feynman brought quantum field theory to the masses"

Figure 6. Julian Schwinger.
From the Nobel Foundation.

the master's lofty language and formulation. Here is how Paul Martin,[15] one of Schwinger's most prominent students from the 1950s and later chair of the Harvard physics department, described what went on in his days. See figure 7.

> "In the dark recesses of the sub-basement of Lyman, where theoretical students retired to decipher their tablets, and where the ritual <u>taboo on pagan pictures</u> could be safely ignored."
>
> Paul Martin
> Physica <u>96</u> (1979), p70

Figure 7. The taboo on pagan pictures: Lyman is the name of one of the ancient buildings housing the Harvard physics department.

Figure 8. Schwinger (left) and Feynman (right) at the Nobel ceremonies in Stockholm, December 1965. People who study body language and its purported link to personality would have a field day with this photo.
© Sebastian Schmittner. CC BY-SA 3.0.

Yes! I revere Schwinger and Feynman much like the gods of Greek mythology, capable of feats almost beyond ordinary mortals in the theoretical physics community. I was fortunate enough to have known both Schwinger and Feynman, albeit as a much lowlier person, the former when I was a graduate student enrolled in his course for four years in a row, and the latter when I was an assistant professor. They were both formidable in their own ways.

Stanley Deser, one of the most distinguished theoretical physicists of our times, a Schwinger student from 1953 and someone who knew both Feynman and Schwinger, wrote an insightful essay[16] marveling at the fact that these two giants came from the same milieu (upwardly mobile middle class New York Jewish) yet were endowed with diametrically opposite personalities, one an extreme extrovert, the other retiring and erudite. See figure 8.

As some readers know, the adulation of Feynman was, and continues to be, stupendous, with an army of idolaters retelling his many exploits (safe cracking during the Manhattan Project, bongo drums, Las Vegas showgirls, Brazilian carnival, strip clubs,[17] etc.), and they were the stuff of legends sweeping physics students off the ground and part of the culture I am immersed in. I believe that, quite understandably, Schwinger felt bitter, and in my opinion, it would run counter to human nature to be otherwise.

There is no doubt, however, that they held each other in the highest respect. Later in life, for a variety of reasons, Schwinger became unhappy with Harvard. After the Turkish American physicist[18] Asim Yildiz taught him tennis, Schwinger wanted to move to sunnier climes, and subsequently accepted a position at UCLA. Reportedly, when Feynman heard about his new "neighbor" in Los Angeles, he made several attempts to contact Schwinger, but they never did meet. Schwinger wrote a memorial essay[19] when Feynman passed away. They both contributed enormously to theoretical physics, even putting aside their historical work on QED.

Notes

[1]The system of units used in electromagnetism is notoriously non-uniform, with different systems used in different areas of physics and in engineering. Quantum field theorists favor the Heaviside-Lorentz system. See *FbN*, appendix M.

[2]The infinite series is apparently of use only if ε is small, like 0.1. In fact it is still correct for ε not that small, like 0.5, for example. In that case, you would have to keep quite a few more terms to get a reasonably accurate answer. $1/(1-\varepsilon) = (1+\varepsilon+\varepsilon^2+\varepsilon^3+\cdots)$. Since $(0.5)^2 = 0.25$, including the second order corrections gives $1+0.5+0.25 = 1.75$, which is already reasonably close to the exact answer $1/(1-0.5) = 1/0.5 = 2$.

[3]*GNut*, pages 207–208.

[4]Note that $24 = 25 - 1 = 25\left(1 - \frac{1}{25}\right)$, so that $\varepsilon = \frac{1}{25}$.

[5]Examples of the ready emission of photons: by electrons rushing about in the sun, by electrons jostling along with the crowd down a wire filament in a lightbulb, or by electrons hopping about as carbon atoms combine with oxygen atoms in a candle flame.

[6]For a glimpse of the history leading up to the revolution in quantum field theory after World War II, see Schwinger's preface to the collection of historic papers he edited, *Quantum Electrodynamics*, ed. J. Schwinger, Dover, 1958. I would not recommend reading the papers contained therein, as the notation and concepts tend to be antiquated, but you could skim them to get a sense of the pervading confusion and excitement at the time.

[7]See, for example, *QFT Nut*, chapter III.7.

[8]I still remember with a shudder my struggles with these complications.

[9]Incidentally, this monumental paper (*Physical Review* 73, page 416, 1948) is exactly one page long, without a single displayed equation or any calculational detail. The result is "buried" inside a long paragraph: "= 0.001162. It is indeed gratifying that ... confirm this prediction." I can only say how times have changed.

[10]The formula for entropy was carved on Boltzmann's tomb.

[11]In J. Schwinger, "Renormalization theory of quantum electrodynamics: an individual view" page 344.

[11]In particular, it takes exactly two pages in *QFT Nut*. See pages 196–198.

[12]Actually, that is not strictly true. Together with a student, Heisenberg calculated photon-photon scattering through coupling to an electron loop. But the calculation is enormously more cumbersome than the later calculation using Feynman diagrams. See D. Kaiser, page 36.

[13]Remarkably, to do things his way, Schwinger had to develop another set of tricks.

[14]Listed in every textbook, for example, *QFT Nut*, page 534.

[15]From whom I learned many body theory.

[16]S. Deser, *American Journal of Physics* 86 885(2018).

[17]I even had the honor of being invited to one such joint, where I enjoyed hearing Feynman holding forth on his theory of aesthetics.

[18]To befriend me, he gave me an elaborately carved Turkish pipe, but surely he realized that as a mere student, my knowledge of quantum field theory was rather skimpy.

[19]It ends with this memorable description: "an honest man, the outstanding intuitionist of our age and a prime example of what may lie in store for anyone who dares to follow the beat of a different drum." See *Physics Today*, vol. 42, 1989. Truly masterful!

The road to gauge theory ▬▬▬▬▬▬

Action requires potential, not force

Open a high school level textbook on physics. You will see forces everywhere, pushing this way and pulling that way. But open a graduate level, or even an undergraduate level, textbook on quantum physics, and you will have some difficulty finding the the word "force." The talk is all about probability amplitude and probability. Force becomes a derived or secondary concept.

In one archetypal class of jokes, the drunkard's wife notices that the probability of finding her husband in the neighborhood tavern is much higher than the probability of finding him in the local library. From this, she could deduce that the tavern exerts an attractive force on her husband, pulling him away from the library. Similarly, in quantum physics, if the probability of finding a particle in one region is larger than that of finding it in another region, we would deduce that the particle's potential energy is lower in one region than in the other. The spatial variation of the potential energy is interpreted as a force pulling the particle toward one region and away from the other.

Even in classical physics, the Euler-Lagrange action formulation signals a movement away from force as a primary concept. It is all about following the path that extremizes the action, an entity defined as the difference between the kinetic energy and potential energy.

In electromagnetism, the same sad ballad about the force is sung as in mechanics. The part of the action describing how the electromagnetic field couples to the charge and current densities has no room for the electric $\vec{E}$ and magnetic $\vec{B}$ fields. As we saw in chapter III.3, Lorentz invariance forces the potentials ϕ and $\vec{A}$ to partner together to form $A_\mu = (\phi, \vec{A})$ which then couples to $J^\mu = (\rho, \vec{J})$, the charge and current densities.

Dumpty's tragic end does not depend on where the wall is

I already asked Humpty Dumpty to demonstrate the action principle in chapter II.1. To explain gauge invariance, I invite Dumpty back to sit on a wall. His potential energy V is given by mgh, that is, his mass m times g, the acceleration due to gravity, times the height h of the wall. (You might recall that g is about 10 meters per second per second on the earth's surface and that the combination mg is known in everyday language as weight.) For future use, we note also that the combination gh is called the gravitational potential.

Recall that the work you've done on an object is equal to the force you exerted on the object multiplied by the distance through which the object was moved. Imagine lifting Humpty Dumpty from the ground up to where he is sitting on the wall. The force needed is mg and the distance moved is h. The work done is converted into potential energy, which is stored indefinitely until it is converted into kinetic energy.[1]

The crucial feature of the children's story is that Dumpty's tragic end does not depend on where the wall is. The wall could have been in a beachside resort or in a mountaintop monastery. Only the difference between the gravitational potential on top of the wall and the gravitational potential on the ground matters. Let's say the monastery is located at an altitude H. Then the gravitational potential on the ground equals gH while the the gravitational potential on top of the wall equals $g(H+h)$. But all that Dumpty cares about is the difference between the gravitational potential on the wall and on the ground: $g(H+h) - gH = gh$. The enormous potential gH is relevant only for someone planning to fall from the monastery all the way down into the beachside resort. Dumpty's end depends only on h, not on H.

But now you see that the seeming frivolity about beach resort and monastery I am yakking about has a point, apparently trivial, but a point nevertheless. We can add a constant to the potential without changing the physics. The math is not difficult: $g(H+h) - gH = gh - 0 = gh$. The point is that gH matters not, as long as we add it to both the potential on the wall, namely gh, and on the ground, namely, 0.

From Dumpty to electric voltage

A similar story applies to the electric potential ϕ, known as voltage in everyday language. Only the voltage drop between two points matters.

Household voltage is measured relative to the ground, defined to have zero voltage. Perhaps you have asked whether an electrical appliance is grounded or not. Or whether a certain person is grounded.* The correspondence

---

*A term now with two connotations in the United States, depending on the age of the person being referred to.

between electrostatics and everyday gravity thus extends even to the term "ground."

Perhaps some readers know that the electric field between two parallel plates maintained at different voltages is given by the voltage difference divided by the distance between the two plates.[2] But this corresponds precisely to the force of gravity (commonly known as weight) being the difference between the gravitational potential energy on top of the wall and the potential energy on the ground divided by the height of the wall: $(mgh - 0)/h = mg$.

So, as was just said, the electric field E between two plates maintained at two different values of ϕ is given by the difference in ϕ divided by the separation d between the plates. More generally, the electric field $\vec{E}$ is given by how the potential ϕ varies in space. A charge q introduced into this electric field would experience a force given by $\vec{F} = q\vec{E}$. The larger q is, the larger the force. (Note that this corresponds to the more familiar example of the gravitational force. The force exerted on a mass m introduced into a gravitational field experiences a force[3] proportional to the mass m.)

Gauge freedom in electromagnetism

Dumpty showed us that we can add to the gravitational potential whatever constant we like without affecting the physics. Similarly, we could shift the potential ϕ on the two plates by the same constant without changing the electric field $\vec{E}$ at all between the two plates. The shift of ϕ by $\phi \to \phi + \text{constant}$ is a particularly simple example of a gauge transformation. That we are free to do this is known as a gauge freedom.

I already mentioned in chapter III.3 the vector potential $\vec{A}$ whose spatial variation determines the magnetic field $\vec{B}$. Similarly, we enjoy a gauge freedom in changing $\vec{A}$ without changing $\vec{B}$.

The situation becomes rather more involved in the presence of both an electric field and a magnetic field: They are intertwined in general into an electromagnetic field. For instance, a moving magnetic field produces an electric field, and vice versa. So the electric field ends up depending not only on ϕ but also on $\vec{A}$. Now, when we gauge transform ϕ, we also have to gauge transform $\vec{A}$ appropriately in order to keep the electromagnetic field unchanged.[4] Not surprising, given that the electric and magnetic fields are intertwined. Incidentally, this point represents a watershed between an undergraduate course on electromagnetism and a graduate course on electromagnetism.

Some jargon for future use. The gauge potentials ϕ and $\vec{A}$ are said to be gauge variant, while the electric $\vec{E}$ and magnetic $\vec{B}$ fields are said to be gauge invariant. Interestingly, although the action is constructed out of the gauge variant ϕ and $\vec{A}$, they are arranged in such an intricate way that the action itself, and hence the path integral, are gauge invariant.[5] Theories of this type are known as a gauge theories, and Maxwell electromagnetism is the granddaddy of them all.[6]

Destined for the dustbins of history, not!

In the late 19th century, there was a heated debate among theoretical physicists whether the potentials ϕ and $\vec{A}$ are even necessary for physics. At one extreme, the influential theoretical physicist Oliver Heaviside thundered that the potentials were destined for the "dustbins of history."

The most forceful argument against the potentials is that, ultimately, experimentalists can only measure the electric and magnetic forces acting on charged particles. True, theorists found it impossible to formulate the action for electromagnetism without ϕ and $\vec{A}$. But still, Maxwell was able to write down his equations using electric $\vec{E}$ and magnetic $\vec{B}$ fields only, without having to mention the potentials ϕ and $\vec{A}$.

At this point you may will ask, so, what is the problem? There seems to be no problem. Experimentalists could continue using Maxwell's equations; they do not need, practically speaking, the action.

The real crunch comes with quantum physics. In all three of the formulations I mentioned in chapter II.2, those of Schrödinger, Heisenberg, and Dirac-Feynman, the potentials ϕ and $\vec{A}$ have to be given the starring roles from the get-go. We could see this most glaringly in the path integral formulation. And as explained earlier, in both the Schrödinger and Heisenberg formulations, force is not a primary concept (and so by extension, neither are $\vec{E}$ and $\vec{B}$.)

The experimentalists, and Heaviside, were wrong.

Spooky: ghostlike presence of the magnetic field

A dramatic demonstration that quantum physics needs ϕ and $\vec{A}$ came in 1959. (Note the late date[7] compared to the founding of quantum mechanics in 1925!) The theoretical physicists Yakir Aharonov and David Bohm proposed a clever arrangement such that (as shown in figure 1) a vector potential $\vec{A}$, but not a magnetic field $\vec{B}$, is present in a region in space. (You might immediately object, how is that possible, since $\vec{B}$ is given by the spatial variation of $\vec{A}$? The secret is that $\vec{A}$ has three components, and each of these components could vary in space in a precisely choreographed way such that their variations cancel each other, not producing[8] a $\vec{B}$. Trust me, Mother Nature's choreography is often a marvel to behold!)

A classical physicist could not care less about this possibility, knowing that all of electromagnetism is determined by Maxwell's equations. A quantum physicist, in contrast, would be puzzled, at least initially. As mentioned earlier, the action has to be formulated in terms of $\vec{A}$, and the action determines the probability amplitude for a particle to go from here to there.

To resolve this puzzle, Aharonov and Bohm considered an experimental arrangement in which a beam of electrons is split into two. See figure 1. The

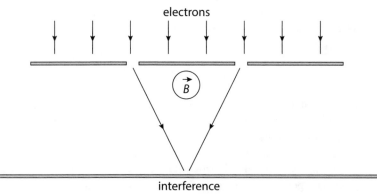

electrons

$\vec{B}$

interference

Figure 1. The magnetic field $\vec{B}$ is restricted to the region indicated by the circle. Electrons travel (from the top of this diagram) through a screen with two slits, interfere with each other, and are then detected on the screen shown at the bottom of this diagram. (The setup is similar to the classic interference experiment demonstrating the interference between waves, as shown in fig I.5.1.) While the electrons never travel through a region with the magnetic field, they could nevertheless feel its ghostlike presence thanks to the mystery of quantum physics.
Modified from Aharonov-Bohm effect in an interference experiment drawn from Sebastian Schmittner. CC BY-SA 3.0.

two beams traverse a region with $\vec{A}$, but without ever encountering a magnetic field $\vec{B}$. They are then allowed to interfere, with the result registered on a detector screen. The prediction is that even though the electrons never come near the magnetic field $\vec{B}$, the field could still influence the interference pattern.

This is such a surprising result that even a quantum heavyweight like Niels Bohr flatly refused to believe it. Sort of spooky that the electron could sense the presence of a magnetic field without actually encountering it. Faraday would be mystified and astonished for sure!

Gauge invariance of the Dirac action

In a relativistic theory, the potentials ϕ and $\vec{A}$ are packaged into a 4-vector A_μ, as we saw in chapter III.3. The gauge freedom we enjoy could be written compactly as $A_\mu \to A_\mu + \partial_\mu \Lambda$, with $\Lambda(x)$ an arbitrary function of spacetime. (As a particularly simple example, the electrician grounding the electric potential by shifting the voltage by a constant is actually setting $\Lambda = V_0 t$, that is, taking Λ to be a linear function of time. The electric potential shifts according to $\phi = A_0 \to A_0 + \partial_0 \Lambda = \phi + \frac{\partial}{\partial t}(V_0 t) = \phi + V_0$. With the appropriate choice of the constant V_0, we can set ϕ to 0 at any location we desire.[9] Note that under this gauge transformation, since Λ does not depend on the spatial coordinate, $\vec{A}$, and hence B, is not changed at all.)

So, we could shift A_μ to $A_\mu + \partial_\mu \Lambda$ without changing the electromagnetic field, and hence without changing Maxwell's action governing the classical dynamics of the electromagnetic field. Let's look at the Dirac action[10] for the electron field ψ interacting with the electromagnetic field A_μ as given in chapter IV.2:

$$S = \int d^4x \; \bar{\psi}\left(i\gamma^\mu(\partial_\mu - ieA_\mu) - m\right)\psi = \int d^4x \; \bar{\psi}\left(\cdots\cdots A_\mu \cdots\right)\psi$$

Never mind all the details, such as γ^μ, ∂_μ, and all that; look at A_μ sitting there, as I have indicated schematically after the second equal sign. If A_μ changes, the action S sure as life is going to change.

Since S also depends on the electron field ψ, the only way the action S could remain unchanged is for the electron field ψ to also change, in precisely such a way as to compensate the change of S caused by A_μ changing. Indeed, Herman Weyl and others pointed out that, if $A_\mu \to A_\mu + \partial_\mu\Lambda$, the electron field has to change as[11] $\psi \to e^{i\Lambda}\psi$. Isn't it funny that Nature works[12] in this way? I find it rather peculiar.

Gauge: a brief etymological note

Weyl named his symmetry gauge symmetry. The term "gauge" comes from low Latin "gaugia," referring to the standard size of casks, and this sense is retained in such modern usage as "railroad gauge" and[13] "gauged skirt." Curiously, the word entered the permanent vocabulary of physics only because Weyl made a serious but justifiable mistake. We now know that the symmetry responsible for electric charge conservation is described by transformations involving quantum fields. Weyl was working before the advent of quantum physics, so he, like everyone else, never dreamed of probability amplitudes and complex numbers, such as $i = \sqrt{-1}$. Instead, inspired by the geometric flavor of Einstein's work, Weyl proposed a transformation in which one changes the physical distance between spacetime points. Weyl was reminded of the distance, or gauge, between two rails—hence the name for his symmetry. He showed Einstein his theory, but they were both deeply disappointed that it failed to describe electromagnetism. When the quantum era began, Weyl's theory was quickly repaired by including an i. Meanwhile, the term "gauge symmetry," although a misnomer, remained.

Notes

[1]Perhaps you also remember from chapter I.3 the heavy box you put in the closet only to use the potential energy you gave it against you. Or the ski lift operator from chapter II.1 whom you paid to provide you with potential energy for you to convert into kinetic energy.

[2]Some readers might know that this arrangement is called a "capacitor" and allows us to store electric charges.

[3]Commonly called "weight." Reportedly, this distinction between mass and weight bedevils high school students of physics. Come on, it

is very simple: Mass has meaning throughout the universe, while weight is a provincial concept specific to the planet the mass happens to be on.

[4]For those readers who know some vector calculus and who are interested in how all this works, see appendix M of *FbN*. Some textbooks try to make gauge transformation sound profound and complicated, but believe me, it is all quite simple and logical.

[5]For further discussions, see for example, *QFT Nut*, chapter II.2.

[6]For the historical roots of gauge invariance, see J. D. Jackson and L. B. Okun, *Reviews of Modern Physics* 73, 2001.

[7]Actually, W. Ehrenberg and R. Siday had independently predicted the effect in 1949.

[8]For those readers who know a bit of vector calculus, $\vec{B}$ equals the curl of $\vec{A}$, and so it is possible to have $\vec{A} \neq 0$ but $\vec{B} = 0$.

[9]In everyday life, that would be the ground.

[10]We do not need to include the Maxwell action governing the electromagnetic field, since that is gauge invariant by itself, as was just remarked.

[11]The reader who knows a bit of calculus could see that, without going to any mathematical detail, that the partial derivative ∂_μ acting on the phase factor $e^{i\Lambda}$ would bring down a term like $\partial_\mu \Lambda$, which is just what we need to cancel the change in A_μ. Meanwhile, $\bar{\psi} \to \bar{\psi} e^{-i\Lambda}$, thus canceling the phase factor coming from ψ. (By the way, a similar cancellation mechanism occurs in nonrelativistic quantum mechanics as well, if we simply reinterpret ψ as the Schrödinger wave function. This particular discussion has nothing to do with special relativity as such. Readers familiar with the Schrödinger equation would know that in the presence of electromagnetism, it contains a term proportional to $\psi^\dagger (\vec{\nabla} - ie\vec{A})^2 \psi$.)

[12]Of course, we could debate whether Nature actually works in this way or whether physicist's description of Nature works this way. Chaired professorship of philosophy at an elite university, here I come!

[13]See the explanation of gauging by the Sewing Academy: https://www.thesewingacademy .com/tag/gauging/ Thus, "Gauging is typically reserved for handling skirt fullness, and develops as a common technique in the early to middle 1840s, when increasing skirt circumferences and fashion preference outstrip stroked gatherings ability to control fullness without increased bulk."

Recap of part IV

Once Dirac promoted the electron to a field and allowed it to interact with the electromagnetic field, the action is no longer quadratic, and the corresponding path integral can no longer be evaluated. Fortunately, the probability amplitude for the two fields to interact is small, so that a perturbative approach works marvelously. This amounts to expanding the path integral in an infinite series in the small quantity e^2 and keeping only the first few terms. Feynman diagrams may be thought of as a pictorial device to keep track of the various processes that appear in the expansion. Schwinger's triumphant calculation of the gyromagnetic ratio of the electron gave physicists confidence in the fundamental soundness of quantum field theory.

The action, so crucial for quantum physics but unnecessary for practical calculations in classical physics, can only be formulated in terms of the electromagnetic gauge potential A_μ rather than the electromagnetic field $\vec{E}$ and $\vec{B}$. This became the prototype of all the gauge theories that were to come in the second half of the 20th century, as will be described in part V.

A well-deserved rest

Congratulations! You made it through the essence of quantum field theory. You have come a long way. Starting from our home village, you have completed an arduous quest and arrived in the fabled northeast quadrant on our map, where special relativity and quantum mechanics reign hand-in-hand.

In part V, we will discuss the application of quantum field theory to the study of the four fundamental interactions in the universe. In part VI, I conclude by placing quantum field theory in its proper place within the intellectual framework of theoretical physics.

Part V is necessarily massive. Each of the fundamental interactions was long shrouded in mysteries, and hence quite readily, each of the chapters in part V could be expanded into an entire book. Think of part V as a book on applied quantum field theory, a book within a book. The term "applied quantum field theory" is however not completely appropriate, since it is not simply a matter of applying what we have learned about quantum electrodynamics to the other three interactions. Many of the concepts in contemporary quantum field theory were developed through the struggle to elucidate these "other" interactions.

At this natural rest stop in our trip, you could perhaps take a well-deserved break. Think of part V as a reward for all your hard work. Now that you understand the essence of quantum field theory, you are in a position to penetrate much further into the physical mysteries of the universe than would normally be possible in a popular book.

Quantum field theory and the four fundamental interactions

Preview of part V

Since quantum field theory was developed to address the wealth of peculiar phenomena that appeared at the confluence of special relativity and quantum physics, its development was naturally intertwined with the development of particle physics, also known as high energy physics. But just as atomic physics is not synonymous with quantum mechanics, particle physics is not synonymous with quantum field theory.

We will start with one of the most striking predictions of quantum field theory, the existence of antimatter. Then we will explore the strong and the weak interactions in turn. The electromagnetic interaction, in contrast to these two, nurtured the growth of quantum field theory and so has already been focused on in the earlier parts of this book. I then explain electroweak unification. This is followed by the exciting possibility of grand unification, by which three of the four fundamental interactions may be unified into a single interaction. Finally, we turn to quantum gravity as a quantum field theory.

Since particle physics is by itself an extraordinarily rich subject, with each of the four interactions easily occupying an entire book, I have to cut even more corners than before, echoing Feynman's warning cited in the preface.

A more casual reader may wish to read only chapter V.1 in detail and then skim the rest of this part to obtain a flavor of how we have gotten to our present understanding of the four fundamental interactions. It is also possible, or perhaps even advisable for some readers, to skip ahead to part VI and come back to this part later at your leisure.

Antimatter!

Undreamed of creatures
in a fabled land

In the quest we embarked on in the prologue, when we journey into the fabulous land of the the northeast, where special relativity and quantum mechanics reign together, we encounter previously undreamed of creatures. Surely among the most astonishing is a tribe of antiparticles.

After Dirac wrote down his equation[1] for the electron in 1928, he noticed that it also described a particle with positive electric charge, opposite to that carried by the electron. Quite excitedly but erroneously, Dirac thought that this other particle might be the proton. However, the proton is almost two thousand times more massive than the electron, while the equation demands that this mysterious particle must have the same mass[2] as the electron. This eventually led to the realization that the Dirac equation does not contain the proton, but a hitherto unknown particle, now called the "positron." All this theoretical confusion was swept away shortly by the experimental discovery in 1932 of the positron, the "antielectron." The electron and the positron are oppositely charged but have exactly the same mass.

We now know that particles and antiparticles come in pairs, for example the proton and the antiproton.[3] Some particles may be their own antiparticles, for example, the photon. (Clearly, these "self antiparticles" cannot carry electric charge.)

Dirac was already revered for his role in the founding of quantum mechanics. With the Dirac equation[4] and the prediction of antimatter, he was catapulted to the very highest pantheon of theoretical physics. Let me now give you a strikingly simple argument, with the help of a Feynman diagram, showing why special relativity and quantum mechanics together mandate antimatter.

The fall of simultaneity

Einstein famously bent the stately flow of time out of shape. In chapter I.3, I dramatized a gedanken experiment he thought up showing that the constancy of the speed of light necessarily alters our cherished and "obvious" concept of simultaneity. In the story of dueling thinkers, one observer could say that Professor Vicious solved a physics puzzle before Dr. Nasty, while another observer said the opposite. Yet they are both absolutely right.

This strange fact presages the existence of antimatter!

Electric charge annihilated and created

As soon as Dirac set the electron free and introduced the electron field ψ, antimatter became obligatory. As a table in chapter IV.2 made clear, if ψ could annihilate an electron, then it has to be able to create a positron. Essentially, this amounts to saying that

$$(\text{create} - 1) \quad \text{is equivalent to} \quad (\text{destroy} + 1)$$

$$(\text{create} + 1) \quad \text{is equivalent to} \quad (\text{destroy} - 1)$$

The field $\bar{\psi}$ does the opposite.

No room for argument there. Still, it would be comforting to have a physical argument[5] that special relativity and quantum mechanics together mandate antimatter.[6]

Special relativity and quantum mechanics together require antimatter: a simple physical argument

Consider the scattering of a photon on an electron, $\gamma + e^- \to \gamma + e^-$, a process that goes on around us all the time. Let's look at the Feynman diagram, shown in figure 1, describing this process in spacetime.

The electron absorbs the photon at the point x, propagates to the point $y = (y^0, y^1, y^2, y^3)$, and emits a photon. The point y occurs later in time than the point $x = (x^0, x^1, x^2, x^3)$, for the simple reason that we don't know what propagating backward in time means. (If the reader knows how to build a working time travel machine, let me know.) In other words, $y^0 > x^0$, that is, $(y^0 - x^0) > 0$.

The separation between x and y, both in space and in time, is too small to be detected and can only be inferred. The experimentalist simply sees a photon bouncing off an electron, not an actual electron propagating from x to y. Indeed, as explained in chapter III.1, between x and y the electron propagates

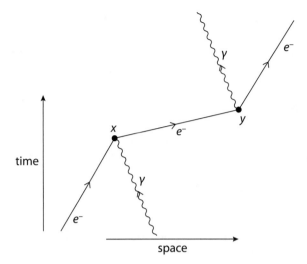

Figure 1. An electron (represented by the solid line) absorbs a photon (represented by a wavy line) at the spacetime point x, then propagates virtually to the spacetime point y, where it emits a photon, and then goes on its merry way. This Feynman diagram depicts the process $\gamma + e^- \rightarrow \gamma + e^-$ as it happens in spacetime.

as a virtual, rather than real, particle, and the separation between the two spacetime points x and y is known as spacelike, not timelike. The jargon means that the spatial separation between the two points is larger than the temporal separation between them.[7]

To make it easier for the reader to see this, let us be definite and take the spatial separation to be along the 1-axis. In other words, set $(y^2 - x^2)$ and $(y^3 - x^3)$ to 0. The spatial separation $(y^1 - x^1)$ could be either positive or negative. Let us choose positive, again just to be definite. Spacelike separation means that $(y^1 - x^1) > (y^0 - x^0) > 0$. In other words, going from x to y, the electron has to move faster* than the speed of light c, a feat forbidden by Einstein in the classical world, but allowed by Heisenberg in the quantum world, albeit only for a fleeting interval of time. (In the units we are using, with $c = 1$ the wordlines followed by light trace out straight lines making an angle of 45° with the horizontal spatial axis, as was explained back in chapter II.1. See figure II.1.3 there. Thus, the angle between the horizontal axis and the straight line joining x and y is depicted as less than 45° in figure 1.)

Note that special relativity hasn't entered yet. Let's go back to the discussion in chapters I.3 and I.4. The crucial point: Einstein tells us that another observer moving by could see x occurring later in time than the point y.

---

*This is just as in everyday life: the speed of the electron is given by the distance traversed $(y^1 - x^1)$ divided by the duration of time $(y^0 - x^0)$, which by assumption is greater than 1, the speed of light in the units we are using.

For the skeptical reader who wants to check this, please look at the Lorentz transformation given in chapter I.4. Let's say we are with Ms. Unprime. One line of high school algebra shows that the time difference seen by Mr. Prime gliding by with velocity u (with c set to 1 as always in this chapter) is given by

$$\left(y'^0 - x'^0\right) = \left(\left(y^0 - x^0\right) + u\left(y^1 - x^1\right)\right)/\sqrt{1 - u^2}$$

Note that the square root $\sqrt{1 - u^2}$ characteristic of special relativity restricts the relative velocity u to be between -1 and $+1$, as usual.

Einstein claims that this time difference could be negative if the propagation from x to y is spacelike. Indeed, just look. For u close to -1, the expression $(y^0 - x^0) + u(y^1 - x^1)$ inside the large parenthesis could indeed be negative.*

The important point is that this reversal of time ordering is not possible if the separation between the two spacetime points is timelike, that is, if $(y^0 - x^0) > (y^1 - x^1)$. The whole point here is that quantum physics allows the separation to be spacelike. Incidentally, this relates back to the discussion in chapter III.2 about the distinction between real and virtual particles.[8]

So, $x'^0 > y'^0$ while $y^0 > x^0$. Mr. Prime sees the field disturbance propagating from y to x, from earlier to later, of course. Since we see negative electric charge propagating from x to y, Mr. Prime must see positive electric charge[†] propagating from y to x. Ta dah, behold the positron!

Nobody noway nohow nowhere sees no nothing going backward in time

Once again, this is the same effect which allows one observer to say that Vicious solved the problem before Nasty, and another observer to say that Nasty solved the problem before Vicious. Note that nobody noway nohow nowhere sees no nothing going backward in time.[9]

Without special relativity, as in nonrelativistic quantum mechanics, we simply write down the Schödinger equation for the electron and that is that. Special relativity allows different observers to see different time ordering and hence opposite charges flowing toward the future. Hence, antiparticles!

---

*For some readers, it may be helpful to pick specific numbers, say, $(y^1 - x^1) = 4$ and $(y^0 - x^0) = 3$ in some unit. In that case, $u = -0.8$, for example, would do the trick: $3 - 0.8 \times 4 = -0.2$ is indeed negative.

†Some readers might realize that this argument, like the more formal argument cited earlier in the text, is also crucially based on charge conservation.

Notes

[1]A legend told around campfires by theoretical physicists: When the extremely chatty Feynman first met the notoriously taciturn Dirac, Feynman was going his usual a mile a minute while Dirac stayed silent. Finally, Dirac said, "I have an equation, do you have one too?"

[2]As was first pointed out by Hermann Weyl in 1931. These days, with the benefit of hindsight, this would be considered "obvious," since a number like $1,836 \simeq m_P/m_e$ "does not grow on trees." For a more precise statement, see endnote 4.

[3]Discovered in 1955.

[4]An unfortunate and persistent confusion around the Dirac equation was generated by some misleading historical concepts, such as "a sea of negative energy states" and "particles traveling backward in time," concepts which mercifully have now been banned from decent modern textbooks and restricted to the web. Heck, if you ever see a negative energy state or a particle actually going backward in time, be sure to contact the Nobel committee pronto.

[5]For those who would like to see a more precise mathematical statement: The Dirac equation for the electron field, with electric charge e, in the presence of an electromagnetic field, reads

$$\left(i\gamma^\mu(\partial_\mu - ieA_\mu) - m\right)\psi = 0$$

(which by the way, we could obtain readily by varying the Lagrangian given in chapter IV.2 with respect to $\bar\psi$.) The mathematical assertion is simply that by complex conjugating this equation (note the two i's!) and rearranging the four components of ψ, we can define a conjugate field ψ^c and rewrite the equation as

$$\left(i\gamma^\mu(\partial_\mu + ieA_\mu) - m\right)\psi^c = 0$$

A sign flips, thus implying that ψ^c carries electric charge $-e$! The field ψ^c describes the positron, with a mass m identical to that of the electron. This offers another example of my favorite dictum that if you see a result in theoretical physics obtained only after many pages of calculation, then it is almost surely not that profound.

[6]The argument given here is adapted from *QFT Nut*, page 157.

[7]Most readers can ignore this endnote; I added it to forestall a potential confusion. Some readers might have noticed that the Feynman diagram shown in figure 1 is exactly the same as the Feynman diagram shown in figure IV.2.2. Simply distort the electron line a bit. But the Feynman diagram in chapter IV.2 looks like the electron propagation from photon absorption to photon emission is timelike. Three responses. (1) I haven't told you what units I am using for space and for time in chapter IV.2. (2) I could have said that the Feynman diagram in chapter IV.2 is in momentum space rather than in spacetime. (3) Feynman diagrams are just sketches of what is happening in the quantum world. Only the topology of a diagram matters.

[8]The use of the word "virtual" has prompted an unfortunate association with the mystical in some people's minds. In reality, it is actually a straightforward and natural concept in theoretical physics best explained with a bit of high school algebra. Einstein demands that the energy E and momentum $\vec p$ of a particle must satisfy (in units with $c=1$) $E = +\sqrt{\vec p^2 + m^2} = +\sqrt{p_x^2 + p_y^2 + p_z^2 + m^2}$, as was explained in part I of this book. (For a particle at rest, $\vec p = 0$, so that $E = m$, which you would recognize as the more sophisticated version of the formula $E = mc^2$ the person in the street is familiar with.) A particle satisfying this condition is real, and said to be "on shell." To understand where this peculiar bit of jargon comes from, you would have to plot (or visualize) E as a function of $\vec p$. To make it easier, set p_y and p_z to 0 and simply plot $E = +\sqrt{p_x^2 + m^2}$. Then visualize putting p_y back in and you would see that the plot resembles a shell.

A particle whose energy E and momentum $\vec p$ do not satisfy Einstein's condition is said to be "off shell." Didn't I just say that being off shell is forbidden by Einstein? But Heisenberg, with his uncertainty principle, assures us that virtual particles could exist in the quantum world, though only for a short duration in spacetime, as was already mentioned in the prologue. So now you also speak the jargon: in figure 1, the electron propagating from x to y, or the positron propagating from y to x, is a virtual particle.

[9]See also *QFT Nut*, page 113.

Too strong and too mean but ultimately free

The strong interaction is strong

The strong interaction is strong, duh; indeed, way too strong for theoretical physicists to handle. Recall, from chapter IV.3, that the coupling strength of electromagnetism, defined by $\alpha \equiv e^2/4\pi \simeq 1/137.036 \cdots \simeq 0.007 \cdots$ is pretty small. (That particular number is easy to remember, eh?) You might also remember why the Greek letter α was carved on Schwinger's tombstone. For comparison, the coupling strength of the strong interaction, namely, the analog of α, is ~ 1.

Consider the series $1/(1 - \varepsilon) = (1 + \varepsilon + \varepsilon^2 + \varepsilon^3 + \cdots)$ we studied in chapter IV.3. For $\varepsilon = 0.1$, the first couple of terms give perfectly sensible results. But for $\varepsilon = 1$, say, the series gives $1 + 1 + 1 + \cdots$ and fails. We could even see why: It is struggling to reproduce the left side $1/(1 - 1) = 1/0 = \infty$. In the same way, for the strong interaction, the perturbative approach to quantum field theory collapses miserably and Feynman diagrams becomes essentially worthless.

To underline this, consider Schwinger's epoch making calculation of the magnetic moment of the electron. With a tiny coupling strength $\alpha \simeq 0.007 \cdots$, only the first order terms need to be kept, corresponding to a small number of diagrams, as shown in figure IV.3.5.

Were you to do the same calculation for the proton, you would have to evaluate an infinite number of Feynman diagrams, poor dear. The incoming proton could have emitted any number of pions (say, 17) before the photon gets there, and these 17 pions could scatter off each other and also continue to interact with the proton. If you ever complete your labor, somebody could always ask you about the process in which the proton emits 47 pions. Furthermore, the quantity of primary interest for the strong interaction is not the magnetic moment of the proton, but rather the coupling of the pion to the proton, and so you should replace the incoming photon by an incoming pion.

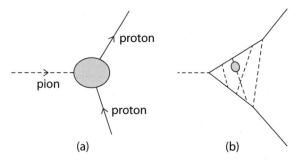

Figure 1. (a) A proton and a pion go into a blob to do their business; afterward, a proton emerges from the blob. (b) One of the infinite number of diagrams that made up the blob.

Just as the incoming photon in figure IV.3.5 could metamorphosize into an electron-positron pair, nothing stops the incoming pion (nor any of the pions emitted by the proton) from metamorphosizing into an proton-antiproton pair (or nine such pairs), with each of them free to emit more pions.

In frustration, you would have to invent the blob, a perfectly academic term in strong interaction circles starting from the late 1950s, to represent the infinite number of Feynman diagrams. See figure 1.

Useful jargon: particles that participate in the strong interaction are known as hadrons, from the Greek root meaning "stout" or "thick," thus hadrosaur, for example. The proton, neutron, and pion are all hadrons. Hadrons also participate in the electromagnetic and the weak interactions, and of course, also gravity. (The electron is an example of a particle that is not a hadron.)

To make things worse, many more hadrons were discovered starting around 1950. The nucleons and the pion turn out to have tens of cousins, with names like the Σ and Λ hyperons, the K and η mesons. Experimentalists kept busy sorting out their properties while theorists wrung their hands. It seemed outrageously profligate to associate a field[1] with each of these hadrons, as Dirac had done with the electron.

A pompous burial for quantum field theory

In the Soviet Union, Lev Landau, one of the most brilliant theorists of the 20th century, declared that quantum field theory was to be buried, but "with pomp" in recognition of its past contributions. In the United States, starting in the late 1950s, the theoretical community was split, roughly speaking, between[2] the east coast and the west coast. A so-called S-matrix school of thought, in direct opposition to quantum field theory, sprang up, centered at Berkeley and

Caltech. Meanwhile, quantum field theory continued to be taught in some places, mostly on the east coast, for instance, by Schwinger at Harvard.

In the fall of 1962, when I was a freshman tackling introductory physics,[3] a fellow freshman and aspiring theoretical physicist excitedly showed me a rather thin book[4] on the S-matrix theory of the strong interaction. Exclaiming that there were hardly any equations inside, he told me that it represented the future of theoretical physics and that we no longer had to learn all that complicated and already passé stuff that was quantum field theory. "All we have to do is master this book!" I was too young to know that a physics monograph that includes so few equations should inspire a healthy dose of skepticism.

Dirac became inessential?

As another indication of the madness of that era, let me imagine showing a physics undergrad nowadays a physics lecture that starts with "Forgetting the Dirac spinors and other inessential details," he or she would likely have thought that this is garbage produced by some crackpot. My generation and younger were raised from childhood to revere Dirac like a god. But no, these words were uttered in 1962 by an esteemed professor at Caltech.[5] Indeed, the catchphrase "ignoring inessential complications such as" was often bandied about in those days.[6]

Fast forward to the fall of 1965, when I had been transformed from an ignorant freshman into a senior thinking about graduate school and when John Wheeler[7] was advising me to go to Berkeley[8] to work for a brilliant rising star named Steve Weinberg. Wheeler said that he already told Weinberg about me and that I should call him. After collecting a huge stack of coins,[9] I found a telephone booth. But Big Steve,* as I have always thought of him since that day, informed me that he was about to leave Berkeley for Harvard.

And thus my fate was sealed.[10] I learned quantum field theory rather than S-matrix theory.

It depresses me greatly to think how many promising young theoretical talents from those days ended up more or less in the dustbins of history. Luckily for me, quantum field theory returned in triumph and fanfare around 1970, just as I was about to be launched into the world. (Indeed, when I arrived at the Institute for Advanced Study in Princeton as a fresh postdoc, a fellow postdoc informed me that I, bereft of the education he had had at Caltech, had not learned the modern stuff. A very helpful fellow, he drew up a deck of flash cards with the properties of the hadrons, such as their masses, lifetimes, and decay channels, etc, written on one side, and their names, such as Σ, Λ, K, and η, etc, written on the other. To educate me properly, he would flash these cards

---

*Sadly, he passed away on July 23, 2021, while this book was being copyedited. As mentioned in the Preface, he read part of this book.

in front of me from time to time. I did start to wonder if quantum field theory was worthless.)

Poles and cuts and the bootstrap

I will pick up the return of quantum field theory later in this chapter and in the next chapter, but for now I want to paint a bit more the desperation the theoretical community felt in facing the monstrous strong interaction. The S-matrix school of thought soon spawned sub-schools,[11] such as polology (as the art of drawing blobs became known as) and dispersion theory (instead of calculating the scattering amplitudes, which was clearly impossible, people studied the location of the poles and cuts[12] of the scattering amplitudes). My graduate school advisor Sidney Coleman quipped that theorists could either group or disperse.[13]

One particularly alluring program is known as the bootstrap. With tens of hadrons on hand, some theoretical physicists decided to throw up their hands and proclaimed that hadron A is made up out of B, C, ..., and Z; that hadron B is then (you guessed it!) made up out of A, C, ..., and Z; and so on, so that finally hadron Z is made up out of A, B, ..., and Y. Nobody is better than anybody else, and everybody is made up of everybody else. Geoff Chew,[14] the father of the S-matrix school, was hailed by one of his followers[15] as "the Thomas Jefferson of nuclear democracy." Self consistency is the name of the game. Pulling on your bootstrap, you are supposedly going to lift yourself up and bring the world of strong interaction into being. You know, levitation and all that.

Dear reader, you have to understand that this is the late 1960s at Berkeley. Riots by day and mellow yellow by night. This "all is one, one is all" manifesto struck many as only a skip and hop away from mystic eastern philosophy, even without any controlled substances being smoked. A popular book equating particle physics with eastern practices became a runaway best seller. You no longer have to bother with the arduous task of learning quantum field theory! Fundamental mysteries of the universe through Buddhist sutras! The appeal is obvious.[16]

American professional football players are often given colorful nicknames. One of my favorites was the aptly named all pro defensive end of the Dallas Cowboys, Harvey "Too Mean" Martin.* In my book *Fearful,* I recollect how my friends and I thought of the strong interaction as "too strong" and "too mean." To avoid dropping out and falling into the dustbins of history, many theorists thought it prudent to work on the weak interaction instead. See chapter V.3.

*He was in the trenches with the likes of "Too Tall Jones" and "Mean Joe Green." Believe it or not, I actually know another one of these giants, Staś Maliszewski, famed for protecting Johnny "Golden Arm" Unitas of the legendary Baltimore Colts.

Gauge invariance

As is often the case in the history of theoretical physics, the developments that finally led to the theory of the strong interaction came out of left field, and not from meditating transcendentally about bootstrap and democracy. I will list some of these strands here.

The notion of gauge invariance in quantum electrodynamics (as already discussed in chapter IV.4) furnishes one strand. For your reading convenience, let me quickly review by focusing on the essentials. First, remember that there was no room for the electric and magnetic fields $\vec{E}$ and $\vec{B}$ in the action, which had to be written in terms of A_μ. Next, as explained in chapter IV.4, we have the gauge freedom to change A_μ to $A_\mu + \partial_\mu \Lambda$, with any function $\Lambda(x)$ of spacetime we like, without changing $\vec{E}$ and $\vec{B}$.

Recall that the action in quantum field theory is just the integral of Lagrangian density over spacetime. Look at the term describing the interaction of the electron with the electromagnetic field in the Dirac Lagrangian density, viz, $\mathcal{L} = \bar{\psi} i \gamma^\mu (\partial_\mu - ie A_\mu)\psi$ with ψ the electron field, in chapter IV.2. It is clear as day that if A_μ changes, $\mathcal{L}$ would change: $\mathcal{L} = \bar{\psi} i \gamma^\mu (\partial_\mu - ie A_\mu)\psi \to \bar{\psi} i \gamma^\mu (\partial_\mu - ie A_\mu - i\partial_\mu \Lambda)\psi$. For $\mathcal{L}$ to remain unchanged, we clearly cannot change only A_μ; we must also change ψ. As was mentioned in chapter IV.4, Weyl showed that at each point x in spacetime, $\psi(x)$ has to be multiplied by a phase factor[17] $e^{i\Lambda(x)}$.

Changing $A_\mu(x)$ and $\psi(x)$ in such a balanced way so that $\mathcal{L}$ remains unchanged is known as a local or gauge transformation, and $\mathcal{L}$ is said to be "gauge invariant."

A small number in the subnuclear world

In 1932, James Chadwick discovered one of the most important small numbers in the history of physics, the mass difference between the proton and the neutron in units of the neutron mass: $(M_n - M_p)/M_n \simeq (939.6 - 938.3)/939.6 \simeq 0.00138$.

Almost immediately,[18] in the same year, Werner Heisenberg[19] and others[20] proposed that the strong interaction is invariant under a set of transformations that turns the proton and the neutron into quantum superpositions of each other. Since the proton is charged while the neutron is not, the electromagnetic interaction was thought to be responsible for the small mass difference. Heisenberg postulated that in a world with electromagnetism switched off, the proton p and neutron n would have equal mass,[21] and we could combine them into a two-component nucleon field $N \equiv \begin{pmatrix} p \\ n \end{pmatrix}$. Since the electromagnetic interaction is much weaker than the strong interaction, we expect that neglecting electromagnetic effects would give an approximate description of the real world. (Incidentally, readers of popular physics books often wonder how physicists could switch various interactions on and off. Where is the switch? But you know better, now that you understand what an action is. Switching electromagnetism off just means crossing out all those terms in the action of the world that have to do with electromagnetism.)

I have often been struck by Nature's kindness toward physicists; it is almost as if we were offered a step by step instruction manual. I mentioned in chapter III.3 that the photon has two helicity states. Well, the electron also could spin in only two directions, conveniently called "up" and "down" by physicists (although this has nothing with terrestrial gravity and the everyday usage of these two words).

So, an electron could exist in two spin states, which transform into each other under a rotation. (The physics and mathematics behind the two quantum states of the photon and of the electron are actually quite different, with the difference traced back to a branch of mathematics known as group theory. Recall that my graduate school advisor told me that I could either group or disperse—now you understand what that cryptic remark means! I chose the former. But in order not to interrupt our story here, I will apply Occam's broom to this rather subtle difference.) The relevant group is known to physicists as $SU(2)$.[22]

Thus, to transform the proton and the neutron fields into each other, theoretical physicists did not even have to learn more mathematics: They already knew it in connection with the electron spin. Happy were they! Hence the name "isospin"[23] for this symmetry Heisenberg introduced. The invariance of the strong interaction under isospin furnished the first example of an approximate symmetry[24] in physics, in contrast to rotation invariance, which, as far as we know, is an exact symmetry.

Prior to isospin, the symmetries of physics, such as translation invariance, rotation invariance, and Lorentz invariance, were confined to the spacetime we live and love in. Heisenberg's profound insight[25] led to the discovery of a vast internal space, the ongoing exploration of which has been a central theme of fundamental physics for close to a hundred years now.

Isospin has a wealth of falsifiable implications. For instance, Yukawa's pion field was expected to transform like a 3-component vector under rotation in this internal space. When we discussed the pion in chapter III.2, we took it to be electrically neutral, and so we now write it as π^0. Isospin demands the existence of two other pions, π^+ and π^-, carrying positive and negative charges, respectively.

From a lifetime living in 3-dimensional space, you know that a rotation requires three angles to characterize: two angles to specify the axis around which we are to rotate and a third angle to specify how much we rotate around that axis. Similarly, an isospin rotation acting on the nucleon field N is specified by three angles, which we will call θ^1, θ^2, θ^3. Mathematically, we write

$N \rightarrow e^{i(\theta^a T^a)} N$ under an isospin rotation. As per Einstein's convention, the sum over the index $a = 1, 2, 3$ is understood. Here T^a denote three matrices acting on the two components of N.

Quarks

I mentioned that in the 1950s, experimentalists were discovering hadrons literally by the dozen. The proton p and the neutron n turned out to have 6 cousins, known as $\Sigma^+, \Sigma^0, \Sigma^-, \Lambda, \Xi^0$, and Ξ^-, with properties similar to p and n. (Allegedly, they were named after the 1911 song "The Sweethearts of Sigma Chi," with a suitable corruption of the sorority's name.) These 8 hadrons are known as baryons (from the Greek root for "heavy;" think of the baritone in music) to distinguish them from the mesons. Well, the pions have cousins, too, in fact 5 of them, altogether forming a set of 8 mesons, namely, $\pi^+, \pi^0, \pi^-, K^+, K^0, \bar{K}^0, K^-$, and η.

Well, do you see a pattern? Yes? A set of 8 is known as an "irreducible representation" in group theory, 8 being equal to $3 \times 3 - 1$. (Theory students these days learn to do such calculations by taking a course in group theory.[26]) The $SU(2)$ of Heisenberg was generalized to the group $SU(3)$ by Murray Gell-Mann and Yuval Ne'eman, with the former proclaiming his approach as the Eightfold Way, in a playful reference to the Eightfold Path of Buddhism. Gell-Mann later said that he was mocking those people who wrote books linking theoretical physics to eastern mysticism. Plenty of the misinformed outside physics have taken such terminology seriously, to this very day!

Leaving such joking allusions aside, are you wondering where the 3 comes from? If you were, you could have been one of the greats in 1964. Yes!

Gell-Mann was asked this very question during a lunch while visiting Columbia University. Reportedly, that very afternoon, he realized that the hadrons are made of 3 species of quarks (as already mentioned in chapter IV.1), which he named up, down, and strange. (The relevant group theoretic calculation is $3 \times 3 \times 3 = 10 + 8 + 8 + 1$. The 8 baryons furnish one of the two 8's.) For instance, the proton, which may be written as uud, is made of two up quarks and a down quark, the neutron as udd and made up of one up quark and two down quarks, the positively charged pion π^+ as $u\bar{d}$ and made up of one up quark and one anti down quark, and so on. By the way, the whimsical names "up" and "down" also represent a joking allusion to isospin's root in "ordinary" everyday spin (but transposed to the quantum world, as was remarked upon earlier in connection with the electron spin.)

Gell-Mann thus brought order to the tohubohu of the hadron world, with its dozens of strongly interacting particles. The zoology of hadrons was reduced to a simple construction of 3 different species of quark. Much rejoicing in the streets as quarks swept away the bootstrap! Hadrons did not pull themselves into existence after all. Since Gell-Mann was at one time a bootstrap enthusiast, he was said to be shrewd enough to have played to both sides of the aisle.

Yes, you guessed it! The group $SU(N)$ transforms N objects into each other. Heisenberg's $SU(2)$ transforms the two nucleons (proton and neutron) into each other, and Gell-Mann's $SU(3)$ transforms his three quarks (up, down, and strange) into one another. Nowadays, students and readers of popular books tend to get the impression that $SU(3)$ and quarks fell from the sky, but in fact they were the end result of a struggle lasting almost 20 years by very brilliant people.[27] For instance, at one time, some people thought that there were two separate isospins, one for the "ordinary" hadrons and the other for the strange hadrons.

Quarks carry fractional electric charge

You are now ready to do one of the most important calculations in theoretical physics of the 1960s, first of the 1960s, first performed by Gell-Mann: determine the electric charge of the various quarks. Can you do it without reading on? Yes, you can. You know that the proton and neutron consists of *uud* and *udd*, respectively, and that electric charges add. So try it!

Denote by $Q(u)$ and $Q(d)$ the charge of the up quark and the down quark, respectively. Since the proton $\sim uud$ and the neutron $\sim udd$ have charge $+1$ and 0, respectively, we have

$$2Q(u) + Q(d) = 1$$

and

$$Q(u) + 2Q(d) = 0$$

Use elementary algebra. From the second equation, we obtain $Q(u) = -2Q(d)$, which when plugged into the first equation gives $-4Q(d) + Q(d) = -3Q(d) = 1$. Thus, we, or rather the incomparable Murray Gell-Mann, find that $Q(d) = -\frac{1}{3}$, and so $Q(u) = +\frac{2}{3}$.

Prejudice, as in human societies, plays a strange role in the theoretical physics community. This astonishing result, that quarks have fractional electric charges, was greeted with fear and loathing. Generations have been taught that electric charge is quantized in integer units, with the proton having charge $+1$ and the electron having charge -1. Some argued that this meant that quarks could not exist. Others concocted schemes to evade this conclusion of fractional charges. With the clarity of hindsight, we now see these reactions as irrational.

Yang-Mills theory

In 1954, Chen-ning Yang and Robert Mills[28] had the idea[29] of combining Weyl and Heisenberg.

Go back a few pages and remind yourself what Weyl did. He showed that the action for electromagnetism is invariant under the local gauge transformation, which multiplies the electron field by a spacetime dependent phase factor $\psi(x) \to e^{i\Lambda(x)}\psi(x)$ and changes A_μ by an additive shift $eA_\mu \to eA_\mu + \partial_\mu \Lambda$. The transformation is called "local" because it varies from point to point in spacetime.

Meanwhile, Heisenberg proposed that the action for the strong interaction is invariant under a global transformation that rotates the two-component nucleon field introduced above according to $N(x) \to e^{i\theta^a T^a} N(x)$. The transformation is called "global" because it does not depend on where we are in spacetime. In other words, θ^1, θ^2, θ^3 are just three real numbers independent of x.

Compare how $\psi(x)$ and $N(x)$ transform. Do you see something curious? The two transformations resemble each other yet are quite different. Two important differences!

Since the nucleon field N contains two components, p and n, and isospin rotates the two into each other, $\theta^a T^a$, which is the analog of $\Lambda(x)$ of electromagnetism for the strong interaction, has to be a two-by-two matrix. In contrast, $\Lambda(x)$ is just a real number.

On the other hand, $\Lambda(x)$ is an arbitrary function of spacetime, while $\theta^a T^a$ does not depend on where we are in spacetime. As a result, there is no analog of A_μ for the strong interaction. The term in the Lagrangian density corresponding to $\mathcal{L}_{\text{electromagnetism}} = \bar{\psi} i\gamma^\mu (\partial_\mu - ieA_\mu)\psi$ is just $\mathcal{L}_{\text{strong}} = \bar{N} i\gamma^\mu \partial_\mu N$.

gauge parameter, $\Lambda(x)$ or $\theta^a T^a$	varies in spacetime	a matrix
Weyl's electromagnetism	yes	no
Heisenberg's strong interaction	no	yes

Why? There is no why

Yang and Mills wanted to force these two Lagrangians to look the same. Why? There is no why. Theoretical physics is not just a bunch of calculations done to agree with observations, as I opined when discussing Dirac setting the electron free. Often, theoretical physicists just play around and try things.

Look at $\mathcal{L}_{\text{strong}} = \bar{N} i\gamma^\mu \partial_\mu N$. As soon as Yang and Mills change the global isospin transformation $N(x) \to e^{i\theta^a T^a} N(x)$ to a local transformation (that is, by making $\theta^a(x)$ dependent on the location x in spacetime), the ∂_μ acting on N in $\mathcal{L}_{\text{strong}}$ will now bring down stuff like $\partial_\mu \theta^a(x)$ describing the spacetime variation of $\theta^a(x)$, in precise analogy with $\partial_\mu \Lambda(x)$ in electromagnetism. Recall that the shift $eA_\mu \to eA_\mu + \partial_\mu \Lambda$ in $\mathcal{L}_{\text{electromagnetism}}$ cancels the $\partial_\mu \Lambda(x)$, as discussed in chapter IV.4. In almost prefect mimicry, we could do the same here! Introduce not one, but three, gauge fields A_μ^a that transform like $gA_\mu^a \to gA_\mu^a + \partial_\mu \theta^a$. (Here g is just a real number characteristic of the strong interaction, the analog of the coupling strength e of electromagnetism.)

Why 3? Because 3 spacetime dependent angles $\theta^a(x)$ are involved. (The index $a = 1$, 2, 3, as mentioned earlier.) A striking characteristic of Yang-Mills theory is that it demands three gauge fields A_μ^a, instead of the single electromagnetic field A_μ in electromagnetism. Each of these three gauge fields is associated with a particle, as described in part IV, just like the electromagnetic field is associated with the familiar photon. These three particles are known as the three Yang-Mills gauge bosons.* You could think of these gauge bosons loosely as the photon's "cousins."

The actual transformation of the three A_μ^a is somewhat more involved than what I just stated. Simply put, the three Yang-Mills gauge bosons also have to be rotated into one another under isospin, just as the proton and neutron are rotated into each other. As a result,[30] these gauge bosons interact with each other,[31] which renders the resulting path integral much more difficult (actually, impossible as of now) to evaluate than the corresponding path integral for electromagnetism. More later.

To summarize, as soon as we opened our eyes after birth, we knew about the photon field A_μ in some sense, but we had to wait for Weyl to tell us that $\mathcal{L}_{\text{electromagnetism}}$ is invariant under a gauge transformation of A_μ, provided that we also transform the electron field accordingly. This inspires Heisenberg to transform the nucleon field N in a way reminiscent of how the electron field transforms, but globally, not locally. Yang and Mills insist that N also transforms locally, not globally. But then mathematical logic forces them to introduce three gauge bosons.

Interestingly, the situations in electromagnetism and in the strong interaction are almost reversed. We already know from classical physics how the electromagnetic potential A_μ transforms, but quantum physics forces the electron field also to transform to compensate. In the strong interaction, we know how the nucleon field transforms. By demanding this transformation to be also local, Yang and Mills require three potentials A_μ^a to exist and play the same role for the strong interaction that A_μ plays for the electromagnetic interaction.

Not even wrong versus wrong wrong wrong

The unsuspecting reader might expect me to say that these three bosons corresponding to the fields A_μ^a were experimentally known in 1954, thus

---

*More about the word "boson" in part VI. For now, just think of gauge bosons as particles with properties similar to those of the photon.

proclaiming a smashing triumph for theoretical physics. Alas no, nothing like them had ever been observed. Even worse, just as gauge invariance forces the photon to be massless, it also forces these three cousins of the photon to be massless. And massless particles are easily produced—the embezzler has to skim off only a vanishing amount. As shown in figure IV.3.2, photons are produced with wild abandon at the slightest provocation. Fires are fearsome to humans and beasts, but a chemical fire corresponds "merely" to electrons rearranging themselves among molecules, with energies involved that are almost infinitesimal in the broad scheme of things. Yet the electrons radiate enough photons to frighten us.

Furthermore, you learned in chapter III.2 that the exchange of massless particles generates infinitely long ranged forces. But the strong interaction is notoriously short ranged, as you also saw in chapter III.2.

Legend has it that when Yang presented a seminar of what we now call Yang-Mills theory at the Institute for Advanced Study in Princeton, Wolfgang Pauli, who was visiting, asked repeatedly if these massless bosons had been seen experimentally. Yang was about to sit down, but Oppenheimer, the director of the Institute at that time, interjected to let the young man continue. The next day Yang found in his mailbox a note from Pauli apologizing for his aggressive[32] behavior.

A senior* physicist who was around in 1954 told me that "everybody"[33] knew that the Yang-Mills paper was somehow important. But how? He kept the paper on his desk for years until it sank to the bottom of some huge pile, out of sight and out of mind.[34]

A widely circulated story about Pauli underlines my repeated assertion that theoretical physics is more than a bunch of calculations done to agree with experiments. Physics journals are full of calculations based on soporific assumptions, invoking tired approximation schemes and tediously fitting parameters to achieve accord with data. The story is that Pauli, when shown such a calculation, dismissed it with a haughty "It is not even wrong," a phrase by now codified into a standard insult in the theoretical physics community. Presumably, in 1954, Pauli regarded Yang-Mills theory as wrong wrong wrong, but intriguingly wrong, far better than not even wrong.

After these amusing stories, I have to reluctantly mention some boring jargon. Electromagnetism is sometimes called an "abelian" gauge theory, while Yang-Mills theory is called a "nonabelian" gauge theory.[35]

Color and flavor

Another difficulty with the Yang-Mills proposal is that the entire scheme is associated with transforming a proton into a neutron and vice versa (or equivalently in modern language, an up quark into a down quark and

*I use the word "senior" to describe those considerably older than I am.

vice versa.) Plenty of strong interaction processes (for example, two protons scattering off each other, viz., $p + p \rightarrow p + p$), have nothing to do with protons transforming into neutrons, at least superficially. Increasingly, isospin was understood to be not an essential ingredient of the strong interaction, but merely due to the fact that the masses of the up and down quarks are small compared to the mass scale characteristic of the strong interaction. (I am glossing over a considerable amount of technical details here that took a decade to unravel. Many stumbled into dead ends applying the Yang-Mills construction to isospin.)

An almost incredibly lucky break for those studying the strong interaction came from a totally unexpected direction—the disintegration of the electrically neutral pion π^0 into two photons. That this decay occurs was not a surprise: all hadrons except for the proton decay. The π^0 is a quantum superposition of two states, one consisting of an up and an anti up quark, and the other of a down and an anti down quark. Focus on the former, and picture an up and an anti up quark rattling around. Since they are electrically charged, the quark and the anti quark are perfectly capable of annihilating each other and in the process emitting two photons. So the decay of π^0 into two photons is understood qualitatively.[36]

The surprise came in the calculation of the probability amplitude for this to occur. Normally, the strong interaction is too strong and too mean, and an infinite number of Feynman diagrams would be involved. The quark and the antiquark could interact repeatedly ad infinitum before annihilating. At first glance, the calculation of the probability amplitude for this decay process appears hopeless, the usual rather depressing situation involving the strong interaction.

But no! Due to an intricate set of circumstances all coming together, an event far beyond the scope[37] of this book to describe, the infinite number of diagrams all cancel except for two diagrams, which are simple enough to be calculated exactly.

This was exciting indeed. Finally, a process involving the strong interaction that could be calculated! But guess what, air out of the balloon: The probability amplitude comes out to be approximately 1/3 that deduced from the experimental measurement. A puzzling result, but definitely in Pauli's category of intriguingly wrong! The kind of wrongness that attracts scrutiny rather than yawns. To exaggerate a bit, if you pass by a seminar on theoretical physics, and the audience is yawning, you could almost be sure that the speaker's work is not even wrong.

Eventually, Gell-Mann and others realized that each quark comes in three copies, which he picturesquely described as each quark having three different colors,[38] let's say for definiteness red, yellow, and blue. We should multiply the theoretical amplitude for π^0 decay by 3, thus bringing it into perfect agreement of experiment. (I should emphasize that the word "color" has nothing to do with how it used in everyday language; it merely reflects Gell-Mann's playful choice of words.)

But then the next big idea eventually dawned on theoretical physicists: Yang-Mills theory is not to be applied to isospin transformation, but to the transformation of the three colors into one another. Thus, for example, in the scattering of two protons, the up and down quarks inside the protons do not change into each other. Rather, the color of the quarks change.

The strong interaction is then described by a Yang-Mills theory based on the group $SU(3)$ acting on the three colors. The theory requires 8 gauge bosons rather than 3, as Yang and Mills had originally thought. Incidentally, you know that the number 8 results from group theory[39] and even how to calculate it from a few pages back. When I see cartoons depicting theoretical physicists at work, wearing lab coats no less, staring at a blackboard filled with complicated expressions looking in part like $\sqrt{\pi + \tan \pi e/3} \int dx \log(e^{\frac{x}{3}} + \sin \sqrt{3}x^2) + \cdots$ (some utter nonsense), I chuckle that the reality was probably more like one physicist showing another that $3^2 - 1 = 8$. It is not rocket science! The resulting theory is known as the dynamics of color, and was named "quantum chromodynamics" (QCD) to parallel quantum electrodynamics (QED).

Incidentally, the attribute that distinguishes an up quark from, say, a down quark is now called whimsically (Gell-Mann again) "flavor." Quarks come in six flavors,* each in three colors.

Confinement

For a long time, Gell-Mann had to fend off critics noting that quarks were not sighted experimentally. Eventually, Gell-Mann proposed that quarks are permanently confined. This also explains why the 8 massless Yang Mills gauge bosons (named "gluons" whimsically by Gell-Mann, because they glue quarks together into hadrons) were not seen. They are also permanently confined, thus in some sense answering Pauli's objection during Yang's talk two decades earlier. For a long time, however, many heavyweight physicists continued to dismiss confinement as just a word, certainly not a theory.

You might think that this particular strand in the development of the theory of strong interaction, from Heisenberg to Yang-Mills to quarks to color, sounds awfully twisted. Be assured that the real history is considerably more so. I glossed over details and omitted mentioning the many blind alleys that numerous smart physicists spent their lives stuck in.

What if there is only one field?

The group $SU(N)$ transforms N objects into each other, and a gauge theory based on $SU(N)$ transforms N fields into each other.

---

*The flavors are up, down, strange, charm, top, and bottom. A bit more about them is given in chapter V.4.

You may be wondering about what happens when there's only one object. That is an exceedingly penetrating question. With only one object, you cannot transform it into something else: All you can do is to multiply it by a phase factor. (Recall the phase factor introduced in chapter II.3.) The corresponding group is called $U(1)$. And guess what? Electromagnetism is a gauge theory based on $U(1)$. You can see that this makes a certain amount of sense. We can imagine a universe with only the electron field and the electromagnetic field. Here and there in this universe, an electron and a positron would stay together for a short time and then annihilate each other, turning into photons. Theoretical physicists like to imagine different kinds of universes, and although this universe, built on the gauge group $U(1)$, is not particularly exciting, it is certainly legitimate and has been much studied.

Running coupling strengths

Each of the four fundamental interactions is characterized by a coupling strength, measuring how strong that interaction is. In quantum physics, the probability that two particles will interact determines the strength of the interaction.

Until 1970 or so, physicists were used to thinking of the coupling strengths as constants, and referred to them as coupling constants. But let us rely on operationalism and consider how one of my experimental colleagues would actually go about determining the coupling strength of the electromagnetic interaction. Well, she would collide two electrons together, for example, and by repeating the experiment many times, determine the probability that the electrons would interact. That probability, in essence, defines the electromagnetic coupling constant. This operational definition makes clear that the coupling constant will depend on the energy with which the experimenter collides the two electrons. Another experimenter repeating the measurement at a different energy will extract a different coupling constant.

Recall that in quantum physics, the wavelength of any particle that we use as a probe decreases as the energy of the particle is increased. Prince de Broglie told us so. Thus, to examine Nature with a finer resolution, physicists have to collide particles at higher energies. This discussion tells us that as we look at Nature with different resolutions, the coupling strengths of the various interactions will vary. The so-called coupling constant is not a constant at all, but varies with the energy scale at which it is measured. This is why, instead of coupling constant, I have been careful to use the more modern term "coupling strength," or simply "coupling."

Around 1970, Ken Wilson,[40] building on earlier work by Murray Gell-Mann and Francis Low, concerned himself with how we describe the world. The reader is familiar, of course, with the fact that the world looks quite different when examined on different length scales. Increase the resolution of the microscope and what appeared as haze crystallizes into detailed structure. In examples drawn from everyday life, our perceptions of the world, with a given resolution, do not tell us very much about what we will see with finer resolutions. However, the logical structure of quantum field theory is so intricate that it can relate a description on one length scale to a description on another. Given a description of the world, physicists can actually say something about how the world would look if seen with a different resolution. The essence of Wilson's work deals with how much the structure of quantum field theory allows us to say.

Speaking picturesquely, physicists say that the coupling strength "moves" as one changes the energy at which the coupling strength is measured. It turns out that the coupling strength moves extremely slowly with energy. Over the entire range of energy that has been studied from the beginning of physics until the late 1960s, the electromagnetic coupling strength has moved by a minuscule, and hence unnoticed, amount. This explains why the coupling strength had always been thought to be constant. With a somewhat twisted humor, theoretical physicists now refer to the concept previously known as coupling constant as a "running coupling."

The inverse of the fine structure constant that measures the strength of electromagnetism and that started this chapter was long believed to equal the integer 137. For many decades, theoretical physicists would receive a steady stream of letters from crackpots claiming that they have somehow derived that mysterious integer from some hocus pocus. The stream eventually slowed to a trickle. The news that coupling strengths run must have spread, not to mention that more precise measurements have long shown that the inverse of the electromagnetic coupling is not an integer.

Asymptotic freedom

In chapter IV.2, I mentioned that a theory with zero coupling is said to be free—the particles in it would be free to move about independently of the other particles. A theory whose coupling strength moves toward zero as the theory is examined at ever higher energies is now called an "asymptotically free theory."

We would like the strong interaction to become weaker at higher energies so that we can deal with it.

The reader may get the impression that the search for asymptotic freedom was motivated purely by wishful thinking. This is almost, but not exactly, true. Around 1970, there was already a hint, at least in hindsight, that the strong interaction may get weaker at high energies. Experimenters had scattered very energetic electrons off protons. In the quark picture, the electron gives one of the quarks inside the proton a good kick. The experimental results indicate that when the kicked quark zips by the other quarks, it barely interacts with them. Asymptotic freedom would be able to explain this phenomenon naturally.

In the winter of 1972–1973, freedom was found. David Gross and Frank Wilczek and, independently, David Politzer, found that Yang-Mills theory is asymptotically free. This stunning news suggests that we may finally have a handle on the strong interaction! Mother Nature made the strong interaction too mean, but then relented, and let the strong interaction become less and less strong at high energies.

The flip side is that the strong interaction remains fiercely strong at low energies. Thus, how quarks are permanently confined to form hadrons has never even been proved,[41] let alone worked out in detail. Particle theorists were spoiled by the golden era of the 1970s, but no principle says that the traditional difficult problems[42] of the strong interaction could be easily solved in the foreseeable future.

A few concluding remarks

So, Yukawa's π meson, which jump-started our understanding of the strong interaction, turns out to be a bound pair of quark and antiquark, "merely" the low energy manifestation of an underlying structure. The strong interaction is mediated by the pion at the level of the nucleons, but by gluons at a deeper level.

To squeeze an account of the strong interaction into this rather short chapter, I have had to suppress a lot of important conceptual developments, most notably the notion of spontaneous symmetry breaking,[43] which in turn led to the Higgs mechanism that is important for electroweak unification, to be discussed in chapter V.3. Historically, why the pion is almost ten times lighter than the nucleons provided the crucial clue. Setting the pion mass to zero in the action, theorists uncovered a previously unknown symmetry.[44] By the way, this apparently crazy approximation made possible the crucial calculation of the π^0 lifetime mentioned earlier in this chapter.

An interesting example of the effect of psychology in physics: Sam Treiman, one of my undergraduate professors and later my collaborator, told me that the generation of physicists who grew up during the war, like him, was awestruck (as the rest of the world was also) by the enormous amount of energy released per nucleon when a nucleus fissions, as in an atomic bomb. The pion mass is more than ten times this fission energy. Thus, to regard the pion mass as approximately zero poses an almost unsurmountable mental barrier for this generation. Imagine, ten times the amount of energy per nucleon released in nuclear fission was later regarded in particle physics as negligible. In physics, whether a quantity is large or small all depends on what we are comparing that quantity to.

What strikes me as truly remarkable about the strong interaction is that the questions regarded by theoretical physicists as important changed dramatically in a short period of time, literally just a few years. At one time, a Nobel-worthy project would be to understand the details of how a proton scatters off a proton, with people dedicating their lives to the problem. Suddenly, nobody gives two hoots. All the attention was shifted to proving quark confinement. I cannot resist mentioning a standard academic joke. Students preparing for a final exam would look up the exams given in previous years. For a while, this approach was useless for a course on particle physics: the questions changed from year to year. In contrast, students in the economics department find that the questions are always the same, but the correct answers change from year to year!

Notes

[1] I already mentioned this in chapter IV.2 in connection with the introduction of a proton field.

[2] In the conclusion of Sakurai's monograph on invariance principles, published in 1964, he worried that quantum field theory might fail at an energy scale several times the proton mass! He then expressed the hope that even if quantum field theory were to fail, the conclusions based on symmetry principles might continue to hold, such as the CPT theorem (to be discussed in chapter V.3). Since Sakurai was at the University of Chicago at the time, his book might represent a rough interpolation between the two coasts.

[3] See the preface of *FbN*.

[4] G. Chew, *S-Matrix Theory of Strong Interactions*, W. A. Benjamin, 1962.

[5] F. Zachariasen, "The theory and application of Regge poles," in *Cargèse Lectures in Theoretical Physics*, ed. M. Lévy; W. A. Benjamin, 1962.

[6] In the end, much of this is now widely dismissed, in contrast to the solid foundational work laid down by the greats such as Dirac, Pauli, Schwinger, and Feynman.

[7] For his influence on me, see Christianson, *Physics Today*, and the preface of *FbN*.

[8] Already known to some as berserk-ely by that time. Berserk comes from "behr sekr," meaning bear skin, referring to the costume of the Norsemen who terrified England and the American Congress.

[9] The significance of which might be lost to the younger readers of this book.

[10] Even though in the end I did not work for Weinberg for reasons not relevant to this book.

[11] With some exaggeration, one might be tempted to say sub-cults.

[12] Never mind what these are.

[13] I chose to group. See, for example, *Group Nut*.

[14] An extremely nice and friendly man, I want to state here for the records.

[15] R. Hwa, of the University of Oregon.

[16] For an account of this era, see chapters 8 and 9 in D. Kaiser, *Drawing Theories Apart: The Dispersion of Feynman Diagrams in Postwar Physics,* University of Chicago Press, 2005.

[17] The term "phase factor" was introduced in chapter II.3. It is a complex number of length 1.

[18] Everything happened so much faster in those days. Chadwick's Nobel Prize came a mere 3 years later.

[19] As I stated in the preface, in a popular book such as this, I opt for a livelier narrative at the expense of a more accurate history. The following brief history of isospin symmetry is taken from *Fearful*. Since the neutron is so close in mass to the proton, Chadwick naturally assumed that the neutron consisted of a proton with an electron stuck on it. Thus, atomic nuclei were erroneously thought to consist of protons and electrons. Heisenberg proposed that the neutron is a particle in its own right and that the nucleus consists of protons and neutrons. He then supposed that strong interaction physics remains invariant if one exchanges the proton and the neutron. Note that this symmetry is considerably weaker than isospin symmetry as we know it, in which one transforms the proton and the neutron into linear combinations of each other. Heisenberg, however, continued to think of the neutron as a proton

with an electron attached. He explained the origin of the interaction between the proton and the neutron as follows: When a neutron gets close to a proton, the electron inside the neutron may hop over to the proton. Heisenberg reasoned that the electron, by hopping back and forth between the proton and neutron, could produce an interaction between the two. In Heisenberg's picture, there is no strong interaction between two protons, since there is no electron around to hop back and forth. The atomic nucleus was erroneously supposed to be held together by the attraction between protons and neutrons. Heisenberg's theory was proven wrong by the experimentalists N. P. Hydenburg, L. R. Hafstad, and M. Tuve, who measured the strong interaction between two protons (following earlier work of M. White) and discovered it to be comparable in strength to the interaction between a proton and a neutron. In 1936, B. Cassen and E. U. Condon, and, independently, G. Breit and E. Feenberg, proposed that Heisenberg's exchange symmetry be generalized to isospin symmetry. (I thank S. Weinberg for a helpful discussion on this point.)

[20]The Matthew principle is in full blast here.

[21]The actual story is more complicated. See A. Zee, *Physics Reports*, 3C, 127, 1972. We now understand that even without electromagnetism, the proton and neutron would not have equal mass, because the up quark and down quark have different masses. Historically, however, Heisenberg was motivated by the near equality of M_p and M_n.

[22]For details, see *Group Nut*, chapters I.4 and I.5. The notation $SU(2)$ stands for something like special unitary transformation of 2 objects.

[23]Replacing the antiquated term "isotopic spin," and even more so, "isobaric spin."

[24]I should have said an approximate internal symmetry; Galilean invariance is an example of an approximate spacetime symmetry, for phenomena slow compared to the speed of light. But that was in hindsight; before special relativity, Galilean invariance was thought to be exact.

[25]For a more accurate historical account, see *Fearful*, pages 333–334. In particular, Heisenberg's original proposal did not involve $SU(2)$ at all. What I present here is known as textbook pseudohistory.

[26]See, for example, *Group Nut*, chapter IV.4.

[27]The history, as with almost all major discoveries in theoretical physics, is confused and convoluted. See, for example, D. Speiser's recollection in *Symmetries in Physics (1600–1980)*, ed. M Doncel, World Scientific, 1988.

[28]A remarkable manifestation of the small world effect: His sister Helen Mills was the design director for my book *Fearful*.

[29]Yang said that he had his idea when he was still a student in China, but was unable to work things out until he shared an office with Mills at Brookhaven in the summer.

[30]For the benefit of some readers, I state without explanation that this fact is a consequence of the preceding sentence. These gauge bosons carry isospin, while the photon does not carry electric charge. See *QFT Nut*, chapter IV.5.

[31]The Yang-Mills action, in contrast to the Maxwell action, contains cubic and quartic terms of the schematic form $\sim AAA$ and $\sim AAAA$. Nasty beasts, in academic slang.

[32]During my time in theoretical physics, I have certainly seen far worse behavior, with never an apology.

[33]I am sure he meant everybody in some small elite circle of theoretical physicists in the know.

[34]Readers having to deal with information overflow surely would appreciate this.

[35]Let R and R' be two rotations that you are going to perform in succession. You could verify easily that RR' and $R'R$ are not the same. Which rotation you perform first matters. See *Group Nut*, page 40, for a pictorial illustration. In contrast, the transformation in electromagnetism involves "merely" multiplication by $e^{i\Lambda}$, a complex number of length 1, and so the order of multiplication does not matter. When the order does not matter, the multiplication is called "abelian," in honor of the great Norwegian mathematician Niels Abel. An irony of the naming convention is that the more intricate case is known as nonabelian rather than the other way around.

[36]Think of the hydrogen atom with the proton replaced by a positron. This state, with an electron and a positron rattling around, attracted to each other via the electric force, is known as a positronium. It also decays into two photons. The calculation of the positronium lifetime, since the strong interaction is not involved, is just a simple exercise, in contrast to the calculation of the π^0 lifetime.

[37] See *QFT Nut,* chapter IV.7.

[38] The necessity for color also comes from hadron spectroscopy. See chapter VI.2.

[39] See *Group Nut,* chapter V.2.

[40] For the story of how Ken Wilson and I spent a month in a basement, see *Fearful,* page 196.

[41] Indeed, this is one of the million dollar Clay prize problems.

[42] Lattice gauge theory, the most useful approach at the moment, more or less follows the procedure sketched in endnote 7 in chapter IV.2 and then applies brute force. Ken Wilson proposed approximating continuous spacetime by a lattice. The uncountably infinite number of degrees of freedom of quantum fields, defined at every point in spacetime, are replaced by a countably infinite number of degrees freedom of variables, defined on each point and link on the lattice. And then let the computer fire away! The reader who knows calculus might be struck that, in some sense, this is reversing the monumental advance of Newton and Leibniz from discrete sums to continuous integrals. Yes, this has produced results, but some theorists, perhaps the more conservative in the community, do not feel that this approach has shed much light on what actually happens in the domain when the strong interaction become too strong and too mean.

[43] For more details, see, for example, *Fearful,* chapter 12.

[44] See, for example, chapter IV.2 in *QFT Nut.*

The weak and the electroweak interactions

Missing energy

The history of the weak interaction began with the discovery that energy went missing when some atomic nuclei relaxed to a state with lower energy[1] by ejecting an electron. A nucleus consisting of Z protons and $A - Z$ neutrons, for a total of A nucleons, is traditionally denoted by (Z, A). Some of these nuclei would eject an electron and "decay" to the "daughter nucleus" $(Z + 1, A)$. (Since the electron carries away 1 unit of negative charge, charge conservation compels Z to increase by 1, while A remains the same.) This process, $(Z, A) \rightarrow (Z + 1, A) + e^-$, is known as beta decay for historical reasons.

Since the experimentalists knew the masses of the mother nucleus (Z, A) and of the daughter nucleus $(Z + 1, A)$ and the mass of the electron, they knew how much mass is "lost"* and thus could use Einstein's $E = mc^2$ (derived in chapter I.4) to determine how much energy was emitted. Measuring the energy of the electron, they found, to everybody's consternation, that energy was missing.

Various eminent theorists championed the suggestion that the sacred law of energy conservation might fail in nuclear decay. We of course now know, in the glare of hindsight, that this isn't so and that energy conservation still stands in nuclear decay. But at the time, with all the weird stuff going on in the quantum domain, this suggestion did not seem so outlandish at all.[†]

---

*Basic accounting: mass lost equals $M(Z, A) - M(Z + 1, A) - m_e$, with the self-evident notation for the masses of the two nuclei and the mass of the electron m_e.

[†]I read that one reason that the great Soviet physics Lev Landau landed in jail at the time was that one of his colleagues informed the KGB (known as NKVD at the time) that he believed in energy nonconservation while Lenin had said that energy was always conserved. Admittedly, Landau had also passed out anti-Stalin leaflets.

Instead, Wolfgang Pauli hypothesized that the missing energy was carried away by an invisible particle, which Fermi called the neutrino, that is, "little neutron" in Italian, to distinguish it from the neutron. Thus, the decay is actually $(Z, A) \rightarrow (Z + 1, A) + e^- + \bar{\nu}$, with the missing energy carried off by an unseen particle called $\bar{\nu}$, with the Greek letter ν (pronounced nu).

That this particle was unseen was explained by theorizing that it participates in neither the electromagnetic interaction nor the strong interaction. It only interacts weakly and gravitationally with other particles, and hence would (in all likelihood) fly straight through the detector. Since the energy carried by the electron is measured to sometimes reach the maximum value $\left(M(Z, A) - M(Z + 1, A) - m_e\right)c^2$ allowed by Einstein's energy-mass accounting rule, it was deduced also that the neutrino must be massless, or nearly massless up to the precision of the energy measurements available at that time. (Incidentally, modern measurements indicate that the neutrino is not strictly massless, but its mass is teeny compared to the electron's.)

Let me slip in a bit of useful jargon here. Particles that do not interact strongly, such as the electron and the neutrino, are called "leptons," Greek for thin, delicate, small, in contrast to "hadrons," particles that interact strongly. Thus, the lepton is the least valuable coin in Greece, and a person having a thin, narrow face is leptoprosopic.

At the risk of being slightly anachronistic, I have followed the standardized naming convention used nowadays and written $\bar{\nu}$: The particle emitted in beta decay is now designated an antineutrino, as indicated by the bar,* and not a neutrino.[2]

The actual detection of neutrino and antineutrino came decades later, proving that they were not merely figments of Pauli's imagination. Incidentally, we moderns living in an age when theorists invent (mostly nonexistent) new particles (that almost nobody cares about) with wanton abandon must appreciate Pauli's boldness in the historical context—it had never been done before. Postulating an invisible particle out of thin air!

The sketchiest of all sketches of the weak interaction: from the nuclear era to the hadron era to the quark era

Later, this process was understood in terms of the more elementary process $n \rightarrow p + e^- + \bar{\nu}$. A neutron inside the nucleus transmutes itself into a proton while emitting an electron and an antineutrino. (The neutron was discovered by James Chadwick in 1932. Up until then, it was generally believed that the

---

*The overbar is used in physics to indicate the opposite. You have already encountered this usage in chapter IV.2, particularly in the table contrasting what ψ and $\bar{\psi}$ can do.

nucleus was made up of protons and electrons. For a while afterward, some continued to believe that the neutron was a bound state of the proton and the electron, until its mass was measured to clearly exceed the sum of the proton mass and the electron mass.)

Still much later, this process was understood, in turn, as due to the even more elementary process $d \to u + e^- + \bar{v}$. A down quark d (I have already mentioned quarks in chapters IV.1 and V.2) inside the neutron transmutes itself into an up quark u while emitting an electron and an antineutrino. By then, the proton was known to be made of two up quarks and a down quark, written as $P = (uud)$, and the neutron to be made of two down quarks and an up quark, written as $N = (udd)$, as was mentioned in chapter V.2. Changing one of the d's in (udd) into a u turns it into (uud).

To summarize, the decay of the down quark generates the decay of the neutron (see figure 1):

$$d \to u + e^- + \bar{v}$$

$$\text{generates} \quad (udd) \to (uud) + e^- + \bar{v}$$

Einstein's urging to make physics simpler and simpler

From $(Z, A) \to (Z + 1, A) + e^- + \bar{v}$ through $n \to p + e^- + \bar{v}$ to $d \to u + e^- + \bar{v}$, physicists have followed Einstein's urging to make physics simpler and simpler, moving from the nuclear era to the hadron era to the quark era.

	the march of physics
1930s	$(Z, A) \to (Z + 1, A) + e^- + \bar{v}$
1950s	$n \to p + e^- + \bar{v}$
1970s	$d \to u + e^- + \bar{v}$

The phenomenology and the accompanying theoretical structure of the weak interaction are extraordinarily rich, and a detailed treatment would fill tomes. I restrict myself to a few aspects[3] with direct connections to the main themes of this book.

A ghost of a particle

The fabled story of the neutrino! You may have heard breathless accounts of the ghostly neutrino, passing through the entire earth without every interacting, cruising "like dustmaids down a drafty hall," as the poet John Updike eloquently put it in his poem, "Cosmic Gall."

NEUTRON DECAY

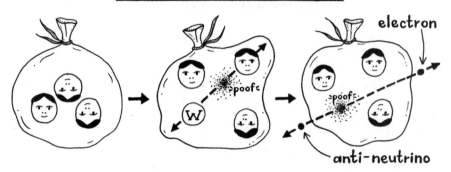

Figure 1. The weak decay $n \to p + e^- + \bar{\nu}$ of the neutron is now understood in terms of quarks. The neutron consists of two down quarks, represented by the upside down faces, and an up quark, represented by a rightside up face, confined (see chapter V.2) within a bag. Suddenly, one of the down quarks emits a W boson (to be explained later in the text) and turns itself into an up quark. The W then disintegrates into an electron and an antineutrino, which, being leptons, escape from the bag. The two up quarks and the down quark left behind constitute a proton.
Redrawn from A. Zee, *Fearful Symmetry: The Search for Beauty in Modern Physics*, Princeton University Press, 1986.

The reason for the neutrino's ghostly behavior, as I mentioned earlier, is that, being an electrically neutral lepton, it interacts neither electromagnetically nor strongly. And so to neutrinos, the earth is indeed "just a silly ball," to use Updike's term, almost a transparent nothing of a ball. Indeed, Pauli soon lamented that he did what physicists should not do, postulate particles that could not be experimentally detected.[4]

Fermi: the quantum field can really create a particle

In 1933, Enrico Fermi proposed his celebrated[5] theory of the weak interaction; to describe neutron beta decay, $n \to p + e^- + \bar{\nu}$, he wrote down the interaction Lagrangian $\mathcal{L}_{\text{Fermi}} = G\,(\bar{e}\,\nu)\,(\bar{p}\,n)$ and added $\int d^4x\,\mathcal{L}_{\text{Fermi}}$ to the action. Here e, ν, p, n denote, respectively, the electron field, the neutrino field, the proton field, and the neutron field. (Particle physicists often use the same letter for a field and the particle associated with it, as already mentioned in chapter IV.1.)

For ease of presentation, I will now write this as it is written in the quark era: $\mathcal{L} = G\,(\bar{e}\,\nu)\,(\bar{u}\,d)$. Now that you have almost mastered quantum field theory, you are initiated (recall chapter IV.2) into reading this secret handwriting: from right to left in $\mathcal{L}$, d annihilates a down quark, $\bar{u}$ creates an up

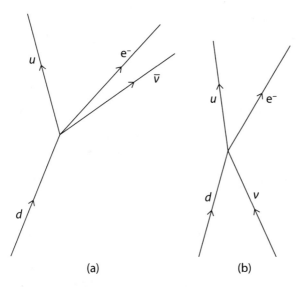

Figure 2. (a) The Feynman diagram describing the process $d \to u + e^- + \bar{v}$, (b) the Feynman diagram describing neutrino scattering $v + d \to e^- + u$; the two processes are related by crossing.

quark, v creates an antineutrino, and $\bar{e}$ creates an electron, thus describing the process $d \to u + e^- + \bar{v}$. The Fermi constant G measures the probability amplitude for the process to occur. (I have simplified by suppressing various "inessential" factors in $\mathcal{L}$. By doing this, I have glossed over several Nobel prizes.[6])

The Feynman diagram describing the process $d \to u + e^- + \bar{v}$ is shown in figure 2(a). As explained in chapter IV.2, crossing symmetry in quantum field theory allows us to bend the outgoing antineutrino into an incoming neutrino, as shown in figure 2(b). Indeed, now that you know how to read physicists' secret handwriting, you could say that the field v in $\mathcal{L} = G \, (\bar{e} \, v) \, (\bar{u} \, d)$ annihilates a neutrino. Thus, Fermi's Lagrangian also describes the process, $v + d \to e^- + u$, neutrino scattering off a down quark producing an electron and an up quark.

Nowadays, one favorite way of studying the weak interaction involves smashing a high energy neutrino beam into a target (in one case, a junked battleship) and then detecting an electron downstream, a possibility undreamed of by Updike's Nepalese lovers in "Cosmic Gall."

Creating the electron: a scandal bringing complete disorder into physics

Nowadays, students of quantum field theory routinely accept that electrons could be created by an electron field. But in 1933, Fermi's theory launched a colossal conceptual breakthrough, which many at the time had difficulty accepting. Senior physicists thought the young turks, Fermi and Heisenberg among them, had gone insane by claiming that the outgoing electron in $(Z, A) \to (Z + 1, A) + e^- + \bar{\nu}$ was actually created during the decay process.

> (I was) criticized very strongly for this assumption by extremely good physicists. I got one letter saying that it was really a scandal to assume that there were no electrons in the nucleus because one could see them coming out; I would bring a complete disorder into physics by such unreasonable assumptions ... [I]t is really difficult to go away from something which seems so natural and so obvious that everybody had always accepted it. I think the greatest effort in the developments of theoretical physics is always necessary at those points where one has to abandon old concepts.[7]

You could literally see them coming out. So they must have been inside the mother nucleus (Z, A) all along. The younger generation is creating havoc in the house of physics!

To say that the electron shooting out of the nucleus was created out of thin air required a fantastic leap of faith. We revere the greats of physics, such as Fermi and Heisenberg, for leaps such as these.

Nowadays, nobody blinks an eye. Simply write down $\bar{e}$ in the interaction Lagrangian.

The range of the weak interaction

If the very short range of the strong interaction came as a surprise after the familiar infinitely long reach of the electromagnetic interaction and of gravity, then the even shorter range of the weak interaction delivered quite a shock. Experimentalists eventually[8] realized that nuclear decay occurs entirely inside the nucleus, which, as you would recall, is teeny, essentially a point compared to the size of the atom. Thus, the range or distance scale over which the weak interaction operates is infinitesimal, much shorter than the separation between the nucleons. Indeed, for a long time, it was believed that the range was zero, so that the interaction occurs at a point, as depicted in figure 2.

Meanwhile, theorists speculated that the range of the weak interaction was 10 times shorter than that of the strong interaction, and then 100 times shorter as experimental measurements improved, and so on. Theorists are flexible and can adapt, which means that they could write many papers. Finally, it was established that the range is about 600 times shorter.

This range implies that the particle mediating the weak interaction interaction, known as the W boson,* is about 600 times more massive than Yukawa's meson[9] for the strong interaction, as was explained in the prologue and in

*This word "boson" is discussed in more detail in part VI.

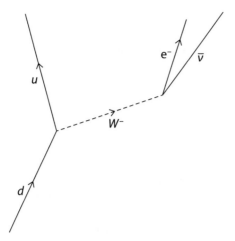

Figure 3. The $d \to u + e^- + \bar{v}$ shown in figure 2(a) occurs via the emission and subsequent disintegration of a virtual W boson.

chapter III.2. An embezzler grabbing a huge amount can't get very far during the short time before the alarm goes off.

In 1983, the long conjectured W boson was finally produced at accelerators. The weak interaction was cracked open, so to speak. In Feynman's diagrammatic language, the decay depicted in figure 2 can now be pulled apart as in figure 3, showing a d quark transforming into a u quark by emitting a virtual[10] W^-, which subsequently morphs into an electron e^- and an antineutrino $\bar{v}$. See also figure 1.

As quantum field theorists, we simply replace the previous interaction Lagrangian $\mathcal{L} = G\,(\bar{e}\,v)\,(\bar{u}\,d)$ by $\mathcal{L} = gW\,(\bar{v}\,e + \bar{u}\,d)$ and add some terms to the action describing how W propagates in spacetime. For instance, the term $gW\bar{u}\,d$ describes the annihilation of a down quark and a W^+ boson, followed by the creation of an up quark, that is, the process $d + W^+ \to u$. In other words, the down quark becomes an up quark upon absorbing a W^+ boson. By crossing, this is equivalent to the down quark becoming an up quark by emitting a W^- boson. Fermi's constant G is determined in terms of the coupling g, which measures the probability amplitude that a W boson is absorbed or emitted.[11]

A modern way of summarizing all this is to write

$$\begin{pmatrix} v \\ e \end{pmatrix} \quad | \quad \begin{pmatrix} u \\ d \end{pmatrix}$$

Simple, right? The weak interaction transforms the neutrino and the electron into each other, and the up quark and the down quark into each other, with the world of leptons and the world of quarks separated by a vertical line. (We will come back to these two separate worlds in chapter V.4.)

I will now show you how readily you could become conversant with the weak interaction. Think of the W^+ boson lifting the electron upstairs to become a neutrino, and the down quark upstairs to become an up quark, while the W^- boson carries the neutrino and the up quark downstairs to become an electron and a down quark, respectively. This could be written, respectively, as

$$e^- + W^+ \leftrightarrow \nu \quad \text{and} \quad d + W^+ \leftrightarrow u$$

and

$$\nu + W^- \leftrightarrow e^- \quad \text{and} \quad u + W^- \leftrightarrow d$$

Indeed, we have already written $d + W^+ \to u$ a couple of minutes ago. The double-headed arrow $\leftrightarrow$ simply reminds us that the process can go both ways. In this example, the up quark turns into a down quark by emitting a W^+.

You merely have to remember that when you move a field from one side of $\leftrightarrow$ to the other, you have to flip it into its anticounterpart, just like in elementary algebra when we move a term from one side of an equation to the other, a plus sign has to be flipped into a minus sign, and a minus sign into a plus sign. You might also have realized that the crossing symmetry explained in chapter IV.2 relates processes involving a W^+ and processes involving a W^-. For example, crossing the W in $e^- + W^+ \leftrightarrow \nu$, we obtain $e^- \leftrightarrow \nu + W^-$, which is also one of the listed processes. Easy peasy. Note that this process is almost foolproof, because electric charge conservation provides a check. For instance, if we had somehow gotten $e^- + W^- \leftrightarrow \nu$, we would know that we had made a mistake, since the left side has charge $-1 - 1 = -2$ while the right side has charge 0.

You can also cross the quark and lepton fields; just remember to flip a field into an antifield. For example, crossing the neutrino field in $e^- \leftrightarrow \nu + W^-$, we obtain $e^- + \bar{\nu} \leftrightarrow W^-$.

Generating various weak interaction processes then amounts to linking these elementary processes together, much like linking various pieces together in a children's construction set. Or, if the reader prefers, it is reminiscent of high school chemistry. As an example, let us generate the process that started, in the early 20th century, this exploration of the weak interaction. Start with $u + W^- \leftrightarrow d$ listed above, and write it as $d \to u + W^-$. Write what we had in the preceding paragraph, $e^- + \bar{\nu} \leftrightarrow W^-$, as $W^- \to e^- + \bar{\nu}$. Link these two pieces, $d \to u + W^-$ and $W^- \to e^- + \bar{\nu}$, together to obtain the process $d \to u + e^- + \bar{\nu}$. Voilà! A neutron changes into a proton by emitting an electron and an antineutrino.

The most intriguing of the three interactions

The weak interaction is so weird that the eminent Japanese American physicist and Nobel laureate Yoichiro Nambu referred to it as "God's mistake?".

I would not go that far, but in many ways, the weak interaction is much more fascinating than its sister interactions, the strong and the electromagnetic. I picture a Victorian novel with three sisters, the weak[12] one being the weirdo, but because of that, the most enchanting, with a convoluted and twisted history. You thought that the history of the strong interaction described in chapter V.2 was involved? Hardly, when compared with the history of the weak interaction.

Since this is not a book on particle physics, I have to restrict myself to a brief mention of some of the most mysterious properties of the weak interaction. Physicists whose education stopped[13] with the electromagnetic interaction don't know what they were missing.

We now turn to one of the weirdest aspects of the weak interaction.

Favoring the left over the right

A priori, the thought that the fundamental laws of physics might distinguish between left and right would seem patently absurd. The principle that Nature has no preference for either left or right is known as parity. Indeed, parity was built into physics as a cornerstone belief.

That was why the violation of parity by the weak interaction was such an almost unimaginable shock to the theoretical physics community, that the Nobel Prize in physics for 1957 was awarded almost immediately to the Chinese American physicists Tsung-dao Lee and Chen-ning Yang for suggesting that the weak interaction would favor the left over the right.[14]

Parity violation is perhaps best explained using the concept of helicity, which describes the spin of a moving particle and which I have already mentioned in connection with the photon back in chapter III.3. As I said there, students are taught to wrap either their right hand or their left hand around the direction the particle is moving, with the thumb pointing in that direction. The fingers are then supposed to point in the direction the particle is spinning. You could see that you have to use either your left hand or your right hand. The particle is said to be left handed or right handed accordingly. All particles known before the neutrino are free to spin whichever way they "like," and so they could be either left or right handed. For instance, the electron and all the quarks are both left and right handed. (By the way, the reader surely recognizes that what physicists call left and right amounts to a mere convention, such as what we call clockwise and anticlockwise, and hardly a profound truth, as was misunderstood by a hoity-toity philosophy professor I tried to explain this to, on a par with some of the eternal verities uttered by, say, Kant. For instance, everyday screws are right handed,[15] but that's just a convention.)

The profundity is that the neutrino was discovered experimentally to be left handed, never right handed! Thus, the weak interaction resoundingly violates parity.

In some sense, what is really strange is not that the weak interaction violates parity, but that the other three interactions respect parity.[16]

You should also understand that the concept of helicity works only if the particle is massless and thus moving at the speed of light. If not, then an observer moving faster than the particle would see the particle moving in the other direction, but spinning the same way, and so its helicity would have flipped sign. Thus, for a massive particle, its helicity would depend on the observer and hence is not a useful concept. Another way of saying this is that to an observer moving alongside a massive particle at the same speed, the particle would appear to be at rest. And so this whole left hand–right hand business simply fails.

Quantum field theory: too accommodating

While parity violation shocked theoretical physicists out of their skulls, it could be readily accommodated by quantum field theory. We simply exclude the right handed component of the neutrino field from the action. Parity violation does not contradict the foundational principles of quantum field theory.

This brings me to another important point: Quantum field theory has been flexible and accommodating. We could easily imagine some experimental discovery that could not be accommodated by quantum field theory and would thus force theoretical physicists to go look for something else (which would, however, contain quantum field theory). Nearly a century has elapsed since the founding of quantum field theory and thus far, no experimental discovery definitively requires quantum field theory to be extended and generalized in the same way that quantum mechanics had to be extended and generalized. Of course, nothing precludes that happening tomorrow.

The neutrino in the mirror

To continue the story, I have to tell you a bit more about parity. Parity actually derives from space inversion, under which we flip the three Cartesian coordinates $\vec{x} \to -\vec{x}$; that is, more explicitly, $x \to -x$, $y \to -y$, $z \to -z$. Consider a rotation by $180°$ around the z-axis, under which $x \to -x$, $y \to -y$, $z \to z$. Combining this rotation with space inversion results in the transformation $x \to x$, $y \to y$, $z \to -z$. This corresponds to reflection in a mirror, if we call the coordinate axis perpendicular to the mirror the z-axis (figure 4).

Trusting rotational invariance to hold, we see that to test for parity, we should simply check whether the fundamental laws of physics we know and love also hold in the world in the mirror. Indeed, a moment's reflection shows that the mirror flips left to right and right to left.

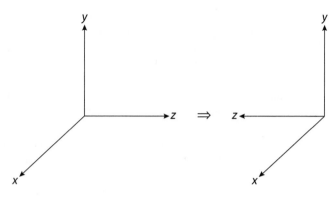

Figure 4. Descartes' coordinate axes reflected (as indicated by the double arrow) in a mirror perpendicular to the z-axis. The x-axis and y-axis are left pointing in the same direction, but the z-axis is flipped.

The neutrino in the mirror, instead of being left handed, is right handed. Hence, the world in the mirror is different from our world: In that world, neutrinos are right handed. This is perhaps the easiest and quickest way to see that the weak interaction indeed violates parity. (I might add, somewhat parenthetically, that Michael Jackson's man in the mirror would have his heart on his right side instead of his left side, but the existence of such a man, although it would make medical news, does not contradict the laws of physics.)

A friendly word of advice: For the purpose of this book, the reader need not absorb all the details I am, and will be, giving you about the weak interaction. I merely want to convey to you the sense of a deep puzzle looming in the weak interaction.

C and CP: symmetry between matter and antimatter

After the shock of parity violation, physicists uncover ever more bizarro properties of the weak interaction. Nambu's "God's mistake"?

Recall that Dirac's discovery of antimatter followed inexorably from combining special relativity with quantum physics, as we saw in chapter V.1. Your first thought might be whether we would have to construct a whole new physics to govern the behavior of antimatter. Physicists subsequently concluded that no, Nature is kind. Antimatter obeys the same laws of physics as matter, a postulate known as charge conjugation invariance, or C for short. (Charge conjugation is jargon for the operation of changing matter into antimatter and vice versa.)

It is important to emphasize that C invariance is not a decree handed down from on high, but a conclusion reached after checking how all the known laws of physics apply to antimatter. For instance, the electromagnetic interaction has been verified experimentally to great accuracy to respect C. In particular, the positron e^+ is obliged by C to have exactly the same mass (as was mentioned in chapter V.1) as the electron e^-, and that is in fact true to some number of significant figures.

By charge conjugation, the properties of an antiparticle are completely determined by the properties of the corresponding particle. One natural question to ask then is how helicity behaves under C. Consider a field described by the Dirac equation. By simple[17] manipulation of the equation, theorists could see that under C, the left handed component of the field is transformed into the right handed component of the conjugate field, and vice versa. This is an incontrovertible consequence of elementary mathematics involving complex conjugation and the properties of the Dirac gamma matrices mentioned earlier.

Let us now go back to the neutrino, and see how its existence continues to mystify. The neutrino is left handed. That was shocking enough, but now, a few pages later, I expect that the reader has gotten over the shock of P violation and is ready for another shock. Ready or not, here it comes!

If C holds for all physical laws, then the neutrino and the antineutrino should spin the same way, and we would expect the antineutrino also to be left handed. So, another shock! The weak interaction also violates C.

I trust that the reader could put two and two together as the idiom[18] goes, and deduce that the weak interaction still respects CP, combining charge conjugation C and parity P. Two minuses equal a plus, and all that high brow mathematics we all learned long ago!

We replace the everyday mirror by a magic mirror that turns matter into antimatter. A left handed neutrino zips by. In the magic mirror, we see a right handed antineutrino go by. The weak interaction violates C and P separately, but still respects the product CP. Well, P interchanges left and right, while C interchanges matter and antimatter. The laws of physics in the world in the magic mirror are the same as those in our world.

T and the microscopic arrow of time

The laws of physics also respect time reversal, T for short, as has been known since the time of Newton. That was what the $\vec{a}$ in $\vec{F} = m\vec{a}$ is all about; the letter a stands for acceleration. Force produces acceleration, not velocity, as Aristotle had asserted. The medieval peasant pushing his heavy cart along a muddy road was devotedly an Aristotelian and thought Newton a loco egghead.

To give a precise definition of T, we abandon our magic mirror and use a more modern technology. Make a movie of a microscopic process, say, $a + b \rightarrow c + d$. Then we run the movie backward. We see the process $c + d \rightarrow a + b$ and ask whether this is allowed by the laws of physics.

A quick digression into a common confusion about acceleration. Everyday language does not distinguish between velocity and speed, but in physics, velocity is defined as speed with direction. In other words, speed is the magnitude of velocity. Saying that a particle is moving at 100 kilometers per second tells me about its speed, but not its velocity; to do that, you need to say 100 kilometers per second due east. Now imagine filming that and running the movie backward. We see the particle zipping by at 100 kilometers per second due west. Thus, under T, velocity flips sign.

Next, suppose the particle is also accelerating due east, so that a second into the future, it is moving at 110 kilometers per second due east. In other words, its acceleration is 10 kilometers per second per second pointing east.

Once again we film and play the movie backward. We see a particle moving due west at 110 kilometers per second, but a second later, it is moving due west at the lesser speed of 100 kilometers per second. The change in velocity equals $(100 - 110) = -10$ kilometers per second pointing west, but this is the same as $+10$ kilometers per second pointing east. What a layperson would call a "deceleration," a theoretical physicist would call an "acceleration in the opposite direction." But an acceleration of 10 kilometers per second pointing east is exactly the same as the acceleration of the particle in the real world being filmed, as noted in the preceding paragraph. We conclude that under T, acceleration remains the same, in contrast to velocity, which flips sign.

Newton's law $\vec{F} = m\vec{a}$ remains invariant, while Aristotle's "law," based erroneously on velocity, does not.

Incidentally, that our friend Humpty Dumpty from chapter II.1 could break into pieces and be made into an omelette, while all the king's scientists could not make a Dumpty out of a poached egg, is a sad consequence of the statistical concept of entropy, which isn't one of "the fundamental laws of physics" as defined here, and so is outside the purview of the present discussion.

Again, time reversal invariance is not an assertion decreed by some authority, but a proposition verified experimentally and theoretically again and again,[19] from Newtonian mechanics through electromagnetism through quantum physics. To cite just one example, with some effort, students could verify[20] that the Dirac equation respects T. Concisely stated, T invariance means that there is no microscopic arrow of time, only a macroscopic arrow, such as aging.[21]

Quantum field theory may be accommodating, but there are limits: the CPT theorem

To summarize, by the late 1950s, charge conjugation C and parity P had been shown to fail in the weak interaction, but CP and time reversal T still stood.

Then in 1964, Jim Cronin[22] and Val Fitch discovered that in the weak decay of the neutral K meson (mentioned in passing in chapter V.2), CP was also violated, for which they later received the Nobel prize. Not only is the world in the mirror not the same as our world as far as the fundamental laws of physics are concerned, but also the world in the magic mirror, the mirror that also reflects particles into antiparticles, is not the same as our world!

I mentioned earlier that quantum field theory is extraordinarily accommodating. Not only was P violation, and C violation, easily accommodated in quantum field theory, but also CP violation. All we have to do is to change the real numbers representing some of the parameters in the action into complex numbers. Easy.

But quantum field theory does have a limit on how accommodating it can be. A theorem was proved, based on the foundational principles that went into quantum field theory, stating that the combination CPT cannot be violated. In other words, if we take a movie of the world in our magic mirror and run that movie backward, that world is the same as our world as far as the fundamental laws of physics are concerned.

Amazing theorem! If we reflect everything in a mirror, turn all particles into antiparticles and vice versa, and reverse the flow of time, then the same fundamental laws still operate. You could violate P, you could violate C, and you could violate T, until your face turns blue, but somehow CPT still stands, as long as the foundational principles of quantum field theory stand.

If we trust the CPT theorem (and there is absolutely no compelling reason not to), then the discovery of CP violation in 1964 would imply that T is violated. The search was on to find an elementary process that would violate time reversal directly (that is, without having to invoke the CPT theorem). Decades, and almost countless number of careers later, T violation was finally observed.[23]

But CPT still stands.

Summary: The weak interaction violates P, C, T, and CP, but in such a way that it still respects CPT! Weird, yes?

Treating the left and the right differently

With our knowledge of parity violation, we can extend the modern way of summarizing the weak interaction given earlier. The field ψ Dirac used to describe the electron (and later used by Gell-Mann to describe the quarks) may be split[24] into a left handed and a right handed field: $\psi = \psi_L + \psi_R$. Parity violation in the weak interaction is accommodated by saying that the W bosons ignore the right handed fields. Thus, behold the modern view of the weak interaction:

$$\begin{pmatrix} \nu_L \\ e_L \end{pmatrix} \quad e_R \quad \bigg| \quad \begin{pmatrix} u_L \\ d_L \end{pmatrix} \quad u_R \quad d_R$$

The subscript L and R denote left handed and a right handed field, respectively.

In contrast to the left handed fields v_L, e_L, u_L, d_L, which are put into doublets, that is, houses with two floors, the right handed fields e_R, u_R, d_R are treated as singlets, that is, they live in one story houses. Hence, the W bosons, whose responsibility is to move fields up and down between floors, ignore the right handed guys totally. In this bizarre way, the shockwave that went through the theory community in 1956 is accommodated in quantum field theory these days by treating the left and the right differently. Some readers might have to agree with Nambu that God made a mistake in creating the weak interaction!

All we have to do is to slap the subscript L on the elementary processes we wrote down before:

$$e_L^- + W^+ \leftrightarrow v_L \qquad \text{and} \qquad d_L + W^+ \leftrightarrow u_L$$

and

$$v_L + W^- \leftrightarrow e_L^- \qquad \text{and} \qquad u_L + W^- \leftrightarrow d_L$$

Compare this with what was written a few pages ago.

Note that the right handed neutrino field v_R does not exist at all in this scheme. Where is that guy? Lost somewhere? Hold that thought for now.

Family problem

Meanwhile, the weak interaction continues to entrance and intrigue. An obese cousin to the electron was discovered,[25] known as the muon, with properties identical to the electron except that it is more than 200 times more massive than the electron.[26] This new lepton, totally unexpected and apparently serving no role in physics, prompted the great Nobel winning experimental physicist Isidor Rabi to quip,[27] "Who ordered that?" The muon, denoted by the Greek letter μ, participates in the weak interaction, much like the electron. In fact, it decays weakly into the electron plus a neutrino and an antineutrino.

In one of the most surprising developments[28] in particle physics, the muon was found to have its own "private" neutrino, denoted by v_μ. The v we have been talking about thus far belongs to the electron, and so we should go back and replace v by v_e. For our purposes here, the one fact you need to know is that the muon decays into its lighter cousin[29] the electron by $\mu^- \rightarrow e^- + \bar{v}_e + v_\mu$. May I test your potential ability as a particle phenomenologist, the subset of particle theorists who relate experimental observation to theory? See if you can draw the Feynman diagram for this decay process. Hint: It is easier than easy.

If you need another hint, look at figures 2 and 3. You're exactly right, all you have to do is to replace the label d by μ^-, u by v_μ, and $\bar{v}$ by $\bar{v}_e$.

Meanwhile, cue ominous organ music here, the plot thickens.

In the 1950s came the discovery of "strange hadrons" (called "strange" because they were not like the familiar proton, neutron, and pion), such as the Σ hyperon and the K meson, as I mentioned in chapter V.2. Gell-Mann eventually proposed that they contained what he called the strange quark s, as also mentioned.

Later, in the 1970s, came the charm quark,[30] denoted by c, as also mentioned peripherally.

After a tremendous amount of (largely) experimental work, particle physicists arrived at this summary:

$$\begin{pmatrix} \nu_{\mu L} \\ \mu_L \end{pmatrix} \quad \mu_R \quad \Big| \quad \begin{pmatrix} c_L \\ s_L \end{pmatrix} \quad c_R \quad s_R$$

I cordially invite you to flip back a page or two and compare this with what we had. The two diagrams are identical with the correspondence $\nu_e \rightarrow \nu_\mu$, $e \rightarrow \mu$, $u \rightarrow c$, and $d \rightarrow s$. A set of quarks and leptons such as that given here is referred to as a "family" or a "generation" in the family. Usage varies. The set with the electron, the up and down quarks (that is, the particles that make up the familiar world of electrons, protons, and neutrons) is known as the first family or the first generation. The set shown above, with the muon, the charm and the strange quarks, is known as the second family.

Even more mysterious sounding music, please! Later, experimentalists discovered yet a third family, consisting of the τ (Greek tau) lepton, and the top and the bottom quarks, as shown here:

$$\begin{pmatrix} \nu_{\tau L} \\ \tau_L \end{pmatrix} \quad \tau_R \quad \Big| \quad \begin{pmatrix} t_L \\ b_L \end{pmatrix} \quad t_R \quad b_R$$

Note that a third neutrino, ν_τ, has to be included.

Why Nature, for no apparent reason at all, repeats the set of leptons and quarks three times is a fundamental mystery. Physics can offer no explanation whatsoever at present.

This is known as the family problem in particle physics. It is worth emphasizing that the gauge bosons, which include the beloved photon, the W and Z bosons, the gluons of the strong interaction, and the graviton, are not repeated. You recognize that these are the mediators of the four fundamental interactions. There is only one photon, not three. What is repeated three times are the quarks and the leptons, sometimes thought of as the matter content of the universe.

The number 3 appears often in fundamental physics, 3 spatial dimensions, 3 colors, 3 families.

Almost separated families lead to unwanted stability

Just as what we did before for the first family, we can now write down how the absorption and emission of the W boson transforms the leptons and quarks of the second family into each other. Ditto for the third family.

I already mentioned that the muon decays. You could have fun constructing the weak interaction process responsible. Flip back a few pages and simply replace the fields in the first family by the fields in the second family. Thus, we have $\mu_L^- \to \nu_{\mu L} + W^-$. Link this to what we had before, $W^- \to e_L^- + \bar{\nu}_{eL}$, to obtain $\mu_L^- \to \nu_{\mu L} + e_L^- + \bar{\nu}_{eL}$. There! The muon decays into an electron, an anti electron neutrino, and a muon neutrino.

How about the quarks in the second family?

You might have noticed that what I have drawn amounts to three almost separate worlds, connected by the photon, the W, and the rest of the "mediators." That the families are almost separate leads to a problem: too much stability in the world!

Look at the strange quark s. The W^+ boson could lift it upstairs to become the charm quark c, thus leading to the process $s \to c + e^- + \bar{\nu}_e$, just like the process $d \to u + e^- + \bar{\nu}_e$ shown in figure 3. Take a look at the Feynman diagram. Simply replace the down quark by the strange quark, and the up quark by the charm quark.

But no! The charm quark c is quite a bit more massive than s. This makes sense if you remember your history: while the strange hadrons have been known since the early 1950s, charmed hadrons (namely, those hadrons containing the charm quark), being more massive, were not discovered until the 1970s. The strange quark s cannot decay for lack of energy. Einstein said no! (Once again, $E = mc^2$: not enough mass, not enough energy.) This process, $s \to c + e^- + \bar{\nu}_e$, is not allowed to proceed. Strange hadrons would be stable.

Similarly, of the two quarks in the third family, whichever is lighter would be stable.

But this poses a problem. Strange hadrons were known to decay rapidly. (Indeed, that was how they were discovered[31] in 1950: Produced in the atmosphere by high energy cosmic ray particles colliding with the protons and neutrons in molecules of air, they disintegrate into the familiar proton, neutron, pion, electron, photon, et cetera, producing what is called a "cosmic ray shower.")

Also, looking around us, we do not see many stable hadrons. In fact, there is only one, the proton. Otherwise, there would be truly exotic atoms formed by electrons bound to the analogs of the proton from the second and third families.

The stability of the proton and the universe

This is a good place to explain why the proton, made up of two up quarks and a down quark, is stable.* Care to try your hand at it?

Very simple. The proton is not massive enough. Recall that we started this chapter by talking about the decay $(Z, A) \to (Z + 1, A) + e^- + \bar{\nu}$. The mother

*I come back to this issue in chapter V.4.

nucleus (Z, A) has to be more massive than the daughter nucleus $(Z + 1, A)$. Otherwise, where would the energy needed to make and eject the electron and the antineutrino come from? It comes from the mass difference between the mother nucleus and the daughter nucleus converted into energy according to the formula $E = mc^2$ that everybody, even the proverbial guy in the streets, knows. For the same reason, the neutron can decay into the proton $n \rightarrow p + e^- + \bar{\nu}$ and have enough energy to make and eject an electron and an antineutrino because it is more massive than the proton.

But then by the same token, the proton can't decay into the neutron via the process $p \rightarrow n + e^+ + \nu$. The proton is not massive enough to decay.

Indeed, the mass of the proton $m_p \simeq 938.2$ MeV, just a tad less than the mass of the neutron $m_n \simeq 939.6$ MeV. (Here MeV stands for million electron volt, the unit used by experimental particle physicists. In case you insist on some silly unit made up by some French revolutionary: $m_p \simeq 1.6726 \times 10^{-24}$ gm versus $m_n \simeq 1.6749 \times 10^{-24}$ gm.)

Here is a fact that you could dazzle your friends with. (At least I was much dazzled when I first read about it in a popular book by George Gamow.) The proton and the neutron appear to differ only in that the proton is electrically charged and the neutron not, as its very name indicates. Since an electric field contains energy (as we all know), physicists expect the proton to be more massive than the neutron. But no, as we've just seen, the proton is a teeny bit less massive than the neutron, a fact that, remarkably, guarantees the stability of the universe. How so?

Consider a hydrogen atom with an electron orbiting a proton. Were the proton more massive than the neutron, it would decay according to the (nonexistent) process $p \rightarrow n + e^+ + \nu$. The positron e^+ and the neutrino ν fly off. With the charged proton replaced by the electrically neutral neutron, the orbiting electron suddenly feels that it is no longer bound by an attractive force and so takes off also. Poof! The hydrogen atom is gone. No more hydrogen gas condensing to form stars, the universe becomes a collection of neutrons, electrons, and positrons, with some neutrinos flying around.

The strange quark yearning to decay

Let us now return to the strange quark s yearning to decay but unable to. At the risk of repeating myself, if s were truly unable to decay, our world would then be quite different—strange, to say the least. There would be all kinds of stuff formed out of the strange quark, in addition to the familiar stuff formed out of the up and down quarks. A stable strange hadron analogous to the proton could form a variety of atomic nuclei together with the neutron and its strange analog, and could electrically attract electrons to form strange atoms. We would truly become strangers in a strange land!

What happens is that the strange quark s, desperate to decay, sneaks into the ground floor apartment inhabited by the d quark; there the W^+ boson could lift it upstairs to become the quark u, so that it could decay according to $s \rightarrow u + e^- + \bar{v}_e$. (The same reasoning as per Einstein: The s quark is more massive than the u quark, and hence has plenty of energy to spare in decaying. This is of course also why strange hadrons were discovered so much later than the non-strange hadrons; they are more massive.)

So the hadrons containing the strange quark s do decay via the weak inter-action, in about 10^{-10} sec. For instance, the Σ^- hyperon, which consists of dds, decays via $\Sigma^- \rightarrow n + e^- + \bar{v}_e$, which is exactly what would happen if s could decay according to $s \rightarrow u + e^- + \bar{v}_e$. In other words, when the s quark inside the $\Sigma^- = dds$ hyperon transforms itself into a u quark by emitting an electron and an antineutrino, we are left with ddu, which you may recog-nize as the good old neutron. In our example, the decay of the Σ^- hyperon is now seen as $(dds) \rightarrow (ddu) + e^- + \bar{v}_e$. To summarize, the decay of the strange quark generates the decay of the Σ^- hyperon:

$$s \rightarrow u + e^- + \bar{v}$$

generates $\quad (dds) \rightarrow (ddu) + e^- + \bar{v}$

(Compare with what I wrote earlier for neutron decay.) By introducing quarks, Gell-Mann has reduced the physics of hadron decays to almost a children's game of building blocks, in which one kind of block could decay into another by emitting electrons and neutrinos.

This time scale of 10^{-10} sec may seem short to you and me, but it should be compared to the characteristic time scale of a strong interaction process, such as the production of a Σ hyperon in the collision of two protons, typi-cally around 10^{-23} sec. These time scales are both so remote from what we experience that, when I was a student, I simply thought of both of them as a zillionth of a second, or more simply as "whatever." And yet that vast dif-ference was what allowed physicists to distinguish the two interactions in the first place. Note that the ratio of 10^{13} between the strong interaction and the weak interaction is the same as the ratio of the age of the universe, around 10^{10} years, to something typical of everyday life, say, 10 hours.[32]

One reader finds this explanation of the observed decay of the strange hadrons "hokey and ad hoc," but please note that no law in physics forbids the strange quark from "sneaking into the apartment" inhabited by the down quark. If you think it is bizarre, then I, and many physicists, would agree with you. Why create two extra families with no apparent role in the universe and then make them decay in this bizarre fashion?

Weak interaction misaligned relative to the strong interaction

I am of course using picturesque language when I talk about the strange quark sneaking into the down quark's apartment. What I actually mean is that what I wrote down earlier, $\begin{pmatrix} u_L \\ d_L \end{pmatrix}$, should be replaced by

$$\begin{pmatrix} u_L \\ \cos\theta\ d_L + \sin\theta\ s_L \end{pmatrix}$$

The tiny angle[33] θ, experimentally measured to be about 13°, is known as the Cabibbo angle.[34] Perhaps a rather imperfect[35] way of describing this is that in that ground floor apartment, we have not only d, but occasionally s would visit when d is out.

The whole concept of the downstairs tenant $\cos\theta\ d_L + \sin\theta\ s_L$ being a "mixture" of d and s is peculiar to the quantum world. The precise statement is that the probability amplitude of finding s there instead of d is given by the ratio[36] $(\sin 13° / \cos 13°) \simeq 0.2$.

Even better, since you and I are both quantum field theorists by now, more or less, we should look at the interaction Lagrangian. The coupling of the W to the relevant quarks is modified from $\mathcal{L} = g\ W\ \bar{u}\ d$ to $\mathcal{L} = g\ W\ \bar{u}\ (\cos\theta\ d + \sin\theta\ s) = g\ (\cos\theta\ W\ \bar{u}\ d + \sin\theta\ W\ \bar{u}\ s)$. Consequently, the W boson can transform not only d to u but also s to u, though with a considerably smaller probability amplitude. If you don't mind an admittedly imprecise analogy, it is as if physicists have modified the famous fairy tale so that at midnight, not only would the carriage turn into a pumpkin, but there is some probability that Cinderella, instead of the carriage, would turn into a pumpkin!

For simplicity, we only talk about what is called "mixing" between the first two families, but this had to be generalized to include the third family after it was discovered. For the same reason, the bottom b quark has to sneak a bit into the apartment occupied by s (and to a smaller extent into that occupied by d) in the simplified scheme I wrote down earlier. Otherwise, hadrons containing the b quark would be unable to undergo weak decay (because it is less massive than the top t quark.)

An eerily mysterious mansion

We don't understand why Nature would want to triplicate the quarks and leptons, and after this senseless repetition, proceed to mix them so that the world is left with only one stable hadron, the proton.

Another way of saying this is that the weak interaction is somehow misaligned relative to the strong interaction. The Cabibbo angle and its generalization measure this (slight) misalignment. Imagine ourselves in a wonderful mansion. Mysteriously enough, one wing is repeated three times, with the same furniture arrangement, except that they are all more massive in the second and third wing. Not only that, but these extra wings are slightly tilted. Perhaps you are more and more inclined to agree with Nambu that the weak interaction looks like "God's mistake."

Finite time problem or not

At this point, I would like to distinguish between two classes of unsolved problems in theoretical physics. (However, I do not want to imply anything about their relative importance or intrinsic theoretical interest.)

Consider, just to give an example, the mechanism for high temperature superconductivity. This is an extremely difficult problem on which many brilliant physicists have worked on for more than 30 years. However, since the interaction between electrons and the nuclei in a superconducting material and the relevant quantum mechanics are known, the theoretical physics community has no doubt that given enough brain power and enough time, the problem will be solved. I propose to call such problems "finite time problems."

The family problem, in contrast, seems to present a different sort of difficulty. We have no clue as to why Nature would want to triplicate the matter content of the universe. According to our understanding, it appears that the universe with only the first family would run perfectly well. After the Big Bang, stars and galaxies would form and everything would proceed according to plan. One indication of this is that the strange quark, the charm quark, the top quark, the bottom quark, the muon, the τ lepton, and their neutrinos, all of this stuff is totally irrelevant to the rest of physics. Physicists working outside particle physics have absolutely no reason to know anything about these particles and fields, and they couldn't care less if all these particles magically disappear from the universe tomorrow. Indeed, even particle physicists may not care.

We have no idea how to even start attacking the family problem. It is not as if we could sharpen our pencils, sit down with a clean pad of paper, and start calculating why Nature wants to have three families. Perhaps there's no explanation at all, or perhaps an explanation will come, who knows, a decade from now, a century from now, or even a millennium from now.

We have no clue whatsoever whether the family problem is a finite time problem or not.

I might mention parenthetically a third class of problems: homework problems for physics students. The knowledge that the solutions exist (and could be found in a finite time measured in hours at most) is of course hugely important and changes the game totally. Furthermore, the methods needed are to be found in the material just covered in the course. This explains why some students who excel in courses fail miserably later as researchers. Getting an A+ in courses is a necessary, but not sufficient, condition for excellence in research, for sure.

Electroweak unification

Over the decades, experimentalists worked hard to determine the properties of the W boson while theorists speculated. Is it associated with a scalar field like

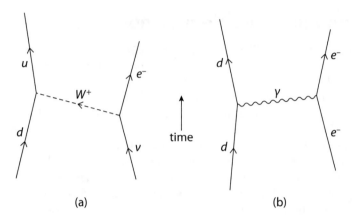

Figure 5. In this figure, (a) depicts a weak process, (b) an electromagnetic process. Note the structural similarity. Time runs upward as always. Here *u* and *d* denote the up and down quark respectively.

Yukawa's φ or with a vector field like Maxwell's A_μ? In other words, does the W boson have one unit of spin just like the photon? After a long struggle to which many devoted their lives, it was ascertained that the W boson is similar to the photon as far as its spin is concerned.

Pull the weak process $v + d \rightarrow e^- + u$ described in figure 2(b) apart so that it becomes figure 5(a). The incoming neutrino converts itself into an electron by emitting a W^+ (the positively charged counterpart to the W^- boson). Note that charge conservation mandates that a W^+, rather than a W^-, is emitted. The W^+ is then absorbed by the incoming down quark d which thus morphs into an outgoing up quark u.

Incidentally, note that this represents the same argument that we used in chapter V.1 to establish the existence of the antiparticle of the electron, the positron. I have drawn the vertex at which the W^+ is absorbed to occur at a slightly later time than the vertex at which it was emitted. Thus, given the existence of the W^-, its antiparticle, the W^+, must also exist.

Now compare this process to the electromagnetic process $e^- + d \rightarrow e^- + d$ depicted in figure 5(b). By now you should know this Feynman diagram like an old friend, first introduced to you back in chapter IV.1. A photon, traditionally denoted by the Greek letter gamma[37] γ, is emitted by the incoming electron and then absorbed by the incoming down quark d.

Did you notice that the two diagrams in figure 6 look structurally the same. Might they be related?

You might begin to suspect that the W bosons, W^+ and W^-, and the photon γ are secretly related. You might even guess that perhaps they are the three gauge bosons of the Yang-Mills theory based on the group $SU(2)$, now that you have learned in chapter V.2 what Yang-Mills theory is about. ($2^2 - 1 = 3$, remember?) If you were around in the mid to late 1950s, you might have been so bold as to propose unifying the weak interaction with the electromagnetic interaction. Such is the power of education (and hindsight): You might have had a shot at a Nobel prize in physics. Indeed, that's exactly what Julian Schwinger suggested to his student Shelly Glashow for his PhD thesis.

But several obstacles might have stopped you cold. While the photon is massless, the W boson is enormously massive. As a result, the electromagnetic interaction is infinitely long ranged, while the weak interaction is short ranged, as you learned in chapter III.3 and in this chapter. Furthermore, while the photon treats left and right equally, the W couples only to the left components of the quark and lepton fields.

This entire discussion is of course in hindsight's bright glare. That the comparison between figure 5(a) and figure 5(b) offered the key for progress came only after decades of dead ends and numerous puzzles.

OK, the W bosons are massive, while the photon is massless. Not knowing anything better, Glashow simply put the mass of the W bosons in by hand. (What this means is that you pick up a pen and write the desired mass term into the action.)

I hope that I conveyed to you in chapter V.2 that the action for Yang-Mills theory must be intricately balanced for gauge invariance to hold. In particular, the gauge bosons are all exactly massless. Remember the question Pauli asked Yang, why the gauge bosons were not seen if they were massless? Just writing in the mass is not acceptable; it breaks the gauge invariance and causes the theory to go haywire. Nowadays, we know that the Anderson-Higgs mechanism[38] (often shortened to the Higgs mechanism) could generate masses for the W bosons spontaneously and softly (that is, without smashing the gauge structure).

Speaking picturesquely, I might describe Glashow's approach as analogous to brutally separating the babies at birth and fattening two of them up, the W^+ and the W^-, leaving the photon massless. The reason the Higgs mechanism is celebrated is because it provides a gentle way of separating the gauge bosons without ruining the intricate balance of the theory.

A bonus feature of the Higgs mechanism is that while generating masses for the W bosons, it could at the same time generate masses for the quarks and leptons. Let me explain how that works for the electron, schematically at the level of this book, merely to give you a flavor. Recall from chapter IV.1 the Dirac Lagrangian $\mathcal{L}_{\text{Dirac}} = \bar{\psi}(i\gamma^\mu \partial_\mu - m)\psi$ for the electron, which I reproduce here. Note that I have not included the coupling of the electron to the electromagnetic field: It is not relevant for the discussion here.

Focus on the mass term $m\bar{\psi}(x)\psi(x)$. Higgs introduced what is now called the Higgs field $h(x)$, and replaced the mass term by $fh(x)\bar{\psi}(x)\psi(x)$, with the coupling strength f measuring how $h(x)$ couples to the electron. (Some terms describing the propagation of $h(x)$ in spacetime also must be added to the Lagrangian, but that is not immediately relevant to our discussion here.) You see that we could produce the mass term for the electron if we simply say that $h(x)$ equals some constant v independent of x, that is, $h(x) = v$. In other words, trivially substituting, we obtain

$$fh(x)\bar{\psi}(x)\psi(x) \to fv\bar{\psi}(x)\psi(x) = m\bar{\psi}(x)\psi(x)$$

with $m = fv$. By the way, v is known as the vacuum expectation value of the Higgs field, but since you are not expecting to receive a physics PhD degree by reading this, you don't actually have to know the jargon. The W boson acquires its mass in a similar fashion, but I will not go into the details here.

If you feel that the Higgs mechanism hardly represents the last word on the origin of masses, almost all physicists would agree with you resoundingly. Surely it will be replaced by something deeper one day.

The Z boson

Coming along some years after Glashow, Abdus Salam and Steve Weinberg both had the idea of adding the Higgs mechanism to Yang-Mills theory. And thus, in this briefest possible account, the weak interaction was unified with the electromagnetic interaction into a single electroweak interaction, for which Glashow, Salam, and Weinberg were awarded the Nobel prize. (Needless to say, I am again skipping over the enormous struggles that actually went on.)

An exciting feature of the theory is that it predicts a variety of hitherto unknown weak interaction processes, generated by an extra gauge boson, the Z boson. I mentioned in chapter V.2 that the number of gauge bosons in a Yang-Mills theory is fixed by group theory. An initial stumbling block to electroweak unification was that a theory with the three gauge bosons, W^+, W^-, and the photon, based on the group $SU(2)$ does not work, rather reminiscent of doing a jigsaw puzzle and realizing that a piece is missing. A fourth gauge boson, the Z boson just mentioned, is needed.

Instead of $SU(2)$, the group has to be $SU(2) \otimes U(1)$. The funny mathematical notation $\otimes$ means that the two groups $SU(2)$ and $U(1)$ have to be loosely joined together. But the $U(1)$ here is not the same as the $U(1)$ Weyl introduced into physics (as was mentioned in chapter V.2), and so theorists were thoroughly confused for a while. A puzzling feature[39] is that the $U(1)$ of electromagnetism, namely Weyl's $U(1)$, sits partly inside the $SU(2)$ and partly inside the $U(1)$ of the group $SU(2) \otimes U(1)$. As it turned out, Nature made it a bit harder for theoretical physicists.

Using the right mass scale

I realize that I am not explaining anything in detail here lest the book end up ten times as thick as it is. For interested readers, numerous textbooks about the weak interaction are available.[40] Instead, I end with a historical footnote underlining the importance in physics of placing physical quantities in the right context. As a simple example, we should not use the same scale to measure the energy of a photon emitted by an excited atom and by an unstable nucleus.

Recall that the Fermi constant G governs the strength of the weak interaction. The value of $1/\sqrt{G}$, which has the dimension of a mass, had long been measured. Given the origin of the weak interaction in the decay of unstable nuclei with the ejection of an electron, physicists naturally compared $1/\sqrt{G}$ with the electron mass. Within this tradition of equating $1/\sqrt{G}$ and the electron mass, the coupling strength of the W boson came out to be teeny compared to the coupling strength e of the photon we talked about in chapter IV.3. Thus, for a long time, theoretical physicists were puzzled that although the weak process and the electromagnetic process in figure 6 look the same, the coupling strengths involved were hugely different.

Around 1950, it occurred to Julian Schwinger that if he used the proton mass instead of the electron mass, the weak coupling strength, though still much less than the electromagnetic coupling strength, is no longer many orders of magnitude smaller. When the young Schwinger mentioned this observation to Oppenheimer, the latter dismissed it as mere numerology. Nowadays, we know that the correct mass to use is yet larger than the proton mass, namely, the W boson mass, and if we do that, the weak and electromagnetic coupling strengths come out to be about the same, indicating that these two interactions could indeed be unified.

Mysteries of the weak interaction

Evidently, I can't possibly survey all aspects of the weak interaction. In contrast to the strong and electromagnetic interactions, the weak interaction, even after almost a century of intense scrutiny, is still shrouded by mysteries. Let me mention just one of these: Are the neutrinos the same as the antineutrinos?[41] This possibility[42] was suggested by the eccentric Italian genius Ettore Majorana, whose mysterious disappearance[43] from a ferry going from Palermo to Naples in 1938 continues to be speculated on by physicists.

The wild child of physics

Finally, a one sentence take-home message from this long chapter: The weak interaction is weird, the wild child of physics, certainly when compared with the relatively bland and straightforward electromagnetic and strong interactions.

Notes

[1] As explained in chapter I.1, many nuclear isotopes are delicately balanced in a perpetual contest between the strong and the electromagnetic interactions.

[2] This naming convention is motivated by later experiments which indicate that the total number of leptons is conserved. Since in the decay $(Z, A) \rightarrow (Z+1, A) + e^- + \bar{\nu}$, the two

nuclei are not leptons, if we call the electron a lepton, then we have to call the invisible particle accompanying the electron an antilepton. Later in this chapter, you would see that in the modern way of writing the theory of weak interaction, this naming convention leads to the electron "living in the same house" with the neutrino, not with the antineutrino.

[3]For more details, see *FbN,* chapter IX.2.

[4]For more details about the neutrino, see *Fearful,* page 37 ff.

[5]Perhaps not surprisingly, his paper was rejected by the journal *Nature.* Too original! For an English translation of his 1934 paper, see http://microboone-docdb.fnal.gov/cgi-bin /RetrieveFile?docid = 953;filename = FermiBeta Decay1934.pdf;version=1. The generally negative reaction to his paper apparently prompted Fermi to turn to experimental work, during which he discovered the activation of certain nuclear processes by slow neutrons, with major impact on world history.

[6]Including the one for parity violation mentioned later in this chapter.

[7]W. Heisenberg, *From a Life of Physics,* page 48, World Scientific, 1989.

[8]"Eventually," used so cavalierly by a theorist, is of course meant to summarize years of experimental work. I am being highly impressionistic here: alpha and gamma decay also occur in the nucleus but they are not short ranged.

[9]Historically, the situation was much more convoluted and confused than could be described in a short book. When he suggested the meson theory for the nuclear forces, Yukawa, also proposed that an intermediate boson could account for the weak interaction. In the 1930s, the distinction between the strong and the weak interactions was far from settled, and it was sometimes not even clear which interaction Yukawa was talking about.

[10]A d quark sitting there, with mass only 1/16,000 that of the W boson, most certainly cannot emit a real W boson. But it sure as heck can emit a virtual one!

[11]A reader to whom I sent the manuscript scrawled "Neat!" on the margin at this point. My sentiments exactly when I first learned this, that the interaction in figure 2(a) could be pulled apart into figure 3.

[12]And in some sense the youngest, being the last to be discovered and studied in detail.

[13]That is, at least 95% of all practicing physicists, but it may well be as high as 99%. A doctor friend who read this remarked that this is like leaving the anatomy class before the kidney was mentioned. I suppose that it would be okay for an ear, nose, and throat guy. Similarly parochial thinking exists in physics also.

[14]On February 11, 2021, the United States Postal Service will finally issue a stamp in honor of C. S. Wu, universally known as Madame Wu in the physics community, one of the leading experimentalists who actually demonstrated this and who many feel should have won a Nobel prize. See also *Fearful,* pages 323–326. https:// www.sciencemag.org/news/2021/02/postage -stamp-honor-female-physicist-who-many-say -should-have-won-nobel-prize?fbclid=IwAR1S F3s6iYm6bnwJsa9MlFSrsXjGcqp4uO2hWoa DGobi9mJEUQw84PrzSjc.

[15]Obviously, by the late 19th century, the industrial nations had to agree on choosing a handedness for screws. See en.wikipedia.org /wiki/Screw_thread#History_of_standardization.

[16]This was explained by Steve Weinberg in *Physical Review Letters* 31, page 494, 1973.

[17]This is by now an elementary homework problem for students. See exercise II.1.9 on page 106 of *QFT Nut.*

[18]Somehow it is never put one and one together in English. Too easy? "The Scripture is plain enough, to proper attention. Any who can put two and two together, to make four, may, and indeed must understand it," according to https://english.stackexchange.com/questions /7734/what-is-the-origin-of-the-phrase-put-two -and-two-together.

[19]For the application of T and P to, for example, classical fluid dynamics, see *FbN,* chapter VI.2.

[20]See, for example, *QFT Nut,* pages 102–104. I say "with some effort," because T is harder to work out than either C or P.

[21]For more about time reversal, see my essay "Time reversal" in the anthology *Mysteries of Life and the Universe,* edited by W. Shore. https://www.kitp.ucsb.edu/zee/research /publications/mysteries.

[22]Professor Cronin ran an evening seminar for undergraduate physics majors, in which he talked about recent discoveries in physics. I still remember that one evening in 1964 the young professor (Jim was 33) told us that he had

discovered something exciting. I mentioned in my book *FbN* (page 303) that I avoided taking a couple of boring physics courses required for graduation, but the professor in charge of undergrads let me go (on condition that I did not tell the other undergrads). Here I can finally reveal that the generous man was Cronin.

[23]A. Angelopoulos et al., *Physical Letters* B 444, 43, 52 1998. Note that after the discovery of P, C, and CP violation, the experimental observation of T violation took several decades. Lots of dedicated people involved!

[24]The interested reader can find this explained on page 100 of *QFT Nut*. A historical curiosity is that Weyl actually had one of these two pieces before Dirac!

[25]In 1936 at Caltech by Carl Anderson and Seth Neddermeyer, as mentioned in chapter III.2.

[26]At first, the muon was confused—Niels Bohr proposed calling it the "yukon"—with Yukawa's π meson, later called the "pion," the alleged mediator of the strong interaction. Indeed, the muon was first named the mu meson, but it was eventually found not be a hadron at all, but a lepton like the electron, as mentioned in chapter III.2.

[27]He was apparently alluding to the group meal the physicists at Columbia University had regularly at a nearby Chinese restaurant.

[28]See *FbN*, chapter IX.3.

[29]That would be something to see at a human family reunion!

[30]I will refrain from saying anything about it. See *FbN*, chapter IX.4 for how its existence was anticipated theoretically.

[31]By V. Hopper and S. Biswas at the University of Melbourne.

[32]There are 365×24 or about 10^4 hours in a year.

[33]Some readers might recall that rotations and the trigonometric functions sine and cosine were mentioned in chapter I.4.

[34]Introduced by the Italian physicist Nicola Cabibbo. It was generalized to three families by the Japanese physicists Makoto Kobayashi and Toshihide Maskawa. By the way, Cabibbo was kind to me when I was young.

[35]It has to be imperfect, because this is an intrinsically quantum phenomenon involving probability amplitudes.

[36]You may recognize this as tan 13°; I didn't write that to avoid confusing those readers less sophisticated than you.

[37]In the early days, nuclear decay in which an energetic photon is ejected was called "gamma decay." Refer to endnote 8 in this chapter. I omit what alpha decay means, since it is not relevant to our discussion.

[38]The mechanism was discovered independently by R. Brout and F. Englert, and by G. Guralnik, Hagen, and T. Kibble.

[39]This is explained in detail in *QFT Nut*.

[40]A brief overview may be found in, for example, *QFT Nut*.

[41]Somehow the crackpots proclaiming that Einstein is wrong never send in what his (and it is invariably a he, at least in my experience) theory has to say about this issue.

[42]See *QFT Nut*, page 102.

[43]No fewer than six hypotheses have been proposed, ranging from suicide to kidnap by Nazi agents.

Addendum to chapter V.3

As was mentioned in the prologue, in April 2021, after the manuscript for this book was submitted to Princeton University Press, the media exploded with articles about the latest measurement of gyromagnetic ratio, that is, the "g factor" of the muon, which was reported to be

$$g_\mu/2 = 1.001\ 165\ 920\ 89$$

Now that the reader has come this far, and has (almost, smile!) mastered quantum field theory, let me explain the significance of this report as a series of bullet points. But first the bottom line: When I heard the news, I promptly consulted with several prominent experimentalists and theorists,[44] and I found that nobody was losing any sleep over this, in spite of the almost hysterical accounts in the press, particularly those on the web.

- The reader recognizes the leading digit 1 in $g_\mu/2 = 1.001\ 16\ldots$ represents Dirac's triumph. Some people prefer to subtract this off to highlight the contribution of quantum field theory, and thus write[45]

$$(g_\mu/2) - 1 = \frac{1}{2}(g_\mu - 2) = 11,659,208.9 \pm 5.4 \pm 3.3 \times 10^{-10}$$

  Because of this utterly trivial[46] elementary school rewriting, this quantity is referred to in everyday conversation among physicists as "g minus two." Incidentally, I have included the systematic and statistical error.
- Flip back to chapter IV.3 and look at figure IV.3.5(b), corresponding to the famous correction to Dirac's result that Schwinger first calculated. The internal photon (labeled by momentum k in figure IV.3.5(b)) could produce a charged particle X and its antiparticle $\overline{X}$, which subsequently annihilate each other to turn back into a photon. See figure IV.3.3. You may recall from chapter IV.3 that this is known as vacuum polarization: A photon merrily moving along can metamorphose into a particle and

an antiparticle and then become itself, happily doing this again and again. This generates a correction of order α to Schwinger's calculation.

- The key point, reflecting the essential nature of the quantum world, is that X could be any particle that the photon cares to couple to. It does not depend on whether we are talking about the g factor of the electron or the muon. The contribution of each and every particle has to be included in the theoretical calculation, up to some desired degree of accuracy. As you might expect, the more massive X is, the less it contributes. For instance, when Schwinger calculated the g factor of the electron, he did not have to worry about the diagram in figure 6(a) (never mind that he did not use Feynman diagrams) with $X = $ muon. This would add to his famous result $\frac{\alpha}{2\pi}$ a term of order $\sim \alpha^2 (m_e/m_\mu)^2$, suppressed not only by the extra factor of α but also by the muon mass m_μ.

- The trouble stems from the fact that X could be a hadron, namely, a strongly interacting particle. In particular, X could be the charm quark c. As shown in figure 5(b), the charmed quark c and the anticharm quark $\bar{c}$ are quite inclined to interact with each other strongly by emitting and absorbing gluons. As was explained in chapter V.2, the strong interaction has always been too strong and too mean. The only known way to calculate its effects to the accuracy needed is through a numerical lattice computation (as was explained in endnote 42 in chapter V.2.) It so happens that there are two competing lattice groups[47] that do not agree with each other! Just to impress the reader with how massive these computational efforts have grown to be, one group involves 172 physicists from 82 institutions spread around the world. I will let the reader draw his or her own conclusion, but I can't help but visualize the young Schwinger all by his lonesome self calculating with pen and paper, racing and beating Feynman.

- Back in 2001, an experiment at Brookhaven National Laboratory in Long Island, New York first claimed that its measurement of the muon $(g - 2)$ disagreed with theory, but ran out of money to continue. This controversial experiment was finally repeated at the Fermi National Accelerator Laboratory in Batavia, Illinois, leading to this "famous and infamous" announcement in April 2021.

- To save money, the experimentalists used the same magnet used in Brookhaven. So, the humongous magnet was floated on a barge all the way from Long Island, around Florida, to the Gulf of Mexico, up the Mississippi River, and then trucked to Illinois at night with road closures. People were naturally concerned whether something could have happened to the magnet en route, but it made for hugely good press.[48]

- After analyzing only 6% of the data collected, the experimental group at Fermi Lab announced that their measurement agreed with the Brookhaven result, and so disagreed with theory. But wait! you cry.

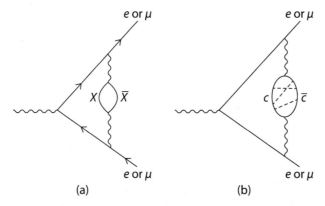

Figure 6. Higher order modifications of figure IV.3.5(b). The external lines could be either the electron or the muon. Photons are denoted by wavy lines, gluons by dotted lines. (a) The internal photon polarizes the vacuum, producing X and $\overline{X}$, which subsequently annihilate each other. (b) The internal photon produces a charm quark c and an anticharm quark $\bar{c}$, which interact by exchanging gluons before annihilating each other.

It was mentioned that the two competing lattice groups disagreed. So which theoretical result are we talking about? Well, somebody's law about maximum confusion here. One of the two theoretical groups agreed with the experimental measurement, and hence no discrepancy at all, while the other disagreed. Oy vay! Also, think what a high school science teacher would say if a student submitted a report based on 6% of the data he or she had collected? But these are big kids, and so they called a press conference. Times have changed.[49]

• In any case, the reported discrepancy, if any, did not even met the statistical criterion physicists routinely impose to rule out statistical flukes. Those readers familiar with data analysis would know that this involves the concept of standard deviation from the mean.

• This frenzied episode generated plenty of shame and embarrassment to go around. Among the hysterical responses of the media, the *New York Times* headline[50] "A Tiny Particle's Wobble Could Upend the Known Laws of Physics" was by comparison among the more restrained. But I am sure that the discriminating reader knows that if one were to believe every headline about physics—laws overthrown, or merely upended, Einstein proven wrong, and so on—the subject would have collapsed long ago.

    Meanwhile, all we could do is to wait for the experimentalists to complete analyzing their data and for the two lattice groups to come to an agreement. You are welcome to set your own betting odds. But already, given what you have learned thus far about quantum field theory, even if the discrepancy turns out to be real, it hardly implies the theory's demise. In all likelihood, it probably just means that we have to include a hitherto unknown charged particle.

Ultimately, the motto of physics should be "nullius in verba," loosely translated as "do not take somebody's word for it." So, we will see.

Notes

[44]Including a founder of the standard theory of electroweak interaction.

[45]As given on the Particle Data Group website.

[46]Nevertheless, the *New York Times* in its first edition gave the wrong reason behind the minus 2, later corrected. See M. Weitzman's remark in figure 7, below.

[47]T. Aoyama et al, https://arxiv.org/pdf/2006 .04822.pdf versus Sz. Borsanyi et al, https:// arxiv.org/pdf/2002.12347.pdf

[48]You could google for some truly astounding photos on the web.

[49]For comparison, see endnote 9 in chapter IV.3.

[50]https://www.nytimes.com/2021/04/07/ science/particle-physics-muon-fermilab-brook haven.html?searchResultPosition=7.

---

41 Responses to *Muon g-2 Result*

Mark Weitzman *says*:

April 7, 2021 at 1:53 pm

I wish the New York Times would have better science reporting – typical error:

"That leads the factor g for the muon to be less than 2, hence the name of the experiment: Muon g-2."

Figure 7. The garbled explanation for the minus 2 in the *New York Times*. From "A Tiny Particle's Wobble Could Upend the Known Laws of Physics" by Dennis Overbye.

Grand unification

After the victory parade

Quantum field theory had two near death experiences, as you have learned. The first was in the late 1940s, when physicists were unable to extract sensible results for quantum electrodynamics. As we saw in chapter IV.3, a new generation of theorists brought it back to life and achieved fantastic agreement with measurements of how the electron spins in a magnetic field, for instance. The second was in the 1960s, when Landau and others buried quantum field theory with the pomp it deserved. Not only was quantum field theory powerless to confront the "too strong and too mean" strong interaction, it also had trouble with the weak interaction.

But almost incredibly, and over a dramatically short period of time around 1970, the dormant notion of nonabelian gauge theory roared to life, and no sooner than you could yell out the names of the gauge groups, the weak interaction was cured of its deficiencies and unified with the electromagnetic interaction, and then with the newly tamed strong interaction into an $SU(3) \otimes SU(2) \otimes U(1)$ gauge theory. (This unified theory of the three nongravitational interactions is known as the standard model of particle physics, but since I dislike this bland and misleading name intensely, I will simply call it the "123 theory.") Happily, I could throw away the thin book that my fellow freshman[1] urged upon me and that I mentioned in chapter V.2.

My thesis advisor Sidney Coleman* proclaimed a victory parade that made the spectator gasp with awe and laugh with joy. See figure 1. After the gasping

---

*I ended up working for him partly because I was told that he was the most intelligent young theoretical physicist in the world. Steve Weinberg referred to Coleman as the "physicists's physicist," just about the highest accolade I have heard a great theoretical physicist bestowed on another. Tragically, he died from a lingering illness that slowly ate his brain. When I visited him a few months before his death, he was literally spouting

Quantum field theory has had two near-death experiences

Late 1940s: inability to produce Lorentz covariant results and cure divergences

- The young people were the (revolutionary) conservatives

Late 1960s: S-matrix school, inability to deal with the strong interaction

- The triumph of field theory = "a victory parade" that made "the spectator gasp with awe and laugh with joy"

Quantum mechanics = (0 + 1) - dimensional field theory
String theory = (1 + 1) - dimensional field theory

Figure 1. Victory parade!

and laughing was over, theorists were seized with grander ambitions and started dreaming about grand unification.

What unification means for the action of the world

Before we grand unify, I would like to say a few more words about the less-than-grand but still-pretty-tremendous unification of the electromagnetic and the weak interactions into a single electroweak interaction. Since you know that the action summarizes the laws of physics, you are actually in a far better position to understand what unification means than the typical reader of popular books on physics.

Recall that the action S is the integral of the Lagrangian density[2] $\mathcal{L}$ over spacetime: $S = \int d^4x\, \mathcal{L}$, and the Lagrangian is a sum of a bunch of terms. For instance, the electromagnetic Lagrangian $\mathcal{L}_{\text{electromagnetism}}(A(x), e(x), q(x))$ is constructed out of the electromagnetic field $A(x)$, the electron field $e(x)$, and the quark fields $q(x)$. (I am running out of letters here; so $e(x)$ is not to be confused with e, which measures the strength of the coupling of the electromagnetic field to the electron field and which I have not shown explicitly here. I show only the fields, and not the various parameters, such as e and m, the mass of the electron. I am using $q(x)$ to denote a bunch of quark fields generically, for example, the up quark field $u(x)$ and the down quark field $d(x)$.)

Also, at this stage, to minimize clutter, I think I can drop the x dependence; by now you know that these guys appearing in the Lagrangian are

---

nonsense. Incredible since I remembered the days when he would chew a seminar speaker alive and spit the bits out. A physicist friend mused about how many laypersons would have even heard of Coleman. Such is the almost laughable distortion in the popular media of the perception of who's who in the theoretical physics world.

all fields. Then $\mathcal{L}_{\text{electromagnetism}}$ could be written somewhat more clearly as $\mathcal{L}_{\text{electromagnetism}}(A, e, q)$.

Similarly, the weak interaction Lagrangian $\mathcal{L}_{\text{weak}}(W, Z, \nu, e, q)$ was constructed out of the weak interaction fields W and Z, the electron field e as well as the neutrino fields ν, and the quark fields q. (Again, I suppress the coupling strength g mentioned in chapter V.3, and various other parameters, such as the mass of the W boson, which was "put in by hand.")

Before[3] electroweak unification, the Lagrangian of the world contains, among other terms, the sum $\mathcal{L}_{\text{electromagnetism}}(A, e, q) + \mathcal{L}_{\text{weak}}(W, Z, \nu, e, q)$. After electroweak unification, these two terms combine into[4] $\mathcal{L}_{\text{electroweak}}(A, W, Z, \nu, e, q, h)$. Notice the arrival of the Higgs field h, mentioned in chapter V.3, which gives mass to W, Z, e, and q.

Schematically,

before electroweak unification	after electroweak unification
$\mathcal{L}_{\text{electromagnetism}} + \mathcal{L}_{\text{weak}}$	$\mathcal{L}_{\text{electroweak}}$

unification has replaced the sum of two Lagrangians by a single Lagrangian.

Furthermore, since we know that this is a Yang-Mills theory based on the group $SU(2) \otimes U(1)$ (as mentioned in chapter V.3), we could write a bit less by denoting the 3 gauge bosons of $SU(2)$ generically by W and the single gauge boson of $U(1)$ by B, and by packaging ν and e into the lepton field l, in parallel to the quark field q. We then have, more compactly, $\mathcal{L}_{\text{electroweak}}$ (W, B, l, q, h).

The reader could probably sense that I am straining to avoid mentioning too many technical details. For those readers who want more, these could be found readily in any number of textbooks, from introductory to more advanced.[5] For those readers who worry, let me assure you that most of these details[6] are not essential for our purposes in this book.

Theoretical physicists want to minimize the number of terms in the action S (and in $\mathcal{L}$). Indeed, even before electroweak unification, writing $\mathcal{L}_{\text{electromagnetism}} = \mathcal{L}_{\text{Maxwell}} + \mathcal{L}_{\text{Dirac}}$ may be considered as an act of unification. I also like to emphasize, if only in passing, how amazing this unification was. Lest we forget, $\mathcal{L}_{\text{electromagnetism}}$ can already account for, if not yet explain, all known physical phenomena except for those involving the strong, the weak, and the gravitational interactions.

The drive toward unification

Physics proceeds on two fronts: to explain every possible observation in the physical world, and to provide a unified understanding of how this world works. For instance, turbulence remains one of the major unsolved problems

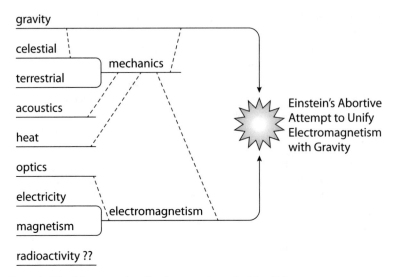

Figure 2. The drive toward unification near the end of the 19th century.
Redrawn from A. Zee, *Fearful Symmetry: The Search for Beauty in Modern Physics*,
Princeton University Press, 1986.

of physics. Yet the extension of Newtonian mechanics to fluids was carried out in the 19th century and is now taught to undergraduates.

Before we grand unify, let's go back in history. Physicists have always dreamed of a unified description of Nature. The story of electromagnetism illustrates well what I mean by the drive toward unity. Electricity and magnetism were revealed to be different aspects of electromagnetism, and optics then became part of electromagnetism. In high school, I read an old physics book that said physics consisted of six parts: mechanics, heat, light, sound, electricity and magnetism, and gravity. In fact, toward the end of the nineteenth century, there were only two parts left in physics: electromagnetism and gravity. The status of the drive toward unity at that time is shown in figure 2.

The drive toward unity may be said to have started with Newton, who insisted that the same laws govern earthly objects and heavenly bodies. Terrestrial and celestial mechanics were unified. Later, sound was recognized as being due to the wave motion of air, and it was realized that sound could be studied with the concepts of Newtonian mechanics. In the nineteenth century, the mystery of heat was finally understood as due to the agitated motion of molecules. The mechanical interaction between objects, such as that due to friction, was traced to the electromagnetic interaction between the atoms and molecules comprising the objects. If we mean by mechanics the description of the motion of particles, then we may say that mechanics has been subsumed into the other interactions.

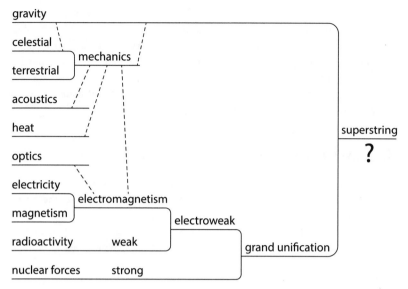

Figure 3. The drive toward unity near the end of the 20th century.
Redrawn from A. Zee, *Fearful Symmetry: The Search for Beauty in Modern Physics*,
Princeton University Press, 1986.

This overview of unification toward the end of the 19th century was eventually replaced by the overview of unification toward the end of the 20th century,
as shown in figure 3 and as we have already seen in chapter V.3.

More is different, yes, but physics is a big tent

Unification reflects the reductionist impulse in theoretical physics, the drive
toward understanding the world in simpler and simpler terms. Yes, I know
about the "more is different" school of thought, advocated forcefully by the
Nobel winning condensed matter theorist Phil Anderson. (Indeed I even played
tennis with Phil when I was young.[7]) People who love to stir up controversy,
some of them outside the physics community, like to set up "more is different" as opposed to reductionism, as if they are inimical to each other. But
nobody is claiming that the reductionist approach is capable of explaining
all of physics, and anybody who claims that is clearly nuts. A knowledge of
quantum mechanics alone hardly enables you to explain the rich and complicated behavior of water in various circumstances. Nor to predict that the
solid it forms suddenly at $0°$ C would be translucent. Understanding that the
proton is composed of quarks will not help elucidate superconductivity, and
certainly even Gell-Mann would not have claimed that. The vitality of physics

is precisely that it is a vast tent capable of accommodating all types, from lion tamers to acrobats to clowns. The reduction of acoustics to the collective motion of air molecules and the collision between them, and hence to mechanics, certainly does not imply in the slightest that we would have to deny the beauty of a soprano voice, nor the progress made in physics. The controversy, if any still exists, is not about physics, but about competition for funding between different areas of physics.

Grand unification

Back to our story. We have achieved electroweak unification and arrived at $\mathcal{L}_{electroweak}(A, B, l, q)$. Meanwhile, as recounted in chapter V.2, the strong interaction is described by the Lagrangian $\mathcal{L}_{strong}(G, q)$ with G denoting the 8 gluon fields. Note no lepton fields. The leptons, such as the electron and the neutrino, are not welcome in the strong interaction. It would be tempting, to say the least, to "grand unify" $\mathcal{L}_{strong}$ and $\mathcal{L}_{electroweak}$.

before grand unification	after grand unification
$\mathcal{L}_{strong} + \mathcal{L}_{electroweak}$	$\mathcal{L}_{grand\ unified\ theory}$

In 1974, theoretical physicists knew[8] that the strong interaction is described by a Yang-Mills theory based on $SU(3)$ and that the electroweak interaction is described by a Yang-Mills theory based on $SU(2) \otimes U(1)$. If you were around at the time, you might have even entertained the thought of combining the two Yang-Mills theories. And that is exactly what Georgi and Glashow (and Pati and Salam independently[9]) did.

Once again, you can see whether you have the making of a theoretical physicist (with the benefit of hindsight, of course). What group should the combined Yang-Mills theory be based on? Well, as you might recall from chapter V.2, $SU(3)$ rotates three objects into each other, and $SU(2)$ rotates two objects into each other.

If you said $SU(5)$, you are absolutely right! That is exactly the group the grand unified theory of Georgi and Glashow is based on. I might even exaggerate a bit and say that $3 + 2 = 5$ is among the most important calculations done in theoretical physics in the 1970s.

You might be puzzled why hundreds of theoretical physicists did not all converge on $SU(5)$. Good question. A few people did come close, but the more accurate answer is that it requires mastery of certain aspects of quantum field theory that most physicists lacked in 1974. Nowadays, of course, even the dullest student in my introductory quantum field theory course would have learned the missing fact. (Rejoice in the progress of physics! What eludes many top theorists around 1974 is now fed to some of the brighter undergraduates.)

Guess what? If you don't quite have this technical mastery, at least you know what it is. In chapter V.3, I mentioned that if a particle is spinning left handed, then the corresponding antiparticle is spinning right handed. And vice versa. Perhaps a bit more picturesquely, a particle that is spinning left handed, when reflected in the magic mirror that turns particles in the antiparticles, would be spinning right handed. This could be extended immediately from particles to fields: the charge conjugate of a right handed field is left handed.

A roll call

The $SU(2)$ group in the electroweak interaction transforms the left handed electron into the left handed neutrino field and vice versa, as explained in chapter V.4. Parity violation is incorporated by having only a left handed neutrino field. The standard electroweak theory does not contain a right handed neutrino field at all.

Now that we have decided on $SU(5)$, let us roll call the quark and lepton fields. Since we don't understand why Nature decided to have 3 families, we simply focus on the first family. Recall from chapter V.4 that the left handed neutrino and electron fields, and the left handed up and down quark fields, form a doublet under the $SU(2)$ in the 123 theory. Keeping in mind that quarks come in 3 colors, we count $2 + (2 \times 3) = 8$ left handed fields. The right handed electron field and the right handed up and down quark fields are all singlets under $SU(2)$, giving us $1 + (1 \times 3) + (1 \times 3) = 7$ right handed fields. In light of what was just said, these could be turned into left handed fields under charge conjugation, namely the positron, the anti up and the anti down quark fields $(e^+, u_r^c, u_y^c, u_b^c, u_r^c, d_y^c, d_b^c)$. Altogether we have $8 + 7 = 15$ fields,

$$(v, \ e, \ u_r, \ u_y, \ u_b, \ d_r, \ d_y, \ d_b, \ e^+, \ u_r^c, \ u_y^c, \ u_b^c, \ d_r^c, \ d_y^c, \ d_b^c)$$

with everybody left handed. By the way, if you think that this list looks blindingly long, keep in mind that quark fields come in 3 colors, and so the list could be abbreviated to $(v, \ e, \ u, \ d, \ e^+, \ u^c, \ d^c)$. Better?

God seems to know a lot of group theory, at least the easy introductory stuff

The group $SU(5)$ transforms 5 fields into each other. Which 5 would you choose from this list of 15? Well, $SU(5)$ contains $SU(2)$ and $SU(3)$. Of the 5 fields, $SU(2)$ transforms 2 fields into each other, but not the other 3 fields, while $SU(3)$ transforms these remaining 3 fields, but not the other 2 fields.

Hey, we are physicists, not mathematicians! Translating from math talk into physics talk, we want 2 fields that listen to the W bosons of the weak interaction but ignore the gluons of the strong interaction, and 3 fields that listen to the gluons of the strong interaction but ignore the W bosons of the weak interaction.

Who are they? These guys are almost handing over their ID cards. Should I pause to let you figure it out?

The two fields that participate in the weak interaction[10] but not the strong interaction have to be, by definition, lepton fields. In electroweak unification, we have precisely two lepton fields, ν and e, transforming into each other.

Who then are the 3 fields who could hang out with ν and e under $SU(5)$? Since they participate in the strong interaction but not the weak interaction, they could only[11] be the 3 anti down quarks (d_r^c, d_y^c, d_b^c), red, yellow, and blue! This set of 5 fields (d_r^c, d_y^c, d_b^c, ν, e) form what mathematicians called the defining representation: The way these 5 entities transform defines the group $SU(5)$.

Now that we have accounted for 5 fields out of the 15 that comprise a family, we have $15 - 5 = 10$ fields left over. Remarkably, the group $SU(5)$ is precisely capable of transforming 10 objects into each other. Why 10? It's a simple consequence of group theory: Any student in my introductory group theory course can see that* $10 = (5 \times 4)/2$. (Are you feeling that I'm trying to entice you to learn a little bit of group theory? Galileo may have said that God knows mathematics. More specifically, God seems to know a lot of group theory, at least the easy introductory stuff that physicists learn, leaving the difficult heavy lifting to mathematicians.)

And thus the dawning of the age of grand unification! I still have a T-shirt somewhere from that era saying[12] "It takes GUTs to be a grand unified theorist."

The universe is stable

Meanwhile, allow me to remind you that the universe has long been known to be stable. Physics should tell us why.

In chapter V.3 on the weak interaction, you saw that in the subnuclear world, particles decay, with the notable exception of the electron and the proton. In the quantum world, what is not forbidden by the laws of physics could proceed with some probability amplitude. For instance, the cousins of the electron, the muon and the tau lepton, decay into the electron plus a neutrino and antineutrino pair. In contrast, the electron is stable. Why?

Einstein with his $E = mc^2$ requires the electron to decay into a particle with a smaller mass, as was repeatedly mentioned in chapter V.3. But all the particles

---

*Imagine drawing 2 cards from a deck of 5 cards, consisting of ace, king, queen, jack, and ten. How many possible hands are there? Five possibilities for the first card, four for the second, five times four and divide by two since the order does not matter. There are 10 possibilities. Yes, it's that easy! See *Group Nut,* pages 230 and 234.

less massive than the electron, namely, the neutrinos[13] and the photon, do not carry electric charge. Thus, charge conservation guarantees that the lightest particle carrying electric charge, namely the electron, must be stable.

The situation with the proton is more controversial. Empirically, the proton must be extremely stable,* for the simple reason that the universe is still here. If the proton is not absolutely stable and disintegrates after a finite amount of time, its lifetime must exceed the age of the universe.[14] But theoretically, there is no particular reason, which drove some people with Nobel prizes to proclaim that this must be so.[†] They formulated a principle that these days would be known as quark number conservation. The universe appears to keep track of the total number of quarks: not one more, not one less.[15] (An antiquark counts as minus one quark, and so it can only pop up accompanied by a quark.)

This principle, which merely amounts to somebody's decree, essentially states that the world of quarks and the world of leptons must be kept separated. Hence the vertical bar in chapter V.3 segregating the quarks and the leptons.

Grand unification risks destroying the universe

But wait, grand unification brings the quarks and the leptons together, by definition.

Recall that the $SU(2)$ gauge theory contains $3 = 2^2 - 1$ gauge bosons and that the $SU(3)$ gauge theory contains $8 = 3^2 - 1$ gauge bosons. All right, how many gauge bosons does an $SU(5)$ gauge theory contain? I'll let you think for a minute.

If you said $24 = 5^2 - 1$, then your mind is quite capable of pattern recognition and generalization.

How many of these gauge bosons are newcomers? That is to say, how many of these were not in the $SU(3) \otimes SU(2) \otimes U(1)$ theory you are trying to unify? Since the 123 theory contains $8 + 3 + 1 = 12$ gauge bosons, there are $24 - 12 = 12$ newcomers.

I remind you that the 5 fields being transformed by $SU(5)$ are $(d_r^c, d_y^c, d_b^c, \nu, e)$. You know the job assignment of the $8 + 3 + 1 = 12$ gauge bosons in the 123 theory: 8 are gluons which transform quarks of different color into each other and are responsible for the strong interaction, while the $3 + 1$ are responsible for the electroweak interaction. So these 12 newcomer

---

*In chapter V.3, we explain that the weak interaction cannot cause the proton to decay into the neutron by emitting an electron and antineutrino, for the simple reason that the proton is less massive than the neutron. The issue here is whether some other, possibly yet unknown, interaction could cause the proton to decay.

[†]In theoretical physics, this is known as proof by authority.

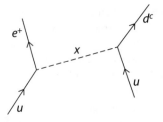

Figure 4. A u quark transforms itself into a e^+ by emitting one of the "newcomer gauge bosons" called X, which when absorbed by another u quark in the proton transforms it into an anti down quark d^c. This results in the process $u + u \rightarrow d^c + e^+$, and hence $(uud) \rightarrow (d^c d) + e^+$, that is, $p \rightarrow \pi^0 + e^+$.

gauge bosons that are not in the 123 theory want to transform the anti down quarks (d_r^c, d_y^c, d_b^c) into either the neutrino or the electron, and vice versa. (The more mathematical readers could chew[‡] on this calculation: $12 = (3 \times 2) \times 2$ And $12 + 12 = 24$.)

You really are ready to read a book on group theory!

In the same way, the 10 fields (u_r, u_y, u_b, d_r, d_y, d_b, u_r^c, u_y^c, u_b^c, e^+) are transformed by these 12 newcomer gauge bosons into each other.

You realize that this is something the world has never seen, never known, and never heard of: gauge bosons scrambling quark fields and lepton fields. Think of these gauge bosons as marauders who are not connected with the strong, the weak, and the electromagnetic interactions.

Are you worried? Yes, you should be. This looks immediately like the mother of all catastrophes. Boys and girls, we might be destroying the universe right here!

The universe goes poof, gone with the gauge fields

Look at the Feynman diagram in figure 4.

The process depicted[*] $uud \rightarrow d^c d + e^+$, would be observed as $p \rightarrow \pi^0 + e^+$. Figure 5 cartoons proton decay. The proton decays into a neutral pion and a positron, and the universe goes poof, gone with the gauge fields.

Executive summary here. Grand unification runs the immediate risk of destroying the universe. So you would think that the theory is ruled out immediately.

[‡]Hint: 3 anti down quarks and 2 leptons, and the extra factor of 2, because each gauge boson comes with its charge conjugate.

[*]By the way, d^c and $\bar{d}$ essentially denote the same thing, one often used for the field, the other commonly for the particle.

PROTON DECAY

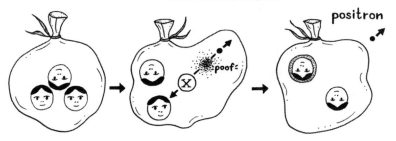

positron

Figure 5. One of the two up quarks emits a newcomer gauge boson, that is, a gauge boson not in the 123 theory, and transforms itself into a positron, that is, an antielectron. The gauge boson is absorbed by the other up quark, which as a result turns into an anti down quark. The positron escapes and the anti down quark settles down with the down quark into temporary domestic bliss as a pi meson. The proton thus decays: $p \to \pi^0 + e^+$. Compare this with the process for neutron decay shown in figure V.3.1. Redrawn from A. Zee, *Fearful Symmetry: The Search for Beauty in Modern Physics,* Princeton University Press, 1986.

Rendezvous à trois

The whole idea of grand unification seems to be dead on arrival. But it still lives! How that is possible came from another consideration that also looms as a disaster for grand unification at first sight: The strong interaction is strong and the weak interaction is weak. How could they be related?

In more precise quantum field theory language (which I claim you now speak), there are three couplings strengths in the 123 theory, g_1, g_2, and g_3, for the gauge group $U(1)$, $SU(2)$, and $SU(3)$, respectively. They are definitely not equal. In fact, g_3 is significantly more than g_2, which is in turn more than g_1; that's what we mean by the strong interaction being stronger than the electroweak interaction. Unification would imply that these couplings should be equal, or at least similar.

How would you, with everything you've learned in this book thus far, resolve these two problems?

Most remarkably, these two difficulties with grand unification manage to knock each other off, so to speak. Remember that in modern quantum field theory, the coupling strengths actually move with the energy scale; they are not fixed constants, as in the elementary textbooks fed to undergraduates. Yes, as we increase the energy scale, g_1 slowly increases, but g_3 steadily decreases, while g_2 also decreases but at a slower rate. So it is possible for the three of them to meet.

Note that quantum field theory fixes whether a coupling strength increases or decreases, and the rate at which it changes. That the whole thing has a

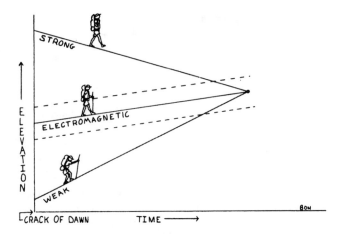

Figure 6. A "parable" of grand unification: At the crack of dawn, a hiker named "strong" starts coming down a mountain while two hikers, named "electromagnetic" and "weak," start climbing up. "Weak" starts out lower than "electromagnetic" and has to move faster to keep up. I have plotted the elevations of three hikers as time passes. Given the starting positions of two of the hikers, the requirement that the three hikers arrive at the same point at the same time clearly fixes the starting position of the third. For instance, if "electromagnetic" starts out too high (the upper dotted line), she will run into "strong" before "weak" catches up. If she starts out too low (the lower dotted line), "weak" will pass her before he runs into "strong." For details, see *QFT Nut*, page 414. Reproduced from A. Zee, *Fearful Symmetry: The Search for Beauty in Modern Physics*, Princeton University Press, 1986.

chance of working out is in itself something of a triumph for quantum field theory. For instance, suppose g_3 increases, g_2 increases at a slower rate, but g_1 decreases. Then the trio would never meet, and grand unified theory goes out the window.

Recall from chapter V.2 that coupling strengths move extremely slowly with energy and indeed, that's why for the longest time theoretical physicists thought that they don't move at all and called them "coupling constants." So if they meet at all, they would have to meet at an extraordinarily high energy scale. The calculation was done by Howard Georgi, Helen Quinn, and Steve Weinberg, who found they meet at the enormous energy of 10^{16} GeV, known as the grand unification scale. (For comparison, recall that the W boson responsible for the weak interaction has mass of order 10^2 GeV, and that the proton mass is about 1 GeV.) Thus, we expect the newcomer gauge bosons to have mass on the order of 10^{16} GeV. Not only is the range of this new interaction responsible for proton decay $10^{16}/10^2 = 10^{14}$ times shorter than the already tiny range of the weak interaction, but furthermore, the reluctance of these enormously massive gauge bosons to move through spacetime renders the interaction to be much weaker than the weak interaction.

The proton lifetime comes out to be around 10^{31} years, which might be compared to the age of the universe, generally given to be about 1.4×10^{10} years. The universe will be around for a long time yet.

Relative strength of the weak interaction and the electromagnetic interaction

We learned as children that two straight lines, if not parallel, would always meet, but in general not three straight lines. Indeed, a child could have told us that requiring the three lines in figure 6 to meet at the same point fixes the ratio of the starting values of g_1 and g_2, the two couplings that determine the coupling strength of the electromagnetic interaction and the weak interaction. Grand unification could predict the relative strength of the electromagnetic and weak interactions, as explained in the caption. That the prediction came out in agreement with experiment gave an enormous boost of confidence in grand unification.

Quantum physics traffics in probability

You might think that the most dramatic predictions of grand unification, that protons decay with a lifetime a factor of 10^{21} greater than the lifetime of the universe, cannot be checked empirically. But we are saved by quantum physics and by our humongous size compared to the proton.

Quantum physics traffics in probability. When you read that the decay life-time of an unstable radioactive nucleus is 10 million years, it's a probabilistic statement, saying that if you have a large number of these nuclei, after 10 million years, something like half of them would have decayed. But a specific nucleus could very well decay within the next hour. So the way to circumvent the very long lifetime for the proton is to gather together a huge number of protons and watch how many of them have decayed after a certain period of time, say, a year. The least expensive way to get a large number of protons is simply to fill a big tank with water and surround it with photon detectors. A dying proton in the water disintegrates according to $p \to \pi^0 + e^+$, producing a neutral pion π^0 (which soon decays into two photons) and a positron e^+ (which eventually annihilates with an electron in the water to produce two photons). These telltale photons are what experimentalists yearning to go to Stockholm hope to find. The entire setup would have to be placed deep underground to shield it from cosmic ray particles crashing in to produce photons.

This humongous experiment is located in a mine near Kamioka, Japan, and known as Hyper-Kamiokande, which succeeds the earlier Super-Kamiokande and is scheduled for operation in 2027. The tank will contain a billion liters of ultrapure water—surely the only experiment inside which experimentalists could paddle around in a rubber boat. (If you remember from high school that a mole of anything contains an Avogadro's number, namely $\sim 6 \times 10^{23}$, of molecules, and that a liter of water weighs about 1 kilogram, you could figure out that this tank contains on the order of 10^{35} protons all waiting to decay.)

Why do the electron and the proton have exactly equal and opposite electric charges?

One of the great attractions of grand unification for me is that it explains a long-standing puzzle about electric charge. We all learned in school that electric charge of the electron and the electric charge of the proton are exactly equal and opposite. How exact? A laboratory measurement[16] in 1962 showed that the magnitude of the electron charge differs from that of the proton by less than 5 parts in 10^{19}. We need hardly quibble over the exact number: Cosmology already tells us that this fantastic agreement must hold. The known universe must be electrically neutral to a high degree of accuracy: Either an excess positive charge or an excess negative charge would have blown it apart. This goes back to the fabulous feebleness of gravity compared to the electrostatic force that we talked about in the prologue.

So why is the electron charge so accurately equal to the exact negative of the proton charge? In elementary textbooks, this remarkable equality is simply stated as a fact. And indeed, before grand unification, there was no explanation whatsoever.

I am now going to show you that with grand unification, given that the charge of the electron, $Q(e)$, equals[17] -1, we could deduce that $Q(p)$, the charge of the proton, must be equal to $+1$. (Be sure not to confuse this with what Gell-Mann did in chapter V.2. Given that $Q(p) = +1$ and $Q(n) = 0$ (the charge of the neutron), he deduced the charges of the quarks. Here, given $Q(e) = -1$, we want to prove that $Q(p) = +1$.)

Group theory mandates that the electric charge of all the entities that transform into each other must add up to zero. This is an ironclad mathematical fact, not open to negotiation. In fact, we already used this fact when we chose the 3 anti down quarks (d_r^c, d_y^c, d_b^c) to accompany the 2 leptons (ν, e) to form a set of 5 fields that transform into each other under $SU(5)$. Given that the neutrino is electrically neutral and the electron carries charge -1, let us denote the charge of the anti down quark by $Q(d^c)$. Then the iron law of group theory says that

$$0 = 3Q(d^c) + Q(\nu) + Q(e) = 3Q(d^c) + 0 + (-1) \implies Q(d^c) = \frac{1}{3}$$

Hence each of the anti down quarks must carry electric charge $\frac{1}{3}$. The down quark has charge $Q(d) = -\frac{1}{3}$.

Since the up quark turns into the down quark upon emitting a W^+ boson, the up quark has charge $Q(u) = Q(d) + 1 = -\frac{1}{3} + 1 = \frac{2}{3}$. Ta da, we conclude that the proton $p = (uud)$, consisting of two up quarks and a down quark, must have charge $\frac{2}{3} + \frac{2}{3} - \frac{1}{3} = +1$, as was to be shown.

You understand that before grand unification, the electron and the proton lived separately in the action without a theoretical bridge between them.[18] That $Q(e) = -Q(p)$ is simply told to students as an empirical fact.

Believers and naysayers

On the issue of grand unified theory, the theoretical community is divided between the believers and the faithless naysayers. I am a believer. For me, the perfect fit of the 15 quark and lepton fields we know and love into a set of 5 and a set of 10 is strong evidence in favor of grand unification, plus the theory passing the stability of the universe test and the deduction of $Q(e) = -Q(p)$.

The naysayers point to the fact that proton decay has yet to be seen. More accurate measurements of g_1, g_2, and g_3 indicate that the three couplings do not quite meet. No question that dark clouds have gathered over grand unification in the minds of many. I feel, however, that given the factor of $10^{13} = 10^{16}/10^3$ or so between the grand unification scale and the highest energy we have attained with accelerators, it may be premature to get down into the weeds. There may be a great deal of physics that we do not know about between these two energy scales. The perfect fit of the group theory and the attractiveness of the general drive of physics toward a unified understanding convince me and others that surely some form of grand unification must reflect the truth.

Yes, during the intervening decades since grand unification was first proposed, numerous attempts to modify the original scheme have been proposed. For example, by adding various hypothetical fields, people could affect the running of the coupling constants and lengthen the proton lifetime. These schemes are almost all too contrived for my taste. We will see.

Grand unifying the grand unified *SU*(5) theory

You might be thinking that splitting the 15 quark and lepton fields into a set of 5 and a set of 10 seems rather unnatural. We are supposed to be grand unifying them. You are absolutely right, and in fact, an extremely attractive possibility exists of embedding the group $SU(5)$ into a larger group known as $SO(10)$. The word "embed" is used here as a mathematical term. You could grasp a flavor of this by noting that a complex number $z = x + iy$ contains two real numbers, x and y. Speaking loosely, we feel that one complex number is worth two real numbers. The group $SU(5)$ is built with 5 complex numbers, while $SO(10)$ is built with $10 = 5 \times 2$ real numbers.[19] In this sense, $SU(5)$ is naturally a child of the larger group $SO(10)$. To repeat, $SU(5)$ is to $z = x + iy$ as $SO(10)$ is to x and y.

Given a group, the number of entities transforming into each other is dictated by the diamond sharp laws of mathematics. For $SU(5)$, the possible numbers are 5, 10, 24, and so on. For $SO(10)$, they are 10, 16, 45, and so on. Yes, 16 has to be 16, not 1 more, not 1 less, neither 15 nor 17 are allowed. Students of group theory would recognize how the calculation[20] goes for $SO(10)$:

$$2^{\left(\frac{10}{2} - 1\right)} = 2^{5-1} = 2^4 = 16$$

So, consider a Yang-Mills theory based on $SO(10)$. The gauge bosons transform 16 fields into each other. The theory demands 16 fields, no bargaining allowed. But we have only 15 quark and lepton fields. No go!

SO(10) theory

But guess what? And this is what sold me totally on $SO(10)$, which by the way I strongly prefer to $SU(5)$. The extra field we need to put into an $SO(10)$ grand unified theory[21] has exactly the properties of a left handed antineutrino field. Believe me, the mathematics is truly miraculous! The properties of this extra field are totally dictated by the mathematics. For instance, it has to be electrically neutral, and it is not allowed to participate in the strong interaction.

Here's an analogy, somewhat imperfect to be sure, but it serves to give you a flavor of what is going on. Imagine a family reunion of 15 family members who have known each other since birth. As the long awaited reunion is about to start, the venue, a private club of some sort with the weird name "Group Theory" that was booked for the party absolutely refuses to open up unless 16 people show up. The family members protest vehemently, stating "we have been 15 for our entire existence." At this venue, the seats are uniquely fitted to various individuals according to weight, height, hair color, and so on. In the design of the universe, this would correspond to fields with or without electric charge, participation in the strong interaction or not, in the weak interaction or not, and so on. You cannot arbitrarily grab somebody off the street to fill the empty 16th seat.

Suddenly, a mysterious stranger shows up with exactly the right characteristics to fill that slot, forming a perfect fit with the other 15 family members. Let me add this long lost antineutrino field ν^c to the list of 15 fields displayed earlier to make a list of 16:

$$(v, \; e, \; u_r, \; u_y, \; u_b, \; d_r, \; d_y, \; d_b, \; v^c, \; e^+, \; u_r^c, \; u_y^c, \; u_b^c, \; d_r^c, \; d_y^c, \; d_b^c)$$

$$= (v, \; e, \; u, \; d, \; v^c, \; e^+, \; u^c, \; d^c)$$

Since I don't believe in miracles, mathematical or otherwise, I just have to believe in $SO(10)$.

Notes

[1]Sadly, he has since disappeared from physics entirely.

[2]I will omit the word "density" henceforth and simply write "Lagrangian." Recall the discussion in chapter IV.2.

[3]Nitpickers, yes, I know that before electroweak unification, the Z boson was not known for sure.

[4]With the coupling strengths e and g, now "related," but again suppressed.

[5]See, for example, $QFT\ Nut$ and FbN. The latter is in fact aimed at undergraduates, hence one step beyond popular books.

[6]One detail (which I am not explaining here) that might amuse some readers is that the pillar of physics, the electromagnetic field, (which, following tradition, I have been denoting by A), is now a linear combination of W_3 and B, and so straddles the two groups $SU(2)$ and $U(1)$.

[7]Here is a difficult problem for which quantum field theory cannot provide a solution: When a young guy plays tennis with a senior professor, should he or she let the old guy win, or beat the hell out of him? In any case, Phil was fond of telling me that he was on the board of the theory institute in Santa Barbara when I was hired.

[8]I don't have space to put in all the ifs and buts. In theoretical physics, you can't wait for all the evidence to come in. By then, others would have done it all. The experimental support for the 123 theory was built up over the years.

[9]J. Pati and A. Salam did not not have $SU(5)$; they used the group $SU(4) \otimes SU(2) \otimes SU(2) = SO(6) \otimes SO(4)$, which turns out to be contained in $SO(10)$.

[10]Again, I'm glossing over one detail, because the Z boson does know about these anti down quarks. I didn't want to confuse some readers

with the mouthful "the weak interaction as understood before the early 1970s."

[11]What about the three anti up quarks, $(u_r^c, \; u_y^c, \; u_b^c)$, you ask? Since the photon is one of the gauge bosons on $SU(5)$, group theory requires that the electric charges of the fields that hang out together must sum to zero. This works for the anti down quarks, $\frac{1}{3} + \frac{1}{3} + \frac{1}{3} + 0 + (-1) = 0$ indeed, but not for the anti up quarks, $\left(-\frac{2}{3}\right) + \left(-\frac{2}{3}\right) + \left(-\frac{2}{3}\right) + 0 + (-1) = -3 \neq 0$. By the way, this is the kind of computation you will be doing when you grand unify.

[12]In American slang, "guts" means courage. I will let you figure out what the acronym GUT stands for.

[13]The neutrinos, long thought to be strictly massless, were discovered to have teeny masses, much less than the electron mass. Thus, in principle, two of the neutrinos could decay into the lightest neutrino by emitting a photon. But such decays have never been observed. In any case, the lightest neutrino would be stable.

[14]Never mind the universe, the distinguished experimentalist Maurice Goldhaber used to quip that the fact that you do not glow in the dark already sets a stringent limit on the proton's lifetime!

[15]This is similar to lepton number conservation mentioned in chapter V.3.

[16]J. C. Zorn, G. E. Chamberlain, and V. W. Hughes, *Physical Review* 129, 2566, 1963.

[17]The choice to define the electron charge as -1 rather than $+1$ is merely a convention, like clockwise versus anticlockwise. Historical note: Benjamin Franklin of course did not know about the electron and the proton, but in his study of electricity, he called one of the two

kinds of electricity as then known "negative," and the other kind "positive." This happens to be an unfortunate choice, since electric currents consist of electrons moving through the wires rather than the sedentary protons. As a consequence, minus signs appear all over the place in electromagnetism and continue to bedevil students in physics and in electrical engineering.

[18]Yes, I know that if the Dirac magnetic monopole exists, electric charge is also quantized. But there is no requirement that the Dirac magnetic monopole exists in quantum electrodynamics. In contrast, grand unified theories, such as $SU(5)$ and $SO(10)$, requires the Dirac magnetic monopole. By the way, the monopole has not yet been seen, but theory does predict that it would be quite massive.

[19]Indeed, the "U" in $SU(5)$ denotes "unitary" and is associated with complex numbers, while the "O" in $SO(10)$ denotes "orthogonal" and is associated with real numbers.

[20]See, for example, *QFT Nut*, chapter VII.7.

[21]Incidentally, $SO(10)$ also contains the gauge theory proposed by Pati and Salam that I mentioned earlier.

Gravity and curved spacetime ▬▬▬

> Einstein was ... one of the friendliest of men. I had the impression
> that he was also, in an important sense, alone. Many very great men
> are lonely.
> Freeman Dyson[1]

Einstein taught us that gravity is a manifestation of curved spacetime.[2] Since
I have devoted a companion* to this book to this epochal story, I will not
repeat it here. I will instead restrict myself to the field theoretic aspects of the
gravitational interaction. Even with this restriction, I still have an enormous
amount to tell you.

But first I have to give you an unreasonably brief overview of Einstein's
theory of gravity.[3] Without this overview, I can't even tell you what constitutes
the gravitational field.

No doubt you have heard that one of the major challenges of fundamental
physics in recent decades has been the difficulty of understanding quantum
gravity. But I have to tell you first about classical gravity, and then about
quantum gravity. Thus, I am splitting the material about gravity into two chapters. First, in this chapter, the easy stuff, Einstein's theory of gravity. Then in
chapter V.6, quantum gravity, which still remains a total mystery.

Gravity and curved spacetime

The clue that Einstein latched onto is the universality of gravity, that all massive objects fall at the same rate regardless of their composition, as was shown
by Galileo dropping cannon balls off the leaning tower of Pisa and in modern
times by astronauts dropping a hammer and a feather on the moon.[4] Imagine flying across the Pacific. You might notice that the plane follows a curved

*G.

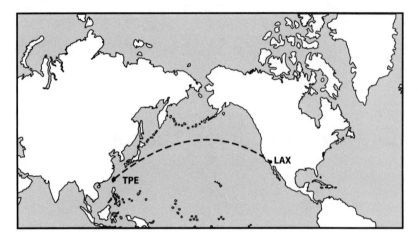

Figure 1. Is the Bering Strait exerting a mysterious attractive force on airplanes flying from Los Angeles to Taipei?
Reproduced from A. Zee, *On Gravity*, Princeton University Press, 2018.

path arcing toward the Bering Strait. Is the Bering Strait exerting a mysterious attractive force on the plane? On your next trip, you try another airline. This pilot follows exactly the same curved path. Don't these pilots have any sense of personality or originality? Not only is the mysterious force attractive, it is universal, independent of the make of the airplane.

Dear reader, surely you are chuckling. You know perfectly well that the Mercator projection distorts the earth, and pilots follow scrupulously the shortest possible path across the Pacific (figure 1).[5]

This little observation means nothing to most people, but it inspired Einstein to think that curved spacetime is masquerading as gravity. Now I have to tell you how to measure the curvature of space and spacetime.

From flat space to flat spacetime to curved space

Consider flat Euclidean 2-dimensional space, also known as a plane, characterized by the distance ds squared between two neighboring points with coordinates (x, y) and $(x + dx, y + dy)$, namely, $ds^2 = dx^2 + dy^2$, as discussed way back in chapter I.3. This could be generalized immediately to flat 3-dimensional space by writing $ds^2 = dx^2 + dy^2 + dz^2$, and indeed, to flat D-dimensional space if you like. Waving a "most valiant piece of chalk" and inserting a clever minus sign, Minkowski extended this to spacetime, by writing $ds^2 = dx^2 + dy^2 + dz^2 - dt^2$, as we discussed back in chapter I.4. Note that we have set $c = 1$.

Fine, but all this is for flat space and flat spacetime. We want curved space-time. Let's start with our favorite curved space, the surface of the earth we dwell on, abstracted as a sphere with radius set to 1. (Just so that we won't have the radius littering our formulas. In other words, we measure distances on the sphere in units of its radius.)

Denote latitude and longitude by the Greek letters θ and φ respectively. Picture a point on the sphere and call it Paris just for ease of reference. Denote the latitude and longitude of Paris θ_P and φ_P, respectively.[6] Consider a place with the same longitude as Paris but a slightly different latitude, namely, $\theta_P + d\theta$. The distance ds between this place and Paris is then given by* $d\theta$. This is because the lines of fixed longitude define "great circles" of radius 1. This place and Paris both lie on the same great circle.

In contrast, lines of fixed latitude do not define great circles, except for the equator. In other words, consider a place with the same latitude as Paris but a slightly different longitude, namely, $\varphi_P + d\varphi$. The distance between this place and Paris is definitely not given by $d\varphi$.

What I just said is that the distance ds between a point with coordinates (θ, φ) and a neighboring point with coordinates $(\theta, \varphi + d\varphi)$ is not equal to simply $d\varphi$, but rather it is equal to $f(\theta)d\varphi$. The distance depends on the latitude θ through a function $f(\theta)$. This function is equal to 1 at the equator, but is considerably less than 1 at the latitude of Paris. See figure 2. As we go north, this function keeps on decreasing, until it vanishes[7] at the north pole.

Thus, on a sphere, the distance ds between two neighboring points, one with coordinates (θ, φ) and the other with coordinates $(\theta + d\theta, \varphi + d\varphi)$ is given by $ds^2 = d\theta^2 + (f(\theta)\, d\varphi)^2$. The key point is that $f(\theta)$ is not a fixed number, but a function[8] of θ, that is, a number that varies depending on θ. You can think of this as a generalization of the Pythagorean formula $ds^2 = d\theta^2 + d\varphi^2$.

From this example, you learned that to go from the flat plane, for which $ds^2 = dx^2 + dy^2$, to a curved surface, we should have written[9] $ds^2 = dx^2 + (f(x)\, dy)^2$, which of course could also be written as $ds^2 = dx^2 + f(x)^2 dy^2$. (The coordinates θ and φ are customary for the sphere, but x and y are still customary for generic curved spaces.)

Enter Bernhard Riemann. He said, following his mentor Carl Friedrich Gauss, "Now that we have inserted a function in front of dy^2, why not insert a function in front of dx^2 also? In fact, why not include $dx\, dy$ and insert a function in front of that also? These three functions could all depend on both x and y!" So, here is the proposal:

$$ds^2 = a(x,y)dx^2 + b(x,y)dx\, dy + c(x,y)dy^2$$

---

*A slight technicality: $d\theta$ might be negative, while distance is usually understood to be positive. This is taken care by all the squares appearing in the generalized Pythagorean formula below, which in this simple case reads $ds^2 = d\theta^2$, so that ds is defined to be the absolute value of $d\theta$.

North pole

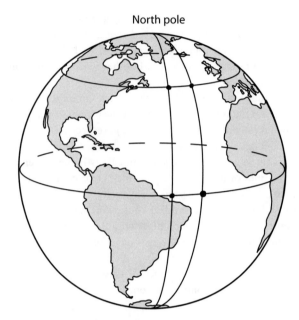

Figure 2. The distance between two nearby points with the same latitude but with longitudes differing slightly by $d\varphi$ is given by $f(\theta)d\varphi$. The function $f(\theta)$ is equal to 1 at the equator, decreases steadily as we move north, and vanishes at the north pole. Reproduced from A. Zee, *On Gravity*, Princeton University Press, 2018.

You specify the three[10] functions a, b, c, and each one of your choices characterizes a curved surface known as a Riemann surface. And thus Riemann started a branch of mathematics known as Riemannian geometry.

To summarize, we've done flat and curved space, and we could immediately generalize to higher dimension by simply introducing more coordinates.

You are ready to curve spacetime!

And then Minkowski, with his valiant chalk, immediately takes us from flat space to flat spacetime, as was explained back in chapter I.4. Simple, just flip a $+$ sign to a $-$ sign. Next, Einstein wants to curve spacetime.

Since you know how to go from flat space to curved space—easy, just stick some functions of space in front of quantities like dx^2—you may be able to show Einstein how to go from flat spacetime, defined by $ds^2 = dx^2 + dy^2 + dz^2 - dt^2$, to curved spacetime. Try it!

Let me show you how by giving you two famous examples, among the most important in Einstein gravity.

First example. Take a function of x, y, z, stick it in front of dt^2, and write

$$ds^2 = dx^2 + dy^2 + dz^2 - (f(x, y, z)dt)^2$$

Yes, it is that easy. Note that, while you have made spacetime curved, space is still flat: Pythagoras still determines the distance dl between two neighboring points in space at a fixed time, namely, $dl^2 = dx^2 + dy^2 + dz^2$. Furthermore, note that at a fixed (meaning that $dx = dy = dz = 0$) location in space, two instants in time are separated by $f(x, y, z)dt$. Voilà! The rate at which time flows depends on where you are, namely, the gravitational redshift you might have read about in the media.

With the appropriate[11] f, Einstein was able to obtain Newtonian gravity as a special case and predict that gravity affects the flow of time. Einsteinian gravity contains Newtonian gravity.

An expanding universe: the simplest possible example

For our second example, we tackle the whole universe. Instead of sticking a function of space in front of dt^2 as we just did, we could stick a function of time in front of $dx^2 + dy^2 + dz^2$:

$$ds^2 = (a(t))^2 (dx^2 + dy^2 + dz^2) - dt^2$$

Again, yes, it is that easy.

Cosmological observations indicate that the universe we live in is fairly well described by this curved spacetime if $a(t)$ is an exponentially growing function of time. Notice that while this spacetime is curved, the space contained in it is flat. At time t, the square of the distance between a point with coordinates (x, y, z) and a neighboring point with coordinates $(x + dx, y + dy, z + dz)$ is given by $(a(t))^2 (dx^2 + dy^2 + dz^2)$, namely, what Pythagoras said it is, multiplied by a factor $(a(t))^2$.

Thus, if $a(t)$ increases with time, this spacetime describes an expanding universe. The distance between any two galaxies, for example, increases by a universal factor $a(t)$, which is therefore sometimes called the "scale size" of the universe.

A friend who has studied quite a few popular physics books and who kindly read an early draft of this chapter wrote to me to insist that the preceding paragraph (which was once an endnote, then later a footnote) must be highlighted, saying: "This is the first comprehensible mathematical representation of an expanding universe that I've (or likely any other lay reader has) ever seen." I have now heeded his advice and also placed these few paragraphs about the universe in a separate section.

Elsewhere, I have spoken quite a bit[12] about Nature's kindness to theoretical physicists. I have often been struck by how Nature keeps it simple at the fundamental level, so that we human physicists are still able to figure Her out. We just came across one of numerous examples: The expanding universe we live in can be described (to first approximation) by one single function $a(t)$ depending on one single variable t.

See how easy[13] it is to learn Einstein gravity! (I caution the reader, however, that what I have given you is a description of curved spacetime, but not the equations governing functions such as $a(t)$ and $f(x,y,z)$.)

Curved spacetimes in general

Not that hard, is it? It does sound a bit too easy. You are right to wonder.* These two spacetimes have particularly simple forms. In general, 4-dimensional curved spacetime requires ten functions to describe. Can you figure out why ten before reading on?

Yes, indeed. In addition to four functions multiplying dx^2, dy^2, dz^2, and dt^2, there could be a function multiplying each of the six combinations $dx\,dy$, $dx\,dz$, $dx\,dt$, $dy\,dz$, $dy\,dt$, and $dz\,dt$. Hence ten functions altogether. Each of these functions could depend on x, y, z, t. Again, not that hard, is it?

A more compact notation

Theoretical physicists are an impressively lazy lot, and so they easily tire of writing ten functions together with ten quantities, such as dz^2 and $dy\,dt$. The index notation, which we have already learned back in chapters I.3 and I.4, rides to the rescue. Recall that instead of writing (t, x, y, z), we now write x^μ, with $\mu = 0, 1, 2, 3$, and $x^0 = t$, $x^1 = x$, $x^2 = y$, $x^3 = z$. So, instead of writing these ten quantities out longhand, simply write $dx^\mu dx^\nu$. As μ and ν separately takes on values $0, 1, 2, 3$, the expression $dx^\mu dx^\nu$ ranges over all ten of these quantities. (For example, $dx^3 dx^3 = dz^2$, and $dx^2 dx^0 = dy\,dt$.)

The metric of spacetime

Using this notation (merely notation, neither physics nor math, nothing profound at all, just bookkeeping), we can then write the most general curved spacetime concisely as

$$ds^2 = g_{\mu\nu}(x)dx^\mu dx^\nu$$

We have already mentioned in chapter III.3 the Einstein repeated index summation convention, implying that the indices μ and ν, ranging over $0, 1, 2, 3$, are to be summed over. In other words, $g_{\mu\nu}(x)dx^\mu dx^\nu$ is shorthand

---

*Indeed, you might wonder why we could get away with such minimal modifications of flat Minkowski spacetime. The general 3-dimensional curved space already requires six functions to describe. In contrast, in each of these two curved spacetimes described here, only one function was needed. That is because these two curved spacetimes are highly symmetric.

for[14] $g_{00}(x)(dx^0)^2 + g_{11}(x)(dx^1)^2 + \cdots + 2g_{01}(x)dx^0 dx^1 + 2g_{02}(x)dx^0 dx^2 + \cdots + 2g_{23}(x)dx^2 dx^3$. Note that, instead of stupidly inventing names for each of the ten[15] functions that appear in front dt^2, $dt\,dx$, $dt\,dy, \ldots$, $dy\,dz$, dz^2, we simply denote them collectively by[16] $g_{\mu\nu}(x)$, known as the space-time metric, and, as the terminology suggests, they measure spacetime.

Flat Minkowski spacetime corresponds to a particularly simple form of $g_{\mu\nu}(x)$. The ten functions actually are not functions, just numbers, and all but four are equal to 0. These four are $g_{00} = -1$, $g_{11} = +1$, $g_{22} = +1$, and $g_{33} = +1$. In other words, $ds^2 = -(dx^0)^2 + (dx^1)^2 + (dx^2)^2 + (dx^3)^2$, which I trust you recognize as flat Minkowski spacetime written using indices rather than using (t, x, y, z).

Since the earth's gravitational field is so weak, the human race has been hanging out in a spacetime that is extremely close to flat Minkowski space-time. Thus, flat Minkowski spacetime is by far the most important spacetime to know and love. Not surprisingly then, theoretical physicists traditionally assign the Minkowski metric, that is, the metric I just described, a special symbol, namely, $\eta_{\mu\nu}$, using the Greek letter η (pronounced "eta"). Nothing profound here: We merely define $\eta_{\mu\nu}$ by specifying that the only nonzero components of η are $\eta_{00} = -1$, $\eta_{11} = +1$, $\eta_{22} = +1$, $\eta_{33} = +1$. In other words, we can write the flat spacetime we wrote down $ds^2 = -(dx^0)^2 + (dx^1)^2 + (dx^2)^2 + (dx^3)^2$ more compactly as

$$ds^2 = \eta_{\mu\nu} dx^\mu dx^\nu$$

Just to give you an example, the metric $g_{\mu\nu}$ of the simple expanding universe mentioned above could be described as follows: $g_{\mu\nu}$ equals $\eta_{\mu\nu}$ except for the three components g_{11}, g_{22}, and g_{33}, which are all equal to the square of a function of time $a(t)$, which measures how large the universe is.

I emphasize that all of this is trivial, just a compact notation to keep track of a large number of quantities needed to describe spacetime. Learning a notation is a bit like learning a language. In the present context, you need it to know what physicists are talking about.

From metric to curvature

Since the metric determines the distance between any two points, once the metric of a curved space (or spacetime) is given, we can deduce all that we need to know, such as the curvature of that space.

Here is an operational procedure a civilization of intelligent mites[17] living on a curved surface would follow to determine how curved their world is. (Remember, they cannot go outside their surface to take a look, any more than we can go outside our universe to see whether it is curved.) As you read the next two paragraphs, please draw your own figure, if only to fix the discussion in your mind.

First, given any two points P and Q, the distance along any path connecting the two points can be determined by cutting the path up into infinitesimal segments and adding up the distance along each of the segments. (This is of course exactly what we do in everyday life: The infinitesimal segments could be the length of your stride as you walk along a path.) By definition, the distance along an infinitesimal segment is given by ds, and hence determined by the metric $g_{\mu\nu}$ evaluated at the location of that segment. Find the path with the shortest distance. Call that path "the straight line" between P and Q. The distance between P and Q is defined to be the distance along that straight line path. (For a sphere, this would be an arc of the great circle connecting P and Q.)

Next, given a point O, find all the points that are located a small distance r away from O. This defines a circle of radius r around that point O. Moving around the perimeter of the circle and adding up the distance between points on the circle infinitesimally separated from each other gives the circumference of the circle. Divide the circumference by the radius r. As r becomes smaller and smaller, if this ratio approaches $2\pi \simeq 6.28\ldots$, then the surface is flat at the point O. If not, then the surface is curved at that point.

Riemann saved us from having to do all this. Given a metric, he found a formula for calculating what is now called the "Riemann curvature tensor." As you could have guessed, it depends on how the metric varies.[18]

The Riemann curvature

Riemann is immortalized in the annals of mathematics for finding what is now known as the Riemann curvature tensor, from which various quantities called the "Ricci curvature tensor" and the "scalar curvature" can be calculated. You do not have to know what these words mean. For our purposes, all you have to know is that Riemann gave us "a machine" into which we can feed the metric, and out comes various measures of curvature.

The formula for the Riemann curvature tensor is rather involved, but nothing that a bright undergraduate[19] could not readily master. It involves two derivatives acting on the metric $g_{\mu\nu}(x)$, thus expressing the notion that the curvature depends on how the metric varies with x, as I just said. A simple example is given by the sphere, which you may recall is described by $ds^2 = dx^2 + f(x)^2 dy^2$. Riemann's formula reduces to some expression involving two derivatives acting on the function $f(x)$.

In general, the scalar curvature, traditionally denoted by R in honor of Riemann, is given schematically by

$$R = (\cdots \partial \cdots \partial \cdots)$$

Here the dots denote various components of $g_{\mu\nu}(x)$ (recall there are 10 of these functions, each depending on the coordinates x), while ∂ denotes differentiation with respect to x. Conceptually simple to understand, even though the actual expression looks rather involved.[20] (These days, my laptop computer could calculate R literally in the blink of an eye.)

Time for an executive summary: Riemann figured out that the curvature of space is determined by the variation of its metric from place to place.

Geometrical concepts coming into physics

Another important geometrical quantity is the volume of an infinitesimal region of space or spacetime. (The word is used generically here. In human languages, the word associated with this concept depends on the dimension of space, so that, for instance, it is called "area" in 2-dimensional space.) Since the metric determines distances, it also determines areas and volumes, as you would expect. For a 2-dimensional curved space, the area dA of an infinitesimal quasi-rectangular region bordered by dx and dy is given by

$$dA = dx\, dy \sqrt{g(x, y)}$$

with g a mathematical expression[21] constructed from the metric. For $g = 1$, this is just the area of an infinitesimal rectangle you learned in elementary school. Indeed, all this highfalutin Riemannian geometry is just a fancier version of what they taught you in elementary school. How could it be otherwise?

We could now take what Riemann did for curved space over to curved spacetime without further ado. After all, we are living more than a hundred years after Minkowski, so while we don't have his valiant piece of chalk, we all have a trusty pen that we could use to flip a sign in the metric.

The metric as a field

This book is about field theory. Boys and girls, do you see a field lurking about? Pause and look for the field in our discussion.

Indeed, $g_{\mu\nu}(x)$, according to the definition given in chapter I.2, is a field, being a function of space and time. If you like, you could say that it consists of 10 fields, but we will refer to the lot as the gravitational field.

Even more, you might have realized that $g_{\mu\nu}(x)$ is what we might call "gauge fields" in the sense described in chapters IV.4. The gauge freedom corresponds to the freedom in choosing the coordinates on the curved spacetime. Instead of the coordinates x, some other physicist could always choose another set of coordinates x', with $x'(x)$ some functions of x. The metric would then change accordingly. With the new set of coordinates, the metric would be given by another set of 10 functions $g'_{\mu\nu}(x')$.

Now that you see these fields all over the place, do you see Planck's quantum lurking around? No! Nowhere in sight. So, Einstein gravity is a classical field theory, just like Maxwell's electromagnetism is a classical field theory.[22]

General coordinate transformation

Two physicists could use two different metrics to describe the same space-time, but they must reach the same conclusion about physics. This parallels the discussion we had about electromagnetism: Two physicists could use two different electromagnetic potentials, $A_\mu(x)$ and $A'_\mu(x)$, to describe the same electromagnetic field $\vec{E}(x)$ and $\vec{B}(x)$.

The freedom to use different coordinates is known as general coordinate transformation, and can be thought of as a type of gauge transformation. In both electromagnetism and gravity, to construct the theory, we are compelled to use quantities A_μ and $g_{\mu\nu}$, respectively, rather than the quantities actually measured, namely, the electromagnetic field $\vec{E}$ and $\vec{B}$ in one case, and the curvature (also known as the gravitational field).

Suppose our colleague, instead of using x and $g_{\mu\nu}(x)$, uses the coordinates x' and the metric $g'_{\mu\nu}(x')$ instead. We live in a free country. The distance ds between two nearby points had better not depend on the physicist! In other words, the ds given in terms of the two different metrics and coordinates must be the same, and hence

$$ds^2 = g_{\mu\nu}(x)dx^\mu dx^\nu = g'_{\mu\nu}(x')dx'^\mu dx'^\nu$$

This relates the two different metrics, $g_{\mu\nu}(x)$ and $g'_{\mu\nu}(x')$, used by the two physicists.

Cartography could convey some of the flavor of this gauge freedom. We are all familiar with the map of the world popularized by Gerardus Mercator (namely, Gerry the Merchant[23]) so that the world could be more conveniently depicted on a rectangular piece of paper rather than on a sphere. The Cartesian coordinates (x, y) on the rectangle are simply related to the (θ, φ) on the sphere.[24] The two metrics[25] $g_{\mu\nu}(\theta, \varphi)$ and $g'_{\mu\nu}(x, y)$ are clearly going to be quite different, but geometrically invariant quantities, such as the area dA of an infinitesimal region, must not depend on the choice of coordinates.

Now recall that I mentioned in passing that a certain quantity determined by the metric and written as $\sqrt{g}$, measures the generic "volume" of the space, namely, the area in this context. Thus, we must have $dA = d\theta d\varphi \sqrt{g(\theta, \varphi)} = dx dy \sqrt{g'(x, y)}$. Even though Greenland looks much larger than China on a Mercator map, we all learned in school that the former is actually much smaller than the latter.[26] Our eyeballs interpret area on the map as $dx dy$ without including the all-important weighting factor $\sqrt{g(x, y)}$.

Hey, just like Molière's bourgeois gentleman, you have been aware of gauge transformation all along! Theoretical physicists, just like cartographers, have the gauge freedom to choose potentials and coordinates as they please.

The Mercator projection has been much criticized in recent decades for predisposing us to exaggerate the importance of Europe at the expense of Africa and South America. There are many other projections other than Mercator's, and each projection amounts to a different choice of the coordinates (x, y) on the rectangle.

"The wretchedness of humanity"

Once Einstein had the profound insight that curved spacetime is masquerading as gravity, he spent ten long years working out the equation of motion[27] governing the dynamics of spacetime, namely, how the metric $g_{\mu\nu}(x)$ varies with x. In contrast to his derivation of special relativity, he struggled and suffered numerous setbacks. I digress slightly to tell the reader a bit of history[28] to underscore for you, once again, the importance of the action, even though, as I mentioned earlier, many practical minded physicists don't even bother to learn it. Einstein, for some reason that I do not fully understand, focuses on finding the equation of motion for the gravitational field, rather than the action[29] governing the gravitational field. I hope that I had impressed on you, in part II, how much simpler the action is compared to the equation of motion. In the case of gravity, we now know that the action is vastly simpler to write down than the equation of motion.

On November 4, 1915, as Einstein neared the end of the struggle he started in 1905, he gave a lecture to the Royal Prussian Academy of Sciences, in which he explained his idea relating gravity to curved spacetime and presented the equation of motion he found. But he got it wrong! Three weeks later, on November 25, he presented to the same academy the correct equation of motion. But meanwhile, Einstein was scooped!

Hilbert, one of the most distinguished mathematicians of the 20th century, heard Einstein's November 4 lecture, found the action for gravity quickly, and on November 20 published it. Einstein understandably boiled over. He wrote to a friend, "The theory is of incomparable beauty. But only one colleague has really understood it, and he is trying, rather skillfully, to 'nostrify' it. ... In my personal experience, I've hardly come to know the wretchedness of humanity better than in connection with this theory." Well, dear reader, nostrification is not only still practiced in theoretical physics, but ever more skillfully.[30]

This historical anecdote brings home the vast superiority of the action formulation. Einstein spent ten years searching for the equations of motion for gravity, while it took Hilbert less than sixteen days to find the action. Of course, knowing that there is an action for curved spacetime to be found is already a huge advantage. (Recall my remark in chapter V.3 about the three classes of problems in theoretical physics.)

Nowadays, physicists call the action for Einstein gravity the "Einstein-Hilbert action," recognizing that the underlying idea is due to Einstein, but still giving credit to Hilbert.[31] Here it is:

$$S_{\text{Einstein-Hilbert}} = \int d^4x \, \frac{1}{G} \sqrt{g} R$$

Trumpet blast, please! The action is amazingly simple, merely the scalar curvature R of spacetime integrated with a volume factor $\sqrt{g}$, divided by Newton's constant G.

Let's summarize by talking what ifs. If Einstein had known about Riemann's work, and if he had more confidence in the action principle—two very big ifs, to be sure—he could have written down this action for gravity many years earlier than he did. Of the various mathematical quantities measuring curvature that Riemann discovered, the scalar curvature R is the only one that could be used in the action. The truly difficult part is of course to recognize that curved spacetime could masquerade as gravity and to identify the metric $g_{\mu\nu}$ as a field. But Einstein had both of these insights early on. He could have saved himself a lot of grief. It didn't happen this way in our civilization but could in some other civilization far, far away.

How to describe a gravitational wave

After the detection of gravity waves[32] in 2015, there was much talk about ripples in spacetime. Dear reader, you who have come so far, you know exactly what this metaphor refers to.

Even better, you may even know what to do concretely. Simply modify flat spacetime by a tiny bit, by inviting ourselves to consider a curved spacetime described by

$$ds^2 = \left(\eta_{\mu\nu} + h_{\mu\nu}(x) \right) dx^\mu dx^\nu$$

In other words, merely add to the bunch of 1's and 0's in $\eta_{\mu\nu}$ some functions $h_{\mu\nu}(x)$, which we are going to regard as small compared to 1. The metric of this (slightly) curved spacetime is given by $g_{\mu\nu}(x) = \eta_{\mu\nu} + h_{\mu\nu}(x)$.

The procedure parallels that followed in numerous situations in introductory physics. Picture water waves on the surface of a placid lake, for example. Let us idealize by supposing that the lake bottom is flat and that the banks are far away from the region we are focusing on. Denote the depth of the water by $g(t, x, y)$. Without any wind, the surface of the lake is flat, and the depth of the water is given by $g(t, x, y) = 1$ expressed in some suitable unit. When a breeze whips up some waves, $g(t, x, y) = 1 + h(t, x, y)$. The surface undulates in space and time.

The equation for fluid dynamics are, notoriously, among the nastiest in classical physics, as would be evident to anybody who has ever watched waves break and swirl on the beach. But if the amplitude of the wave is small, then we may treat $h(t, x, y)$ as small compared to 1. Elementary school children know that a small number, small meaning smaller than 1, multiplied by another small number gives an even smaller number; this simple fact forms the basis of perturbation theory in physics, as I explained in chapter IV.3. So, after we plug $g(t, x, y) = 1 + h(t, x, y)$ into the equation, if we are allowed to throw away all the terms in which h gets multiplied by h in the resulting mess, then the equation simplifies to one that almost any physics undergrad can solve.

Surely it has not escaped your notice that the form of the metric of spacetime $g_{\mu\nu}(x) = \eta_{\mu\nu} + h_{\mu\nu}(x)$ is structurally[33] the same as $g(t, x, y) = 1 + h(t, x, y)$.

Einstein gave us a set of equations* for determining $g_{\mu\nu}$. When we plug $g_{\mu\nu} = \eta_{\mu\nu} + h_{\mu\nu}$ into these equations, if we are allowed to treat $h_{\mu\nu}$ as small, things simplify enormously, leaving us with equations for determining $h_{\mu\nu}$ that are only marginally more complicated than the equations for electromagnetic waves.[34]

During the excitement over the detection of gravity waves, some misleading reports in the media proclaimed that gravity waves has something to do with quantum gravity. Well, no, detecting gravity wave has little to do with quantum gravity, not any more than tuning into your favorite radio station means that you are probing quantum electrodynamics. To be sure, your receiver depends on quantum physics, whether transistors in the old days or integrated circuits on chips now. But the electromagnetic wave you are detecting is very much classical.

Kaluza, Klein, and higher dimensional spacetime

I might mention in passing that under a change of coordinates, the metric $g_{\mu\nu}$ transforms in a way that may look quite involved to the uninitiated. As you would expect, however, things simplify enormously if (1) $g_{\mu\nu}$ deviates from the flat metric $\eta_{\mu\nu}$ by only a tiny amount, as characterized by a small $h_{\mu\nu}$, and (2) if the two sets of coordinates differ by a small amount, as characterized by a small $\varepsilon^\mu(x) = x'^\mu(x) - x^\mu$. In that case, the gravitational field $h_{\mu\nu}$ transforms as $h_{\mu\nu} \to h_{\mu\nu} + \partial_\mu \varepsilon_\nu + \partial_\nu \varepsilon_\mu$. Recall that under a gauge transformation, the electromagnetic field $A_\mu \to A_\mu + \partial_\mu \Lambda$. Compare this with how $h_{\mu\nu}$ transforms. Between us friends, what is an extra index flying around?

---

*Since you know the action for Einstein gravity from our discussion above, you could in principle vary that action to obtain these equations.

The intriguing resemblance between how the gravitational field and how the electromagnetic field transform is striking to say the least, and almost begs for some kind of unification. Indeed, this is what led Kaluza and later Klein to propose[35] a 5th dimension in spacetime into which electromagnetism could be neatly tucked,[36] so to speak. I will show the curious reader how this works in an endnote.[37]

Most readers have heard by now that string theory has to be formulated in 10 dimensional spacetime. This apparent extravaganza all started with the basic observation given here, that the gravitational field and the electromagnetic field transform in mysteriously similar ways, thus setting off the mad rush into higher dimensional spacetime.

Einstein's quest for a unified field theory

Inspired by the Kaluza-Klein theory, Einstein devoted the latter part of his scientific life to a quixotic and futile quest for a unified field theory, a quest that some biographers view as tragic. To understand this, we have to remember that during Einstein's formative years, the only fundamental interactions around were electromagnetism and gravity. The new fangled discoveries around the turn of the century, such as atomic nuclei, radioactive decays, and so on, were characterized by empirical rules and phenomenological descriptions, which, while intriguing and exciting, could hardly be compared with the magnificence of Maxwell's theory of electromagnetism and of Einstein's own theory of gravity, both based on gauge symmetries. In the preceding chapters, we were able to tell a coherent story of the strong and weak interactions only with the benefit of hindsight.

So Einstein chose to pursue the unification of electromagnetism and gravity, and dismiss the strong and weak interactions as epiphenomena that he expected would fit into the theory eventually. To his contemporaries, Einstein's quest appeared boneheaded and misguided. They felt that it was absurd and ignorantly old fashioned to insist on the unification of electromagnetism with gravity when the world contained two other mysterious interactions that appeared to have nothing to do with gauge symmetries.

Laughing at Einstein's futile labors, Pauli once quipped, "Let no man join together what God has put asunder." But Einstein had the last laugh on Pauli. In some sense, grand unification realizes Einstein's impossible quest. Physicists have joined together what God has only appeared to put asunder. While it is true that unification of the other three interactions, leaving gravity out, was quite different from what Einstein had in mind, his vision of a unified design inspired the grand unifiers of the 1970s and continues to inspire us today.

Now that the strong, weak, and electromagnetic interactions have been grand unified into a gauge theory generalizing electromagnetism, the search for unification with gravity once again amounts to the unification of two geometric theories, as Einstein had wanted. In this ironic twist of history, Einstein turns out to be right in spirit, if not in detail.

An interim summary before going quantum

Let's pause for an interim summary. Riemann showed us that by differentiating the metric twice, we could determine the curvature of a curved space. Minkowski with his chalk showed us how Riemann's result could be extended to spacetime by flipping a sign. Einstein understood that the universality of gravity must be due to the curved spacetime we navigate in. Astonishingly, the Einstein-Hilbert action was uniquely determined in terms of Riemann's scalar curvature R. Given this action, Euler and Lagrange (remember them from chapter II.1) showed us how to obtain the equation of motion for the gravitational field.

So far, all of this stuff is classical Einstein gravity, stuff that nowadays many undergraduates routinely learn.

You have surely read a lot of the wild stuff about warped spacetime, gravity waves, expanding universes, black holes, crazy scenarios such as falling behind the horizon never to get out again, with time slowing down and even standing still. Please appreciate that all that and more comes from the classical physics contained in the action S given above. Thus, you might say that all these amazing phenomena are baby stuff compared to quantum gravity. Since this book is about quantum field theory, and, by extension, the quantum field theory of gravity, I will not go into details about all this mind boggling, but "merely" classical, stuff. To put things into perspective, however, I will mention a few salient points.

Much of the classical manifestation of Einstein gravity deals with spacetimes that are almost flat, that is, for $g_{\mu\nu}(x) = \eta_{\mu\nu} + h_{\mu\nu}(x)$ with $h_{\mu\nu}$ small compared to 1. Plugging this into the action, we obtain an infinite series with the schematic form $S = \int d^4x \, \frac{1}{G} \, \partial\partial(hh + hhh + hhhh + hhhhh + \cdots)$. This schematic notation means that the two differentiation operators ∂ are to act on an infinite series of ever higher power of h. (Just as an example, the term with h to the fifth power equals $hhh\partial h\partial h$ schematically. Understand that this is highly symbolic. Since $h_{\mu\nu}$ carries two spacetime indices and ∂_λ carries one index, the 12 indices in $hhh\partial h\partial h$ are to be summed over in all possible ways.)

Newtonian gravity, gravitational waves, black holes, and all that

Hello, Newtonian gravity must be contained in here somewhere. It appears as a small piece of $\partial\partial hh$ if we make some further simplifying restrictions.[38] For instance, suppose we do not allow $h_{\mu\nu}$ to depend on time. Newton's gravitational potential ϕ that we talked about in chapter I.2 turns out to be just $\phi = \frac{1}{2}h_{00}$. In other words, the other nine $h_{\mu\nu}$'s do not appear in Newtonian gravity at all! I mention this to impress on you how rich Einstein gravity is compared to Newtonian gravity.

As another example, to study how gravitational wave propagates through spacetime, you only have to keep the term $\partial\partial hh$ quadratic in h, with the $h_{\mu\nu}$ oscillating in time and in space. The resulting equations are slightly more involved than the corresponding equations for the propagation of electromagnetic waves, for the simple reason that $h_{\mu\nu}$ carries one more index than A_μ. Hence, this derivation is just a homework exercise these days for advanced undergrads in a course on Einstein gravity!

The weird stuff about black holes happens when $h_{\mu\nu}$ get to be so large that we could no longer expand S as an infinite series in h. I showed in chapter V.2 the infinite series $1/(1 - \varepsilon) = (1 + \varepsilon + \varepsilon^2 + \varepsilon^3 + \cdots)$ going haywire when ε approaches 1. The same mathematical phenomenon occurs around a black hole!

Imagine solving Einstein's equation for the metric of the curved spacetime outside a spherically symmetric object of mass M, such as a star or a black hole. (Not very difficult at all; most physics undergraduates can learn to do it.) In fact, denote by r the radial distance from the center of the star or black hole; then the radial component of the metric comes out to be

$$g_{rr} = \frac{1}{1 - \frac{2GM}{r}} = 1 + \frac{2GM}{r} + \left(\frac{2GM}{r}\right)^2 + \cdots$$

(It is even the same series as the one I gave in chapter V.2.) The quantity $2GM$, that is, twice Newton's constant times the mass M, is known as the Schwarzschild radius.[39]

Far away, with r much larger than the Schwarzschild radius, the deviation $h_{rr} = \frac{2GM}{r}$ is small. But when r approaches the Schwarzschild radius, g_{rr} blows up.[40]

Einstein's equation also determines the time component of the metric g_{tt} to be[41] $g_{tt} = -1 + \frac{2GM}{r}$. At the Schwarzschild radius, g_{tt} vanishes, hence all that mind boggling stuff about time standing still!

Notice also that as you move inside* the Schwarzschild radius, g_{tt} and g_{rr} both flip sign, and according to Minkowski, time becomes space and space becomes time. Weird!

---

*Some readers might be wondering that this discussion seems to apply to a garden variety star as well as a black hole. What is the difference? Excellent question! A black hole is defined as an object whose Schwarzschild radius is larger than its actual radius. A star's Schwarzschild radius is much smaller than its actual radius. You cannot move inside the Schwarzschild radius of a star by definition. In other words, the solution given here for g_{tt} and g_{rr} is restricted to those values of r larger than the actual radius of the object we are studying. Once inside the star, we have to include the physics of the hot gas that the star is made of.

Weird though all this stuff may be, I think that you could see that it is, at least mathematically, all fairly straightforward and could be, and is, safely fed to undergraduates without causing harm. Nothing like quantum physics! Allegedly, nobody understands quantum physics, but everybody understands Einstein gravity.

Notes

[1]F. Dyson, in *The New York Review of Books*.

[2]For an intuitive introduction to why this must be so, see G, chapter 8. See also my talk at the Royal Institution on YouTube.

[3]Unfortunately, also known as general relativity.

[4]Google "Hammer vs Feather—Physics on the Moon" for the demonsration on YouTube.

[5]This story is taken from G.

[6]Of course, the French had insisted that φ_P should be set to 0, but unfortunately for them, the Brits were more powerful when these things were determined.

[7]Why? Think about this for a moment. It is because longitude ceases to be defined at the north pole.

[8]For the mathematically sophisticated reader, $f(\theta) = \cos\theta$, with θ defined to be $\pi/2$ at the north pole. A trivial matter: In physics and mathematics, θ is defined to be 0 at the north pole and $\pi/2$ at the equator, so that $\theta_{physics} = \frac{\pi}{2} - \theta_{everyday}$ and in physics books, $f(\theta) = \sin\theta$.

[9]If you are at all into math, you would have fun figuring out the properties of the spaces described by various metrics. For example, consider $ds^2 = (dx^2 + dy^2)/y^2$ with $y > 0$. The space it describes is called the "Poincaré half plane" and has some weird properties. See *GNut*, page 67.

[10]Note that $dy\,dx$ is the same as $dx\,dy$ and should not be counted separately.

[11]That's the hard part, to determine the appropriate f, but still not that hard. It is easily mastered by undergrads. I should know, since I have taught it to undergrads on numerous occasions.

[12]*Fearful, QFT Nut,* and *GNut.*

[13]Seriously. I kid you not: way way easier than learning quantum mechanics. The math involved only goes a bit beyond what is discussed here.

[14]There are ten terms altogether, but I have not bothered to write them all out; the ones I did not write out are indicated by dots.

[15]The mathematically inclined readers might enjoy counting. For d-dimensional spacetime, the number is $d + d(d-1)/2 = d(d+1)/2$. Thus, for $d = 4$, the metric has $4 \cdot 5/2 = 20/2 = 10$ components. If you have trouble with the counting, try doing it for $d = 1$ and $d = 2$.

[16]Let me forestall a potential confusion here. The notation $g_{\mu\nu}(x)$ is shorthand for $g_{\mu\nu}(x^0, x^1, x^2, x^3)$. The letter x is used to denote x^0, x^1, x^2, x^3 collectively.

[17]See *GNut,* pages 6 and 27.

[18]The mathematically sophisticated reader who has read endnote 8 might even see that Riemann's expression of the curvature must involve two spatial derivatives. Refer to the earlier endnote and consider a point on the equator, that is, a point with $\theta = 0$. At that point, $f(\theta) = \cos\theta \simeq 1 - \frac{1}{2}\theta^2$. Indeed, the second derivative of f with respect to θ gives the curvature. See *GNut,* or at a lower level, *FbN.*

[19]If you feel that you are comparable to a bright undergrad at a large U.S. state university, then you can for sure learn how to derive the Riemann curvature tensor. I have experimentally established this fact. I mention this to encourage you.

[20]For instance, just so some curious readers could see what R might look like, for the expanding universe described earlier, $R = 6\left(\frac{\dot{a}^2}{a^2} + \frac{\ddot{a}}{a}\right)$. Here $\dot{a} \equiv \frac{da}{dt}$ is just the first derivative of $a(t)$ with respect to time. Similarly, $\ddot{a}$ is the second derivative. See, for example, *GNut,* page 609.

[21]For those readers who know what a determinant is, think of $g_{\mu\nu}$ as a matrix; then g is the determinant of the metric.

[22]Indeed, the famous textbook by Landau and Lifshitz on electromagnetism and gravity is titled *The Classical Theory of Fields*.

[23]Actually, Mercator had never been a merchant; his Latinized scholarly name was a direct translation of his family name Gerard Kremer.

[24]By $x = \varphi$, $y = -\log \tan \frac{\theta}{2}$ (here I use the (θ, φ) customary in physics rather than everyday life; see an earlier endnote), as was first derived by the English mathematician E. Wright. The mathematically sophisticated reader would notice that y actually ranges from $-\infty$ to $+\infty$ and so needs to be cut off for practical use.

[25]A trivial remark almost not worth making: To avoid confusion, I use the indices $\mu\nu$, even though I am talking about space and not spacetime here.

[26]See G, page 65.

[27]There are actually 10 equations, but for ease of exposition, I will simply use the singular. Why 10? Yes, you are right. The metric has 10 components.

[28]See *GNut*, page 396.

[29]Even though Einstein indisputably knew and understood the action principle.

[30]And I might add, aided and abetted by sycophants and idolaters and the popular media—not to mention authors of popular books.

[31]For more details and historical references, see notes 6, 7, and 8 on page 299 of *GNut*.

[32]On "gravity wave" versus "gravitational wave," see G, page 5.

[33]I say "structurally" because there are clearly some difference in the details. For one thing $g_{\mu\nu}(x)$ consists of 10 functions, instead of the one function $g(t, x, y)$. For another, x is now a compact notation denoting (t, x, y, z), but that

is just because we live in 3-dimensional space, while the surface of the lake is 2-dimensional.

[34]Needless to say, this is a simplified first description. In real life, the spacetime around the two black holes merging could hardly be taken to be flat Minkowski spacetime. But once the gravitational wave leaves this region, then the description given should be more or less adequate, except for the fact that the universe has expanded some during the 1 billion years or so that the wave took to reach us.

[35]For more details, see *GNut*, chapter X.

[36]The need for an extra dimension has to do with $g_{\mu\nu}$ having one more index than A_μ.

[37]Denote the index for 5-dimensional spacetime by $M = (0, 1, 2, 3, 4, 5) = (\mu, 5)$. In other words, in addition to the values the μ runs over, the index M could also equal 5. In this spacetime, the gravitational field h_{MN} transforms as $h_{MN} \to h_{MN} + \partial_M \varepsilon_N + \partial_N \varepsilon_M$. But this implies that $h_{\mu\nu} \to h_{\mu\nu} + \partial_\mu \varepsilon_\nu + \partial_\nu \varepsilon_\mu$, and also that $h_{\mu 5}$ transforms as $h_{\mu 5} \to h_{\mu 5} + \partial_\mu \varepsilon_5 + \partial_5 \varepsilon_\mu$, which becomes $h_{\mu 5} \to h_{\mu 5} + \partial_\mu \varepsilon_5$ if we assume that ε_μ does not depend on the x^5, the 5th coordinate, so that $\partial_5 \varepsilon_\mu = 0$. Now simply rename: $h_{\mu 5} = A_\mu$, $\varepsilon_5 = \Lambda$. We recover $A_\mu \to A_\mu + \partial_\mu \Lambda$ and hence electromagnetism! A genius observation on Kaluza's part.

[38]These are detailed on pages 302–303 in *GNut*, for example.

[39]You might have seen it as $2GM/c^2$, but I have set $c = 1$ mostly everywhere in this book.

[40]Some readers are also worried about the series blowing up term by term as $r \to 0$, but that is another story I am not getting into here.

[41]In light of what I said above, you could even see Newtonian gravity emerging for r large: Newton's potential is just $\phi = \frac{1}{2} h_{tt} = \frac{GM}{r}$.

Quantum gravity: The Holy Grail of theoretical physics?

Quantum gravity

Now that we have "mastered" Einstein's classical theory of gravity, are you ready to blast off to the quantum field theory of gravity?

Guess what, dear reader, you are ready to quantize gravity! You simply follow the instruction manual I gave you back in chapter III.1. For your reading convenience, I repeat here the four steps you have to go through in the path integral formulation: (1) Identify the relevant fields. (2) Find the action governing these fields. (3) Disturb the fields. (4) Evaluate the path integral.

Step (1), check. The fields are the metric $g_{\mu\nu}(x)$.

Step (2), check. Einstein gave us the action.

Step (3), check. Just like the source of the electromagnetic field is the distribution of charge and current in spacetime, the source of the gravitational field is the distribution of energy* and momentum in spacetime.

Easy peasy thus far! (This is of course a bit deceiving; Newton, Einstein, Dirac, Schwinger, Feynman, and many others did all the work for us.)

It is step (4) that brought us to grief. As I explained in chapter IV.2, we can evaluate path integrals only if they are quadratic or Gaussian. But, it is meaningless to even talk about whether Einstein's action, $S = \int d^4x \ \sqrt{g}R/G$, is quadratic or not, unless we expand in the field $h_{\mu\nu}$. When expanded, we have $S = \int d^4x \ \frac{1}{G} \ \partial\partial(hh + hhh + hhhh + hhhhh + \cdots)$, as shown earlier. I mentioned before that Yang-Mills theory contains terms cubic and quartic in the gauge field A_μ, in contrast to the much easier Maxwell theory, which is merely quadratic in A_μ. This allows us to calibrate how difficult quantum gravity would be: S contains not only terms cubic and quartic in h, but also quintic, sextic, ad infinitum.[1]

---

*Remember that in Einstein's world, mass is equivalent to energy.

Figure 1. The cubic, quartic, quintic, ... vertices in Feynman diagrams controlling the self interaction of gravitons. The graviton is represented by a straight line.

	terms in the action
Maxwell	quadratic
Yang-Mills	quadratic, cubic, quartic
Einstein	quadratic, cubic, quartic, quintic, sextic, ..., ad infinitum

One simple way of understanding this is that the electromagnetic field couples to electric charge, but the photon itself does not carry any electric charge. In contrast, the gravitational field couples to energy and momentum. But everything carries energy and momentum, including the graviton itself. Thus, the graviton couples to itself, and once that starts, there is no end to the self couplings. So, even in a quantum universe devoid of anything else but the gravitational field, things could get out of hand (figure 1).

Here is another way to calibrate the difficulty. I mentioned in chapter V.5 that Einstein gravity treated classically already produced a truckload of mind boggling phenomena: warped spacetime, black holes, time standing still, expanding universe, and so on. But classical physics "merely" corresponds to finding and examining the extrema of the Einstein-Hilbert action S. But for quantum gravity, we have to take this amazingly rich action, put it in the exponential, and then integrate over all possible metrics, that is, all possible spacetimes. When faced with having to evaluate $\int dg \, e^{iS}$, theorists tremble with trepidation. In truth, people are still arguing about what these symbols mean and how best to define the integral.

Perturbative quantum gravity

The astute reader might ask, is gravity not much more feeble than the electromagnetic force (as was explained in the prologue)? Hence, can't we do perturbation theory by simply following what Schwinger and Feynman did for quantum electrodynamics? Yes indeed.[2,3] We could calculate, for instance,[4] quantum corrections to Newton's law. Classically, the gravitational potential between two massive objects separated by distance r is given by $V(r) = -Gm_1 m_2 / r$. But in the quantum world, the uncertainty principle already tells us that r is fluctuating, so right off the bat there has to be a correction to Newton's law.

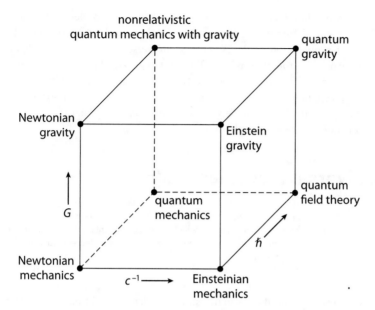

Figure 2. The cube of physics. Notice that the square of physics in the prologue forms the bottom face of the cube.
Redrawn from A. Zee, *Einstein Gravity in a Nutshell*, Princeton University Press, 2013.

Recall chapters III.2 and III.3. Newton's law comes from the exchange of a single graviton between the two masses; this is accounted for by the term $\partial\partial hh$ quadratic in h. You could even see how the cubic term $\partial\partial hhh$ may come into play. One of the masses emits two gravitons, the other mass emits one graviton, and the three gravitons could then interact via the cubic term.

But these perturbative calculations of minuscule corrections do not touch the heart of the challenge posed by quantum gravity, thus leaving most theoretical physicists dissatisfied. Our experience with the strong interaction showed us that there are plenty of fascinating physical phenomena beyond the reach of perturbative treatments.

From the square of physics to the cube of physics

We started our quest with a map in the prologue showing two "directions" leaving our "home village," one based on the speed of light c, the other based on quantum uncertainty $\hbar$. But with gravity in this chapter, we introduce a third direction based on Newton's constant G. The square of physics in the prologue has to be extended to the cube of physics; the original square now forms the bottom face of the cube (figure 2).

From our home village we turn on G to climb up to Newtonian gravity. Turning on c^{-1} from 0 to a finite value (in other words, changing the infinite speed of propagation of the Newtonian universe into the speed of light c), we go from Newtonian gravity to Einsteinian gravity. From there, we would like to turn on $\hbar$ to arrive at the "Holy Grail" of quantum gravity. Quantum gravity, seemingly so near yet so far, sits at the opposite and far corner of our starting point in physics, namely, Newtonian mechanics. That is the challenge of our time for theoretical physicists!

Universal Planckian units

The ... constants ... offer the possibility of establishing a system of units for length, mass, time ... which are independent of specific bodies or materials and which necessarily maintain their meaning for all time and for all civilizations, even those which are extraterrestrial and non-human.
Max Planck

To do physics, we need units to measure mass, length, and time with. One way to see if the units we use make universal sense is to imagine communicating with an extraterrestrial being about physics. "What? You measure mass with some lump of alloy[5] kept in a place called Paris[6]?" The need for universal units far transcends the petty, even pathetic, human level argument about metric versus imperial.

As soon as Max Planck introduced his constant $\hbar$, he was intelligent enough to see that, with three fundamental constants, G, c, and $\hbar$, we could construct three fundamental units meaningful throughout the universe. Shortly after launching quantum physics, Planck wrote a celebrated paper establishing what fundamental physicists now call Planckian, universal, or natural units, defined by the Planck mass $M_P = \sqrt{\frac{\hbar c}{G}}$, the Planck length $l_P = \frac{\hbar}{M_P c} = \sqrt{\frac{\hbar G}{c^3}}$, and the Planck time $t_P = \frac{l_P}{c} = \sqrt{\frac{\hbar G}{c^5}}$. With three fundamental constants, you can form three combinations having dimension of mass, length, and time respectively, which could be used as units understood throughout the universe wherever they have reached a certain level of theoretical physics. For example, if you want to tell an extraterrestrial the size of our home planet, you could give it as a multiple of the Planck length. (As you can see, once the Planck mass is defined, the Planck length is just the de Broglie wavelength corresponding to that mass, $l_P = \frac{\hbar}{M_P c}$, and the Planck time is the time it takes light to traverse the Planck length $t_P = \frac{l_P}{c}$.)

By the way, you do not need to know what these three Planckian combinations are, just that they exist, but do note that they all go away in the pre-Planckian classical world with $\hbar = 0$.

A whiff of why quantum gravity is so difficult

I can now give you some hints of the extreme difficulty quantum gravity poses.[7]

The Planck mass measures the natural mass or energy scale at which quantum gravity kicks in, and it works out to be about 10^{19} times the proton mass m_p. That humongous number 10^{19} is responsible for the Mother of All Headaches plaguing fundamental physics today. (The much publicized Large Hadron Collider operates at an energy equivalent to order 10^4 m_p, almost pitiful compared to what quantum gravity theorists traffics in.) Thus, we have no hope of ever directly confronting quantum gravity experimentally.[8]

Note that in units with $\hbar$ and c set to 1, the Planck mass is given by $M_P = \frac{1}{\sqrt{G}}$. That M_P is so gigantic compared to the masses of known particles can be traced back to the extreme feebleness of gravity: G tiny, so M_P enormous.

As the Planck mass is huge, the Planck length and time are teeny. If you insist on knowing how these length and time units, understood throughout the universe, are related to artificial and arbitrary human-made units, t_P comes out to be $\simeq 5.4 \times 10^{-44}$ sec and $l_P \simeq 1.6 \times 10^{-33}$ centimeter. Almost unfathomably remote from human experiences!

The distance scale probed by a spinning electron in a magnetic field, which gives the fabulous agreement between quantum field theory and experiment that I started this book with, and the much smaller scales probed by the strong and weak interactions, while incredibly short compared to human distance scales, are still humongous compared to the Planck distance scales.*

The extreme smallness of the Planck length and time shows us how difficult quantum gravity truly is: It is almost incomprehensibly remote from any conceivable experiment! Over a distance l_P during a time t_P, spacetime is expected to fluctuate wildly out of control and takes on all sorts of nasty topologies, with holes all over the place, so to speak.[9] The traditional notions about space and time would no longer make any sense. Compared to these savage fluctuations, the warping of spacetime envisioned by Einstein would be like a gentle breeze on the "breath of God." We are experiencing a gently curving spacetime smoothed out over these fluctuations.

So, it is not merely that we have a difficult path integral to evaluate, but we have almost no idea what that integral even means!

---

*Back in chapter V.3, we talk about how remote the time scales characteristic of the weak interaction and of the strong interaction, 10^{-10} sec and 10^{-23} sec, respectively, are from everyday experiences. Compare them with the Planck time characteristic of quantum gravity. In the time it takes the strong interaction to act (and the phrase "lightning fast" fails utterly to describe it), 10^{21} Planck epochs have elapsed!

Much of contemporary work on quantum gravity is focused on playing with Planck sized black holes. I would like to clarify a common confusion. There has been considerable press coverage of the likely fact that at the center of every galaxy lives a black hole, including our home galaxy the Milky Way. It is important to distinguish these astronomical black holes from the black holes studied by quantum gravity theorists. Black holes occupy the extreme opposite ends of the mass scale, from the Planck mass to tens or hundreds of solar masses. The astronomical black holes have little, if anything, to do with quantum gravity as such. Even if you're as close as a few light years to one of these, the deviation from flat spacetime would still be rather small.

Colliding gravitons

So, as you could imagine, several schools of thought about quantum gravity suggest themselves.

One school, championed by Weinberg and others, asserts that the gravitational field should be treated like any other field, and the graviton like any other particle cruising through spacetime. Just to indicate to the reader how heated theoretical physicists could get, I was shocked by one distinguished quantum gravity theorist who told me that he would not allow Weinberg's textbook on gravity into his office. Yikes, agkkah mukkah!

Quantum gravity exhibits many distress symptoms.[10] Here I mention one such within the perturbative framework. Imagine colliding two gravitons with energy E together. (Theoretical physicists like to perform so-called thought experiments, as we have already seen in chapter I.3. Never mind that we have no way to carry out the actual experiment.) The probability amplitude $\mathcal{M}$ for this process is expected to be $\sim G$ for the simple reason that if G were equal to zero, there will be no gravity and no scattering.

Now calculate to order G^2. We obtain, by definition, $\mathcal{M} \sim G + G^2 X = G(1 + GX)$ for some X, unknown to us, well, because we didn't calculate it, duh. But still, we could figure out what it must be. Recall that in relativistic units, mass and energy are equivalent, and $G = \frac{1}{M_P^2}$. For the expression $(1 + GX)$ to even make sense, $GX = \frac{X}{M_P^2}$ must be dimensionless, that is, a pure number, and hence the unknown X must be some mass or energy squared. But since the only energy around is E, we conclude that $\mathcal{M}$ must have the form

$$\mathcal{M} \sim G\left(1 + a\frac{E^2}{M_P^2} + \cdots\right)$$

with a some numerical factor.

As we crank up the energy E up to M_P, the correction term $a\frac{E^2}{M_P^2}$ becomes comparable to the leading term 1, and the scattering amplitude threatens to

Already, in non-relativistic QM, photon (electromagnetic field) treated as a field but not the electron $\Rightarrow$ Jordan, Heisenberg, Dirac, ...

All particles are excitations in some field

(graviton just a particle like any other, an excitation in the gravitational field (e.g. S. Weinberg's textbook on gravity) but somehow also responsible for the spacetime arena in which all fields work and play —

Is it somehow different?

Quantum gravity?

Cosmological constant?
(With gravity, cube of physics)

Figure 3. The graviton is different from the other particles.

blow up.[11] This behavior offers another indication that quantum gravity kicks in only at the Planck energy.

An avant garde theater

In the perturbative framework, the gravitational field could be described in terms of gravitons. But wait! The graviton is also responsible for constructing the curved spacetime arena in which the other fields work and play. See figure 3.

Imagine yourself stuck in some avant garde theater. It soon becomes apparent that there's no stage. Then you eventually realize that one of the actors is the stage. The other actors are literally walking around on him. The graviton is special: it is the stage as well as the actor.

Should gravity even be quantized?

A dissenting attitude, perhaps articulated most forcefully by Freeman Dyson, is that gravity should not be quantized at all. I will let Dyson speak for himself.

> If you try to detect individual gravitons by observing electrons kicked out of a metal surface by incident gravitational waves, you find that you have to wait longer than the age of the universe before you are likely to see a graviton. If individual gravitons cannot be observed in any conceivable experiment, then they have no physical reality and we might as well consider them non-existent. ... Einstein ... was happy to get rid of the ether, and I feel the same way about gravitons. According to my hypothesis, the gravitational field described by Einstein's theory of general relativity is a purely classical field without any quantum behavior.[12]

Note that Dyson is dismissing quantum gravity because of its weakness, but gravity is weak precisely because Mp is so huge. So it is basically the same symptom mentioned above, that the sickness does not show up until the Planck energy becomes relevant. If you were offered a chance to "buy" quantum gravity, you could always say, "Thanks but no thanks, I am already quite happy with Einstein's low energy calculation of, say, the bending of light."

On Fridays, Saturdays, and Sundays I subscribe to the Dyson view, but during the work week I am less sure. At least superficially, it appears bizarre to have the world partly quantum and partly classical. I believe that it is not consistent.[13]

Quantum physics holds up amazingly well over many orders of magnitude, from the atomic energy scale of a few electron volts (and even lower in some other areas of physics) all the way to the energy scale probed by high energy physics, of order $10^4 m_p$. But does it hold unmodified all the way to the Planck energy of $10^{19} m_p$? I'm not so sure. It troubles me that the problems in this difficult marriage between the quantum and gravity are always blamed on gravity, but not on the quantum. Perhaps quantum physics as we know it would be modified at some energy scale far above what we have explored, but still far below the Planck mass.

A more radical speculation may be that Einstein's theory holds only if the metric does not fluctuate too far away from the Minkowski metric. In other words, large h is not allowed. In particular, what would it mean for h to be so large that the metric goes negative? Indeed, as you have learned, at the horizon of a black hole, two components of the metric flip sign. More mysteriously, if the time time component g_{tt} flips sign and thus undoes Minkowski's magic chalk, spacetime could morph into plain 4-dimensional space, a world without time. Or, consider the other extreme, in which g_{rr}, say, flips without g_{tt} flipping, we could end up locally with two time coordinates. What does a two timing universe even mean? The possibilities for wild, and somewhat irresponsible, speculations are almost endless. I could join the mob, but I won't.

Even though I make my living partly through quantum field theory, I would be loath to see the standard approach to quantum field theory work for quantum gravity. A true theorist always hopes for something dramatically (and conceptually) new, rather than the same old same old.

Refusal to dance

There are actually two separate issues with the problem of gravity in theoretical physics. As some readers might know, Einstein spent the last few decades of his life fruitlessly trying to unify electromagnetism and gravity. (He chose to ignore the more recently discovered strong and weak interactions, regarding them to be "epiphenomena" that would resolve themselves once electromagnetism and gravity are unified.) You read in the preceding chapters that the

Figure 4. Gravity resists being quantized and refuses to join in the dance of the other three interactions.
Reprinted with permission from A. Zee, *Einstein's Universe: Gravity at Work and Play*, Oxford University Press, 2001.

three non-gravitational interactions, the strong, the weak, and the electromagnetic, have now been harmoniously joined together in a single grand unified interaction.

So, not only does gravity resist being quantized, it also stubbornly refuses to be unified with the other three interactions. Referring back to the four sisters in the Victorian novel I alluded to in chapter V.3, we could envisage the oldest sister, oldest in the sense of being the first to be introduced to us in this story of physics, sitting by herself and spurning her three sisters in their dance.

The alluring promise of string theory, which we will describe briefly in the next section, is that it could solve both problems at once. That remains to be seen, as many readers have no doubt heard, but it does seem to many physicists that the two issues are likely related.

String theory and the pants diagram

Surely anyone reading this book is aware that over the past several decades, a theory of quantum gravity in the form of string theory has been in the limelight. Yes, you read that correctly, past several decades, not past several

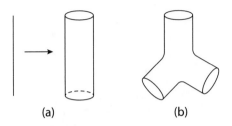

(a)   (b)

Figure 5. The graviton, treated as a point particle in quantum field theory, appears as a closed loop of string in string theory. (a) A graviton propagating in spacetime, described by a line in quantum field theory, is revealed to be a tube in string theory. (b) The inverted capital Y in figure 1 becomes a pair of pants.

years; the struggle has gone on for a long, long time. Evidently, this is neither the place nor the time to talk about string theory in detail. We also do not know what the future holds for string theory.

I would like to mention, however, one attractive feature of string theory. Look at the cubic interaction of gravitons shown in figure 1: two gravitons merging into one graviton as in an inverted uppercase Y. Topologically, this letter is singular at the bifurcation point. In string theory, the graviton appears as a closed loop of string. As a point particle moves through spacetime, it sweeps out a line. In contrast, as a loop of string moves through spacetime, it sweeps out a tube. See figure 5(a). Thus, the upside down Y is replaced by what is known as the pants diagram. See figure 5(b). The surface is smooth, without any singular points. Consequently, graviton scattering is better behaved in string theory than in quantum field theory.

The optimism of two giants and the execution of a young physicist

In 1929, after Heisenberg and Pauli quantized* the electromagnetic field, they opined rather airily that "the quantization of the general relativity field ... may be carried out without any new difficulties by means of a formalism fully analogous to that applied here." Ha! Even quantum electrodynamics was not so easy, let alone quantum gravity.[14] Moral of the story: Do not take the pronouncements of the giants of physics, whether true giants or those anointed by the popular media, too seriously.

Remarkably, not long after, a brilliant 28 year old Russian physicist Matvei Bronstein outlined in an obscure paper the essence of what we now understood

---

*As the reader of this book has learned, this early attempt at quantum electrodynamics was afflicted by infinities and various inconsistencies, difficulties that were not cleared up until the late 1940s by the generation consisting of Schwinger, Feynman, Tomonaga, and others, as described in chapter IV.3.

about quantum gravity. Unfortunately, he was purged and executed[15] at the age of 31 in 1938.

Cosmological constant and the vacuum in quantum field theory

One tremendous advantage you now have over the typical reader of the typical popular physics book is that you know what an action is. Furthermore, you understand that physicists can add and subtract terms to and from the action, within certain guidelines (such as symmetry considerations and the behavior of the theory at high energy or short distances).

For Einstein gravity, the action has to be invariant under coordinate transformations. But I've already mentioned that the integral $\int d^4x \sqrt{g}$ measures the volume of spacetime. The volume, being an intrinsically geometric concept, cannot change when we change coordinates, just as the area of Greenland cannot change as we go from Mercator to some other projection. The area is a property of Greenland, not of the map. Thus, we can add to the Einstein-Hilbert action the term $\Lambda \int d^4x \sqrt{g}$ with an overall constant Λ.

Well, the constant Λ represents the fabled cosmological constant that you may or may not have heard of. Einstein, having been raised on the 19th century conception of the static universe, was alarmed that his equation of motion for gravity led to an expanding universe. In response, he added the cosmological constant Λ to his equation and adjusted its value to precisely cancel the expansion.

When the astronomers showed that the universe was indeed expanding, he dropped the cosmological constant and was seen as adjusting his theory to conform to astronomical observations.

Einstein thus missed another world-shaking prediction that he could have made, but in some sense, Einstein has the last laugh. The observed dark energy, which is causing the expansion of the universe to accelerate,[16] may well represent the cosmological constant, at least that's what many theorists are betting on.

Incidentally, contrary to what is often reported in the popular media, Einstein never said the introduction of the cosmological constant was his greatest blunder.[17]

The real difficulty with Einstein's attempt to stop the universe from expanding is that adjusting Λ to counter the expansion is akin to balancing a pencil on its point. It can be done, but it is unstable. Einstein could do this because he was working within classical physics, but quantum fluctuations would make this balancing act untenable.

Quantum field theorists can easily estimate the value of the cosmological constant produced by the fluctuations of all the other fields in the universe. The so-called cosmological constant paradox[18] is that this comes out too large by

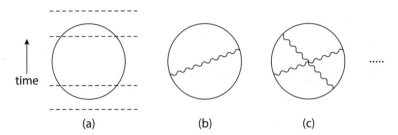

Figure 6. The fluctuating vacuum in quantum field theory.

an almost unthinkable factor of about 10^{120}. (Again, this is large, because M_P is so large compared to other relevant energy scales. This estimate is regarded as trustworthy, because it does not involve quantum gravity as such, but rather depends only on the fluctuation of the other quantum fields in the universe.)

From nothingness to nothingness

Now that you have almost made it through this book, you could refer back to the prologue where I talked about the state of nothingness known as the quantum field theoretic vacuum. The Feynman diagrams depicting the fluctuating vacuum are shown in figure 6. The circle corresponds to an electron and a positron popping out of nothingness, and then annihilating into nothingness.

I will now help you to read this "mystic" circle, going on everywhere in the universe at every instant. I have superposed on it four dotted lines. At the time indicated by the bottom dotted line, there was nothing. As time passes, moving ever upward, we come to the next dotted line. At that instant, we see two solid lines, namely, two arcs on the circle, corresponding to the electron and the positron, respectively. They are seen to be moving apart. Later, at the instant indicated by the next dotted line, they are coming toward each other, ready for an embrace. At the time indicated by the topmost dotted line, there is again nothing. From nothingness to nothingness.

The next Feynman diagram, figure 6(b), a circle with one wavy line moving across, corresponds to the process with the electron and the positron exchanging a photon before they annihilate. Slicing time with dotted lines as I did in figure 6(a), you see that for some duration in time, there was a photon in addition to the electron and the positron. The photon was emitted and then absorbed.

Next, we draw in figure 6(c) a circle with two wavy lines, corresponding to the electron and the positron exchanging two photons before they annihilate. And so on ad infinitum. You should congratulate yourself for coming so far in your understanding of Feynman diagrams. Ever more complicated diagrams depicting the constant morphosis of nothingness into nothingness!

The point is that, while we may not be aware of this roiling and boiling going on in the quantum nothingness, the graviton, which couples to any bit of energy in the universe, is perfectly aware and is affected by it, thus generating the Λ term in the Einstein-Hilbert action.

One of the difficulties facing quantum field theorists at the moment is to explain why the cosmological constant, if indeed it corresponds to dark energy, is so tiny. But not only is it tiny, the amount of dark energy per unit volume is comparable to the amount of energy per unit volume accounted for in matter, whether dark or luminous. We have reasonable estimates of how much dark matter is present, and we have fairly reliable estimates of how much luminous matter (that is, stuff contained in stars and other electromagnetic wave emitting objects) there is.

The cosmic coincidence puzzle

Now comes yet another mystery about the physical universe! As the universe expands, the density of matter steadily decreases: The amount of matter is (to first approximation) conserved and stays constant, while the volume of the universe expands. The dark energy per unit volume, in sharp contrast, stays constant. This assertion is based on observations, since astronomers can look back at the universe in earlier epochs. It also follows if dark energy indeed corresponds to Einstein's cosmological constant, which, as I had mentioned, is just a constant in the action.

So, this is the mystery[19] of the cosmic coincidence problem! If, as the universe evolves through the eons, the density of dark energy stays constant while the density of dark matter* decreases steadily, why are the two comparable just at this particular moment in the universe's evolution? Surely, the universe does not care that humans are around at this instant in its billions and billions of years of expansion to wonder about this mystery.† Or does it?

I might add that, logically, these two mysteries, why the dark energy is so tiny on the Planck scale and why it happens to be comparable to dark matter just during this tiny flash of the anthropocene era, may or may not have anything to do with our difficulty quantizing gravity.

The ongoing cosmic struggle

A cosmic struggle is going on, right now as we speak, between dark energy and dark matter. (Luminous matter is being neglected simply to enable a smoother narrative.) For your amusement, I show in figure 7 the fate of the universe according to the relative proportion of dark energy and dark matter

---

*Luminous matter is negligible compared to dark matter. The relative amount is not yet, but should be eventually, understood. But that is not the issue being discussed here.

†Just to elaborate a bit on the coincidence puzzle: A billion years ago, there was unlikely to have been any life forms losing sleep over this mystery. A billion years from now, who knows? But by that time, the density of dark matter will be much less than what it is now.

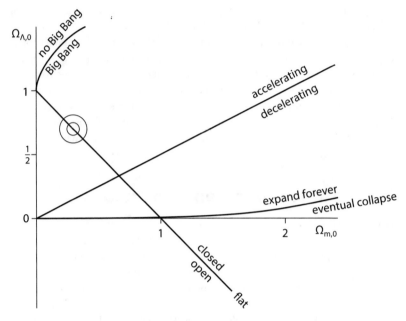

Figure 7. The fate of the universe: The variables $\Omega_{m,0}$ and $\Omega_{\Lambda,0}$ on the two axes are dimensionless measures of how much dark matter and how much dark energy, respectively, the universe contains at the present epoch. How the universe evolves depends on these two variables as indicated. According to observation, our universe is located inside the circle, the size of which indicates uncertainty in the data. We see that it is flat and accelerating. As another example, a universe with $\Omega_{m,0} \simeq 2$ and $\Omega_{\Lambda,0} \simeq \frac{1}{2}$ is closed and will expand forever but at an ever slower rate.
Redrawn from A. Zee, *Einstein Gravity in a Nutshell*, Princeton University Press, 2013.

you put in. I am following the default assumption in contemporary cosmology that the dark energy is simply Einstein's cosmological constant Λ (as is indicated by the notation on the diagram.)

Notes

[1]This strongly suggests that this is not the right way to approach quantum gravity.

[2]To be sure, there are some technical difficulties that were not surmounted until the 1970s. Mainly, people did not know how to deal with the gauge freedom.

[3]Some readers might be puzzled by where G appears in the action. The resolution is that we could define a field $k_{\mu\nu}$ by $h_{\mu\nu} = \sqrt{G}k_{\mu\nu}$. Then $S = \int d^4x \ \partial\partial(kk + \sqrt{G}kkk + Gkkkk +$

$\sqrt{G^3}kkkkk + \cdots)$, giving an expansion in $\sqrt{G}$. This is known as a field redefinition in quantum field theory. See *QFT Nut*, page 68.

[4]N. E. Bjerrum-Bohr, J. F. Donoghue, and B. R. Holstein, arXiv/0212072 and references therein.

[5]I am speaking intentionally a bit behind the times to the emphasize the absurdity of human made units. Away with the king's foot! The platinum-iridium cylinder kept under two bell

jars in Paris was finally abolished on May 20, 2019. The Système International of units used by scientists and engineers is now based on the spectroscopy of the cesium atom. The new system, based on c, $\hbar$, and a particular atom, almost reaches Planck's natural units but not quite. The reason is that G, ironically the first fundamental constant known to physics, could only be determined to about one part in 10^5 at present, far below the precision to which c and $\hbar$ are known. See W. Ketterle and A. Jamison, *Physics Today* 73, page 33, 2020. By the way, can you figure out why G is so poorly known?

[6]The new system meant that Paris lost its position of eminence. Any laboratory with the know-how to isolate some cesium atoms could now establish the standards independently.

[7]For a collection of heuristic thoughts about quantum gravity, see *GNut*, chapter X.8.

[8]Hence physics by pure thought. See *GNut*, chapter X.8, for example.

[9]John Wheeler was fond of talking about spacetime foam. One day, when I was an undergraduate, he gave me a dollar, sent me to buy a kitchen sponge, and had it photographed. There in his book, spacetime foam!

[10]See, for example, *GNut*, chapter X.8, and *FbN*, chapter IV.4.

[11]Indeed, this is analogous to the series $1 + \varepsilon + \varepsilon^2 + \cdots = (1 - \varepsilon)^{-1}$ in chapter IV.3 blowing up as ε approaches 1 from below, a mathematical fact I already mentioned in this chapter in connection with the Schwarzschild radius.

[12]F. Dyson, *New York Review of Books,* May 13, 2004.

[13]See *GNut.*

[14]Statements like that made prematurely by Heisenberg and Pauli are known cynically in academia as "staking out territory."

[15]See *FbN*, pages 155–156.

[16]*GNut*, chapter VIII.2, for example.

[17]That conversational remark was attributed to him by George Gamow, who may or may not be reliable.

[18]*GNut*, chapter X.7, for example.

[19]See, for example, *GNut*, page 751.

Recap of part V

Perhaps the most stunning consequence of combining special relativity and quantum mechanics is the unexpected appearance of antimatter. A real particle cannot travel between two spacelike separated points A and B in spacetime: Einstein forbids particles from traveling faster than the speed of light. But a quantum particle can, at least for a short while.

Suppose A occurs before B. Consider a quantum particle carrying electric charge +1 traveling from A to B. At A, a unit of electric charge has been carried away (that is, lost), while at B, a unit of electric charges is gained. But with the fall of simultaneity, another observer could see B occurring before A, and hence a quantum particle going from B to A. By charge conservation, the unit of electric charge +1 gained at B must be balanced by a negative unit of charge carried away by the quantum particle going from B to A. Thus, this observer sees a quantum particle carrying electric charge −1.

Hence, for every particle that carries a positive unit of charge, there must be a particle that carries a negative unit of charge, with exactly the same mass. Voilà, antimatter!

I understand that after this quick argument showing the existence of antimatter, the rest of this rather massive part V presents a long slog for the reader. Let me summarize.

The generation of Schwinger and Feynman succeeded in turning Maxwell's theory of electromagnetism into a quantum field theory. It was then the task of the following generation to come up with a quantum field theory of the strong and weak interactions. After many twists and turns, they triumphed and celebrated with a virtual victory parade.

Contrary to what Einstein thought, electromagnetism is not to be unified with gravity, but with the weak interaction, which Einstein could barely be bothered with. The resulting electroweak interaction may very well already be unified with the strong interaction to form a grand unified theory.

Sadly, gravity stubbornly resists being unified with the other three interactions. One of many mysteries: The graviton propagates in curved spacetime just like any of the other particles, but then it is also the one responsible for generating the curved spacetime.

All four of the fundamental interactions are now to a large (or at least some) extent understood using the language of quantum field theory. Of the three non-gravitational interactions, the weak interaction is by far the weird-est, presenting us with numerous puzzles yet to be resolved. In contrast, I think that most physicists would say that we now understand the electromagnetic interaction in considerable detail, and the strong interaction in broad outline. Of course, how they behave in specific situations has yet to be worked out in detail. But the same could be said of Newtonian mechanics. While the low energy aspects of the strong interaction cannot be calculated analytically, they are under control numerically using a lattice gauge approach.

Understand that much of what the public hears about black holes involves Einstein gravity as a classical field theory. The big problem is to reconcile Einstein gravity with quantum physics. For several decades, some people have been hopeful the string theory will finally provide us with a quantum theory of gravity, but the optimistic initial hope has faded somewhat, at least for the time being. It is perhaps worth emphasizing that whether or not string theory turns out to be the answer, it is known to reduce to quantum field theory. So the quantum field theory you have learned here will endure.

We humans finally realize that quantum fields are all around us, that quantum fields pervade the universe.

Quantum field theory is more intellectually complete than quantum mechanics ▬▬▬▬▬▬▬▬▬▬

Preview of part VI

In this last part of the book, I wrap up. But instead of telling you about one more attempt to master quantum gravity, or about the latest gee whiz razzmatazz, all reported breathlessly on the web but more likely than not of merely ephemeral interest, I would like to place quantum field theory in its proper place in the development of physics.

Physics is a vast subject. Circumstances, inclinations, and abilities determine which area a physicist specializes in. I am aware that some go into physics to make a living, but believe me, there are easier and softer ways. Still, many of us went into physics to understand the physical universe to the greatest extent that the human mind is capable of. Most are by nature inclined to stay away from academic subjects in which authority figures force feed the acolytes with received wisdom stated as facts. Rather, with a burning need to understand, they want to penetrate to the deepest rather than to be told a series of "Because I told you so, that's why" stories.

Yet I am here to tell you that in many areas of physics, important theoretical principles appear to simply rain down from the sky, with neither rhyme nor reason.

In this part, we will chat about the intellectual completeness of different areas of physics. Necessarily, some areas are more intellectually complete than others.

A question of identity

Distinguishable versus identical

Consider two billiard balls. They may look and feel perfectly identical, painted precisely the same color. But you know, just know, deep down in your heart, that they are not identical. The two balls are made of different chunks of wood. They may be indistinguishable in practice, but surely not in principle. If only you had a more precise scale, you could show that their masses are slightly different. Or, if you had a powerful enough microscope, you could see that their surface textures are different.

But when physicists venture into the quantum domain, they have to confront the issue of whether different hydrogen atoms are distinguishable or not, and whether two electrons are distinguishable or not. Even in principle, we have no way of telling two electrons apart. Electrons are all identical. In the quantum world, individual identity does not exist.

The issue at hand, that of identity, is not a matter of idle philosophical hyperventilation. In particular, in the statistical physics championed by Boltzmann and others toward the end of the 19th century, to calculate the thermodynamic properties of a gas, physicists have to count the number of possible states the gas could be in. This counting is embodied in the concept of entropy, much bandied about in the early 21st century and familiar, at least vaguely, to the intelligentsia. Meanwhile, experimentalists are able to measure various properties of gases at lower and lower temperatures. To obtain results in agreement with experiments, theorists[1] soon learned that they have to treat atoms in gases as indistinguishable, even in principle.

Interestingly, at high enough temperatures (say, room temperatures), the results of these calculations approach the well known results of classical physics, as they should and must. This agreement is well understood by now: at higher temperatures, the atoms are moving with larger momentum,

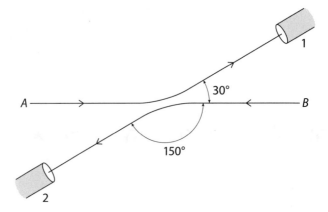

Figure 1. The source A shoots out a helium-4 nucleus toward the right, while the source B shoots out a helium-3 nucleus toward the left. The two helium nuclei scatter off each other, with the helium-4 nucleus headed to detector 1, and the helium-3 nucleus headed to detector 2.

corresponding to smaller de Broglie wavelength. The atoms behave less like quantum wave packets and more like billiard balls.

The claim that electrons are identical is of course ultimately based on experiment, like everything else in physics, and so is falsifiable. Some experimentalist measuring some property of the electron to be different from the accepted value would cast doubt on this amazing identity. Or far more likely, on his or her credibility.[2]

A series of five scattering experiments to explore distinguishability and identity

Indistinguishability has astounding consequences in the quantum domain.[3] Consider the experimental setup in figure 1. The source A shoots out a helium-4 nucleus toward the right, while the source B shoots out a helium-3 nucleus toward the left.

Let me mention that a helium-4 nucleus consists of two protons and two neutrons, while a helium-3 nucleus consists of two protons and one neutron. The only nuclear physics you need to know is that a helium-4 nucleus and a helium-3 nucleus are distinguishable: one contains four nucleons, while the other contains three nucleons. Hence their names.

Also, for simplicity, assume that the nuclei are shot out with the same momentum. The nuclei collide in the middle and are scattered into two detectors, as shown in figure 1. The detectors beep when they are hit by a nucleus; well, that's their job in life.

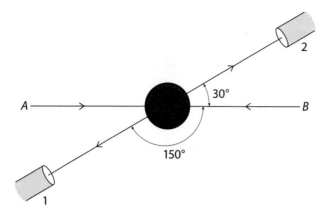

Figure 2. The same experimental setup as in figure 1 but with the two detectors interchanged. The black circle covering the collision region serves to emphasize that the details of how the scattering occurred is not relevant to the discussion here.

As a mere matter of convention, physicists define the scattering angle as the angle between the direction that the particle was initially moving in and the direction it moves off in after the collision. Thus, in the figure, a collision with the scattering angle equal to $30°$ is shown. Now, suppose that detector 1 is sensitive only to helium-4 nuclei, and detector 2 sensitive only to helium-3 nuclei. Denote the probability amplitude for scattering at $30°$ by $A(30°)$. The probability P for the detectors to beep is then the absolute square of the probability amplitude $A(30°)$: $P = |A(30°)|^2$.

It is worth emphasizing that we are not interested in nuclear physics here, that is, how to calculate the probability amplitudes for the helium nuclei to scatter at different angles. Thus, we will also assume that the collision occurs at such a low energy that the two nuclei are not broken up or excited in any way. Indeed, the strong interaction, being short ranged, is not even involved: The collision is entirely due to the electric repulsion between the two protons contained in the helium nucleus, be it helium-3 or helium-4. The repulsion keeps the two nuclei farther apart than the range of the strong force. The helium nuclei just bounce off each other without even "making contact." What we are interested in is the question of distinguishability and identity, not nuclear physics.

Good. Now let's do a second experiment. See figure 2. Interchange the two detectors. For a helium-4 nucleus to arrive at detector 1, it now has to scatter by $180° - 30° = 150°$, which occurs with probability amplitude of $A(150°)$. (Typically, $A(150°)$ would be significantly smaller than $A(30°)$, but that is not relevant here.)

Incidentally, the lines tracing the trajectories are merely to guide the reader's eyes. In fact, we may cover the collision region by a black circle, since the details of the collision are not relevant for the following discussion.

Very good. Time for a third experiment. We replace the two detectors with two cheaper and less discriminating detectors, that would beep regardless of whether a helium-4 nucleus or a helium-3 nucleus hits them. What is the probability P now for the detectors to beep? Care to try to figure it out?

Well, the detectors would beep when the helium-4 nucleus has scattered off by 30°, with probability $P = |A(30°)|^2$, or when the helium-4 nucleus has scattered off by 150°, with probability $P = |A(150°)|^2$, with the helium-3 nucleus going off in the opposite direction by momentum conservation. In either case, both detectors would beep, since they don't care whether a helium-4 nucleus or a helium-3 nucleus hits them. Hence the probability for the detectors to beep is just $P = |A(30°)|^2 + |A(150°)|^2$: we add the two probabilities. Adding probabilities is what casino gamblers do all day and all night.

That's indeed the correct answer.

Indistinguishability has astounding consequences: physicists were totally surprised

Ready to take the game to another level? We perform a fourth experiment by setting up both sources, A and B, to emit helium-4 nuclei. The detectors beep.

What are the detectors telling us? They have each detected a helium-4 nucleus. But, was it a helium-4 nucleus from source A that got to detector 1, or a helium-4 nucleus from source B that got to detector 1?

The point is that we cannot tell, not in practice and not in principle. All we can say is that two helium-4 nuclei, one from source A and one from source B, collided, with one scattered into detector 1 and the other scattered into detector 2.

In the quantum domain, we cannot tell whether a helium-4 nucleus from source A scattered by 30° into detector 1, or scattered by 150° into detector 2, (and in each case accompanied by another helium-4 nucleus from source B scattering into the opposite detector.)

Recall the rules of quantum physics given in chapter II.3. We are to add the probability amplitudes associated with all the possible histories. Thus, the probability amplitude A for the detectors beeping equals $A(30°) + A(150°)$, namely, the sum of the amplitudes for the two histories. The probability P of the two detectors beeping is then given by the absolute square of this amplitude: $P = |A(30°) + A(150°)|^2$.

This is emphatically not the same as $P = |A(30°)|^2 + |A(150°)|^2$. (Well, even second graders know that $(2 + 1)^2 = 9$ is not the same as $2^2 + 1^2 = 5$.)

Let us now perform yet another experiment. Replace the two sources, A and B, by sources that emit helium-3 nuclei. Go through the same discussion as before. Guess what, and this is an experimental fact, not some theoretical fantasy: The probability P of the two detectors beeping is now given by $P = |A(30°) - A(150°)|^2$. Surprise, surprise! (Even a second grader understands that $(2 - 1)^2 = 1$ is yet another number, equal to neither 9 nor 5.)

A trivial comment. Evidently, 30° is just for for definiteness in the exposition. It could of course be any angle θ you like.

The executive summary:

	probability detectors beep	possible experiment
distinguishable	$P = |A(\theta)|^2 + |A(180° - \theta)|^2$	helium-4 on helium-3
indistinguishable	$P = |A(\theta) + A(180° - \theta)|^2$	helium-4 on helium-4
indistinguishable	$P = |A(\theta) - A(180° - \theta)|^2$	helium-3 on helium-3

Voodoo quantum magic

Readers of this book have surely read a few expositions[4] on the weirdness of quantum physics. The scattering discussed here gives another example of the voodoo magic of the quantum world.

Picture the helium-4 nucleus as a black bag containing two protons and two neutrons. The helium-3 nucleus is a black bag containing one fewer neutron. But the neutrons are not "doing anything" in the scattering process. Let me emphasize this important point at the risk of repeating myself.

We learned in chapter III.2 that the strong interaction has a very short range compared to the electromagnetic interaction. The scattering of the two nuclei off each other is entirely due to the electric repulsion between the protons. Just let the momentum of the nuclei coming out of the two sources A and B be so low that they barely get near each other, as I have already said. To emphasize again, the question of identity and the scattering phenomenon I'm talking about have nothing to do with nuclear physics as such.

Now that we have set up the experiment, we just sit back and listen to the beeps. That's incontrovertible, beep beep beep: the rate at which the detectors beep tells us about how likely the nuclei are scattering off each other at the specified angle. The force between the nuclei is the same in each case, whether we are scattering helium-4 on helium 3, or 4 on 4, or 3 on 3.

Yet each nucleus somehow knows whether it is scattering off another nucleus identical to itself or not, even though it is too far away to discern whether the other guy contains 3 or 4 nucleons. Sort of spooky, no?

Interchange in the path integral formulation and in Feynman diagrams

In the nifty table summarizing our five scattering experiments, we see that for helium-4 on helium-4 we are to add the two possible probability amplitudes and then absolute square to obtain the probability $P = |A(\theta) + A(180° - \theta)|^2$, while, in contrast, for helium-3 on helium-3 we are to subtract the two possible probability amplitudes one from the other and then absolute square to obtain the probability $P = |A(\theta) - A(180° - \theta)|^2$.

Here we are going to examine this bizarre quantum happening in the Dirac-Feynman path integral formulation, and in the process show that adding and subtracting the probability amplitudes represent the only two possibilities.

We could even keep the discussion a bit more general. Let two indistinguishable particles start from two initial positions, labeled x_{1i} and x_{2i}. After some specified time, they end up at the final positions x_{1f} and x_{2f}. But we cannot know, even in principle, whether the particle that ended up at x_{1f} is the particle that left x_{1i} or the particle that left x_{2i}.

According to Dirac and Feynman, we are to find all the paths, or better in this context, histories, that satisfy the given initial and final positions. Sum up the probability amplitude for each of the possible paths, and call the resulting amplitude $\mathcal{M}(1, 2)$ with the corresponding probability $P = |\mathcal{M}(1, 2)|^2$. (Note that I am using a different symbol to distinguish it from the amplitude A in the preceding discussion.)

Now, let's ask what $\mathcal{M}(2, 1)$ might be. At first sight, this appears to be a really senseless question. Didn't we say that the particles are identical? How could $\mathcal{M}(2, 1)$ be different from $\mathcal{M}(1, 2)$?

Ah, here lies a quantum subtlety. True, the probabilities are equal, that is, $|\mathcal{M}(1, 2)|^2 = |\mathcal{M}(2, 1)|^2$. But since this is an equality between the absolute square of the two probability amplitudes but not the probability amplitudes themselves, we should allow for a more general possibility and write $\mathcal{M}(2, 1) = \xi \mathcal{M}(1, 2)$, with ξ some as yet unknown complex number. Interchanging 1 and 2 could multiply the amplitude by this unknown overall factor ξ. Let us will now try to determine ξ.

The requirement that the probability remain unchanged (namely, that $P = |\mathcal{M}(1, 2)|^2$ must equal $P = |\mathcal{M}(2, 1)|^2 = |\xi \mathcal{M}(1, 2)|^2 = |\xi|^2 |\mathcal{M}(1, 2)|^2$) implies that ξ has to be such that $|\xi|^2 = 1$.

Okay. Now let us interchange again. We have $\mathcal{M}(1, 2) = \xi \mathcal{M}(2, 1) = \xi\big(\xi \mathcal{M}(1, 2)\big) = \xi^2 \mathcal{M}(1, 2)$. Interchanging twice brings us back to what we started with, and thus we must have $\xi^2 = 1$.

In case you got confused, let us summarize. That the probability must be the same upon interchange tells us that $|\xi|^2 = 1$. That interchanging twice is effectively no interchange at all gives us the stronger condition that $\xi^2 = 1$.

Should we ask a mathematician to help us solve the equation $\xi^2 = 1$?

A profound discovery in mathematics showed that there are two (and only[5] two) solutions: $\xi = +1$ and $\xi = -1$.

Familiarity[6] may breed contempt, and you may not think that this is so profound. Yes, it is profound: $(+1)^2 = 1$ and $(-1)^2 = 1$.

In conclusion, there are indeed only two possibilities allowed, corresponding to the two entries $P = |A(\theta) + A(180° - \theta)|^2$ and $P = |A(\theta) - A(180° - \theta)|^2$ in our nifty table. In the former, $\xi = +1$ upon interchange, and in the latter, $\xi = -1$.

Profound is as profound is: quantum particles choose either +1 or −1

A profound statement in mathematics is often linked with a profound statement in physics.

The quantum world allows two possibilities.[7] Almost incredibly, Mother Nature loves this piece of mathematical truth and exploits both possibilities. All known quantum particles are either bosons or fermions.

Bosons, named in honor of the Indian physicist Satyendra Bose, choose the + sign and act like the helium-4 nucleus. Fermions, named in honor of the Italian-American physicist Enrico Fermi, choose the − sign and act like the helium-3 nucleus. The photon and the graviton are bosons, while the electron, the nucleons (namely, the proton or the neutron), and the quarks are all fermions. This terminology, of bosons and fermions, was coined by Paul Dirac in[8] 1945.

We now understand the difference between helium-4 and helium-3 nuclei. The helium-4 nucleus contains four nucleons (that is in fact what the "4" means), two protons and two neutrons. Interchanging two helium-4 nuclei amounts to interchanging 4 nucleons, and each time we interchange two nucleons, we get a − sign, and so we get the − sign 4 times: $(-1) \times (-1) \times (-1) \times (-1) = (-1)^4 = +1$. The helium-4 nucleus is a boson. In contrast, the helium-3 nucleus contains three nucleons, and interchanging two helium-3 nuclei gives us $(-1) \times (-1) \times (-1) = (-1)^3 = -1$. The helium-3 nucleus is a fermion.

In general, an assembly of an even number of fermions is a boson, of an odd number, a fermion. In fact, we can now explain the fact that nucleons are fermions by saying that they are made of three quarks and that quarks are fermions.

Scattering at 90°

For the most dramatic demonstration of what I am tempted to call "identity politics," set θ to 90°. Since $A(180° - 90°) = A(90°)$, we could simply read off the various probabilities that the detectors would beep from the "executive summary" table given earlier in this chapter. The probability for the scattering of helium-4 on helium-3 (the boson on fermion case) equals

$$P = |A(90°)|^2 + |A(90°)|^2 = 2|A(90°)|^2$$

The probability for the scattering of helium-4 on helium-4 (the boson on boson case equals)

$$P = |A(90°) + A(90°)|^2 = |2A(90°)|^2 = 4|A(90°)|^2$$

The probability for the scattering of helium-3 on helium-3 (the fermion on fermion case) equals

$$P = |A(90°) - A(90°)|^2 = |0|^2 = 0$$

The result, that the probability of scatterings in the three cases equals, respectively, 2, 4, and 0 times $|A(90°)|^2$, as summarized in the following table.

	relative probability of scattering at 90°	
distinguishable	2	boson on fermion
indistinguishable	4	boson on boson
indistinguishable	0	fermion on fermion

The relative probabilities, 2, 4, 0, reflect the bizarre quantum world, pure and simple.

To repeat the mathematics involved, $(1 + 1)^2 = 4$ versus $(1 - 1)^2 = 0$!

Two identical bosons are extra happy to scatter off each other at 90°, but two identical fermions absolutely refuse to do so. This extra eagerness for two bosons to interact with each other, amazingly enough, underlies the phenomenon of superconductivity.

Electron electron scattering: test your skill at drawing Feynman diagrams

That fundamental particles are offered this almost bizarre choice between + and − could also be illustrated by Feynman diagrams. Consider two electrons with momentum p_1 and p_2, respectively, scattering off each other by exchanging a photon, and emerging with momenta p_3 and p_4. Thus, in a self evident notation, $e(p_1) + e(p_2) \rightarrow e(p_3) + e(p_4)$ (this is the sort of thing physics students have to learn to write). By now, you know enough to draw the relevant Feynman diagram for the process. Please go ahead and draw it.

Ha ha, did I trick you? There is not one diagram, but two! Did you draw only one?

The point is that we have no way, not even in principle, of knowing whether the electron with momentum p_3 "was once" the initial electron with momentum p_1 or the initial electron with momentum p_2. Read that again, no way no how, not even in principle! Thus, the quantum world demands two diagrams (see figure 3). Do we add or subtract the two diagrams?

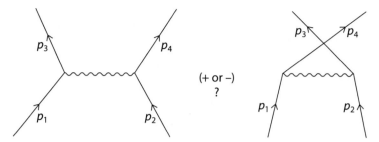

Figure 3. The two Feynman diagrams should not be added together, but subtracted from each other.

This is an experimental question. If the electron were a boson, we add, and if a fermion, we subtract. By the time electron electron scattering could be calculated, it had long been known from atomic physics (see chapter VI.2) that the electron is a fermion. Thus, we subtract and then absolute square[9] to obtain a probability to compare with experiment. If we had added, our result would not agree with experimental observation. Indeed, that is one way physicists painstakingly find out whether a specific particle chooses + or −.

The Creator offered each quantum particle a choice: go with Bose or go with Fermi.

Scattering cold neutrons off crystalline materials

Scattering low energy neutrons (known as cold neutrons) off various crystalline materials is a process once at the frontier of quantum physics but has by now developed into a relatively routine industrial technique.[10] Cold neutrons can penetrate deep (due to their lack of electric charge) inside the sample material and scatter off the atomic nuclei inside. The scattered neutrons are then detected on some kind of screen, as shown schematically in figure 4.

Since we could not tell, even in principle, which of the roughly 10^{23} or so atomic nuclei the neutron scattered from, the rules of quantum physics instruct us to add the amplitudes for scattering on each atomic nucleus. A huge sum, by the way, with order 10^{23} terms! We are then to take the absolute square to obtain the probability that a neutron would arrive at a particular spot on the screen. The measured intensity displays a pattern with peaks characteristic of a diffraction phenomenon, long familiar from 19th-century physics from studies of water waves and light waves (as explained in chapter I.5.) The separation between the peaks then tells us, via an undergraduate level calculation, about the spatial separation between the atomic nuclei in the sample.

Some readers might want to see the preceding written out mathematically. Others could safely skip this paragraph. Denote by A_j the probability amplitude that the jth nucleus scattered the incoming neutron. Here the index j

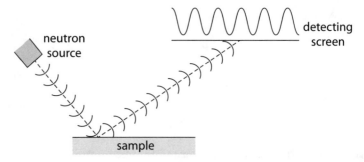

Figure 4. A beam of cold neutrons is scattered off a sample material. The detecting screen shows the characteristic interference pattern of the kind observed in everyday life with water waves.

labeling the nucleus runs over, just to be definite, say, $N = 10^{23}$ possible values. Since we could not tell, even in principle, which nucleus was the "culprit," we are instructed to add the probability amplitudes and then take the absolute square to obtain the probability that the neutron arrives at a specified location on the screen: $P = |A_1 + A_2 + \cdots + A_N|^2 = |A_1|^2 + |A_2|^2 + \cdots + |A_N|^2 +$ an enormous number of cross terms[11] like $A_k^* A_l + A_l^* A_k$. Since the distances from nucleus k and nucleus l to the specified location on the screen are slightly different, the two probability amplitudes A_k and A_l have slightly different phase angles. It is this huge number of cross terms that generates the observed interference pattern.

What if we could tell which nucleus scattered the neutron?

So far so good. I have just sketched for you the physics underlying neutron diffraction. Now the amazing issue of identity arises.[12]

To continue, we need a few more basic facts. The neutron, just like the electron, spins. Its spin could be either up or down, defined relative to some direction chosen by the experimentalist.

Some atomic nuclei do not spin, but many others do. For the sample, use a material whose atomic nuclei spin and arrange[13] for them all to be spinning down. At the source, where the neutrons are emitted, select only those neutrons with spin up.

Most of the time, a cold neutron scattering off an atomic nucleus would just lightly bounce off, keeping its spin up. (This is because we are scattering at very low energy.) But once in a while, in what is called a "spin flip process," the neutron would go off with its spin down, while the nucleus, originally with spin down, ends up with spin up.

In principle versus in practice: Mother Nature knows and that's enough for Her

Now that we have learned these relevant facts, we can start our experiment. Let us spend our research funding on expensive detectors that can tell whether the incoming neutron is spinning up or down. Set things up to detect only spin up neutrons. Then the diffraction pattern would show peaks and valleys as before. But now let's detect only spin down neutrons.

Surprise, surprise! The peaks and valleys disappear! We just get a dull, more or less constant "background" (that is, background to the spin preserving scattering). Why? Because now we can tell, in principle, which one of the 10^{23} atomic nuclei the neutron scattered off. How? By putting all these nuclei in a police lineup and checking which guy has his spin flipped up.

Never mind that we have absolutely no way of doing this in practice. Mother Nature could care less if humans could or could not find out which of the 10^{23} atomic nuclei flip the neutron spin. She knows and that's all She cares about!

The rules of quantum physics now instruct us not to add the probability amplitudes for the neutron scattering off each of the 10^{23} atomic nuclei and then absolute square. Instead, we are told to add the probability of the neutron scattering off each of the 10^{23} atomic nuclei (since it could be any one of them that flipped), and each one of these probabilities is just a positive number between 0 and 1. The sum is some dullsville of a number. There is no interference between the 10^{23} quantum waves, and so no diffraction pattern.

Again, for those readers who want to see the preceding written out mathematically, let us denote by S_j the probability amplitude that it is the jth nucleus which scattered the incoming neutron. The letter S reminds us that the spin has flipped. Since we could tell in principle, though assuredly not in practice, which nucleus was the "culprit," we are now instructed to add the probabilities, not the probability amplitudes, to obtain the probability that the neutron arrives at a specified location on the screen: $P = |A_1|^2 + |A_2|^2 + \cdots + |A_N|^2$. No cross terms and hence no interference and diffraction pattern.[14]

The quantum world is shockingly different from our familiar classical world. The distinction between "in principle" and "in practice" can be determined experimentally. A reader so inclined can no doubt pen tomes about the metaphysics of "in principle" versus "in practice" and earn a good shot at obtaining an endowed chair in philosophy at an elite university.

By the way, these types of experiments, to probe at the foundational rules of quantum physics, involve low energy and are relatively inexpensive. This is in contrast to building giant accelerators that dwarf entire cities in the hope of detecting some hitherto unknown massive particle that might shed some light on some as yet unknown physical law. Thus far, the rules of quantum physics as laid down in the mid 1920s, bizarre though they are, have continued to hold, unshaken.

Bose and indistinguishability

I close with a bit of history. One day, Einstein received a letter from India written by someone he had never heard of, describing an alternative derivation of Planck's result (which ushered in the quantum era) for a gas of photons. Einstein, being Einstein, saw the worth of Bose's derivation, translated it into German, and arranged to have it published. Fortunate for Bose! He himself likened his career to a meteor: after that flash of brilliance, he never did do much else in physics. He also wrote to the right person. (Incidentally, Bose was among the first in India to refuse his wife's dowry and to reject the caste system.)

German physicists wanted to have Bose explain his derivation at an important conference but confusedly invited a different Bose.[15] It has been jokingly said that to Europeans at that time, Indians named "Bose," just like bosons, were also indistinguishable.

Notes

[1]Already in the late 19th century, Josiah Willard Gibbs, the very first American theoretical physicist, recognized that to determine the entropy of a gas with N molecules correctly, a factor of $1/N!$, now known as the Gibbs factor, had to be introduced. This offered the first hint of indistinguishabilty in the brave new world of physics yet to come.

[2]But not before a flood of articles with sensational titles about the death of quantum field theory get published. Some decades ago, physicists complained about the media's neglecting advances in physics, but now almost every day we read that the foundations of physics have been shattered.

[3]I follow Feynman's discussion in his freshman *Lectures on Physics,* vol. 3, page 4-3. Since this issue of indistinguishability has been understood for around a hundred years, the reader should not suppose that Feynman was the first to consider the effect of identity on scattering experiments.

[4]I recommend T. Bub and J. Bub, *Totally Random: Why Nobody Understands Quantum Mechanics,* Princeton University Press, 2018.

[5]Those who love jargon might say that the group Z_2 has only one irreducible representation besides the identity representation. See, for example, *Group Nut.*

[6]Many elementary school kids understand this mathematical fact, because they know what a good deal is. If we sell an item for $1 above cost, our profit is $1 \times 1 = 1$ dollar. If we sell an item for $1 below cost, our profit is $1 \times (-1) = -1$ dollar, that is, we have a loss. Buying is the opposite of selling. So, if we buy an item for $1 below cost, we get a good deal. Effectively, our profit is $(-1) \times (-1) = 1$. The key is that (many if not all) little kids recognize a good deal when they see one. Humans are hardwired to reach for the good deals. (Just like the classical particles studied by Euler and Lagrange?)

[7]If space were 2-dimensional rather than 3-dimensional, other possibilities exist. But this subtlety lies far beyond the scope of this book. See, for example, *QFT Nut,* chapter VI.1.

[8]Surprisingly late, considering that these two words are now in the everyday vocabulary of physics.

[9]See, for example, *QFT Nut,* page 135.

[10]To detect unexpected stresses and strains in aerospace and automotive components, for example.

[11]Allow me to remind you that when you square the sum of two numbers you obtain $(w + z)^2 = w^2 + z^2 + 2wz$. The term $2wz$ is known as the cross term. Since quantum physics deals with complex numbers, this has to be slightly modified. Given a complex number $z = re^{i\theta}$, its conjugate is defined as $z^* = re^{-i\theta}$, that is, the phase angle is reversed. The absolute value squared of z is defined as $|z|^2 = z^*z = r^2$, in contrast to the ordinary square $z^2 = zz = r^2 e^{2i\theta}$. Note that information about the phase angle θ is lost in $|z|^2$ but preserved in z^2. Thus, for w and z two complex numbers, $|w + z|^2 = |w|^2 + |z|^2 + w^*z + z^*w$. The cross terms $w^*z + z^*w$ carry information about the phase angles of w and z, unlike $|w|^2 + |z|^2$.

[12]Again, I am following Feynman here.

[13]By using a magnetic field. Magnetic fields affect how elementary particles spin, as indicated by the fantastic agreement between quantum field theory and experiments on the magnetic moment of an electron that opens this book.

[14]Referring back to the table given in the "executive summary," you see that this case corresponds to the first line in the table, while the experiment involving the less expensive detectors that do not measure spin corresponds to the second line.

[15]Instead of S. Bose, they invited D. Bose, who was totally puzzled when asked about Bose statistics. Historians cite two possible explanations. Einstein had the paper published as authored by Doctor Bose, which is abbreviated in German as D. Bose. Alternatively, D. Bose had spent some time in Berlin and might have met Einstein, thus confusing the latter.

Exclusion, inclusion, and quantum statistics

Atomic spectroscopy

In the early days of quantum mechanics, Niels Bohr postulated that the orbits of electrons in an atom are quantized,[1] that is, only certain orbits are allowed. Electrons are said to leap from one orbit to another, either with higher or lower energy. (The reader might have heard of the terms "quantum jump" and "quantum leap," which have to some extent entered our vocabulary.)

If the state the electron jumps to has lower energy, a photon (namely, a quantum of light) carries away the excess energy. In sharp contrast, in classical physics, the electron is expected to radiate electromagnetic waves (typically light) continuously, thus losing energy and spiraling into the nucleus.

According to Max Planck, the energy of the photons that comprise the emitted light is directly proportional* to the frequency of the light. The light emitted by a heated vapor consisting of one type of atom or another could be separated into light of different frequencies by passing it through a prism, as Isaac Newton and others had studied and as seen in everyday life (such as light diffracting through water droplets to form a rainbow, or light reflecting off a thin film of motor oil on a puddle of water). Because only certain orbits are allowed, only photons corresponding to certain frequencies are emitted, and these are recorded on photographic plates as "discrete lines" instead of a continuous spectrum. In another type of experiments, white light is shone through a heated vapor, and various frequencies are absorbed by electrons jumping from one orbit to another with higher energy. Thus, the spectrum of light emitted and absorbed by vapors consisting of atoms of a specific kind tells us about the characteristic quantum structure of that type of atom. This

---

*Indeed, the proportionality constant is called "Planck's constant," as already mentioned in the prologue.

field of study is known as atomic spectroscopy and was essential for the early development of quantum mechanics.

Exclusion principle

As quantum mechanics developed, theorists became increasingly confident that they could calculate the energies of the allowed orbits. The energy spectrum of the hydrogen atom, with a single electron orbiting a proton, was almost completely understood. But surprise, surprise! Detailed spectroscopic studies of atoms with more than one electron showed that many expected frequencies were missing. This puzzling fact was eventually explained by Wolfgang Pauli and others.[2] According to what is now known as the Pauli exclusion principle, two electrons cannot occupy the same quantum state.

The bottomline is that some of the quantum states expected to exist in an atom are forbidden by the Pauli exclusion principle, and hence some of the lines, corresponding to quantum jumps between forbidden states, do not occur. This fact turned out to be in triumphant accord with experimental observation of atomic spectra.

The importance of the exclusion principle

> There is no one fact in the physical world which has a greater impact on the way things are than the Pauli Exclusion Principle.
> I. Duck and E.C.G. Sudarshan[3]

Indeed, the importance of the exclusion principle and its later generalization to the spin statistics rule (as explained in this chapter) can hardly be overstated. From the microscopic structure of atoms to the macroscopic structure of neutron stars, a dazzling wealth of physical phenomena would have been incomprehensible without this spin statistics rule. Much of condensed matter physics (for instance band structure, Fermi liquid theory, superfluidity, superconductivity,[4] and the quantum Hall effect) follows as consequences of this rule.

The vast majority of the human race has hardly lost sleep over the proposition that matter is stable, which a few theoretical physicists worked long and hard to prove. By now you probably have realized that a crucial ingredient of the proof has to be the Pauli exclusion principle, that each and every one of the electrons in a lump of matter insists on having its own private space.

Perhaps you once wrestled with the periodic table, as I did. Since the chemical affinity between elements is determined by the inclination of electrons to jump from one type of atom to another, much of the periodic table follows from the exclusion principle.

The plus and minus signs again

"Aha," you say, "we talked about bosons and fermions in the preceding chapter. The Pauli exclusion principle is just stating that electrons are fermions."

Excellent! Indeed, one easy way to understand the exclusion principle is to go back to the Feynman diagrams for the scattering of two electrons $e(p_1) + e(p_2) \to e(p_3) + e(p_4)$ shown in figure VI.1.3. The two diagrams, which differ only with p_3 and p_4 exchanged, are to be subtracted one from the other. But if $p_3 = p_4$, the two diagrams are the same, and we get a big fat zero. Yes, we learned it in school, $x - x = 0$! The two electrons refuse to scatter into the same state. (I am glossing over a technical detail regarding the spin of the two electrons.)

Another way of seeing this is as follows. Characterize a quantum state a by saying that if a particle is in that state, then the probability amplitude for the particle to be at the location $\vec{q}$ is given by* $\psi_a(\vec{q})$. Next, suppose we have two distinguishable particles, say, Jack in state a and Jill in state b. Then the amplitude for Jack to be at the location $\vec{q}_1$ and Jill to be at the location $\vec{q}_2$ is given by $\psi_a(\vec{q}_1)\psi_b(\vec{q}_2)$. End of story.

Now suppose the two particles are indistinguishable. We cannot tell, not even in principle, which particle is in state a and which is in state b.

If they are bosons, then we learned in chapter VI.1 that when we interchange $\vec{q}_1 \leftrightarrow \vec{q}_2$, the probability amplitude must remain the same. But under the interchange, $\psi_a(\vec{q}_1)\psi_b(\vec{q}_2) \leftrightarrow \psi_a(\vec{q}_2)\psi_b(\vec{q}_1)$. (Note carefully the subscripts 1, 2, a, b!) We conclude that the probability amplitude must equal

$$\mathcal{A}(\vec{q}_1, \vec{q}_2) = \psi_a(\vec{q}_1)\psi_b(\vec{q}_2) + \psi_a(\vec{q}_2)\psi_b(\vec{q}_1)$$

If they are fermions, then when we interchange $\vec{q}_1 \leftrightarrow \vec{q}_2$, the probability amplitude has to flip sign. So the probability amplitude equals

$$\mathcal{A}(\vec{q}_1, \vec{q}_2) = \psi_a(\vec{q}_1)\psi_b(\vec{q}_2) - \psi_a(\vec{q}_2)\psi_b(\vec{q}_1)$$

Verify for yourself that with the minus sign, we indeed have $\mathcal{A}(\vec{q}_1, \vec{q}_2) = -\mathcal{A}(\vec{q}_2, \vec{q}_1)$.

All fine and good. As I said, that's what we learned in the preceding chapter.

Now, see for yourself what happens to the probability amplitude for bosons and fermions if a and b are the same state, that is, if $a = b$. The probability amplitude for two bosons to be in the same state gets doubled, but the probability amplitude for two fermions to be in the same state becomes[5] zilch! Hence the Pauli exclusion principle. (See that? If $x = -x$, then $x = 0$. We learned that in school also.)

---

*Some well informed readers might know that $\psi_a(\vec{q})$ is also called the "wave function" in the Schrödinger formulation. See chapter II.3.

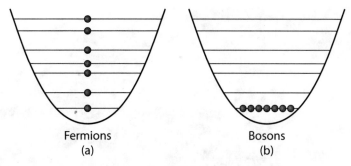

Figure 1. Quantum particles in a quantum potential well. The horizontal lines represent the energy levels in a quantum potential well, and the solid dots represent particles. We can think of this as an apartment building with many floors. Each fermion insists on occupying an entire floor and practices extreme social distancing. Meanwhile, the bosons all crowd into the ground floor apartment to party party party. But strangely, two fermions together act like a boson.

And thus, all those nagging puzzles in atomic spectroscopy[6] I mentioned at the beginning of this chapter got cleared away with the supposition that the electron is a fermion. In particular, the periodic table, and hence much of chemistry, could then be understood.

Real estate crisis in the quantum world

To summarize, quantum particles exhibit two extreme personality types. Fermions (such as electrons and quarks) obey the Pauli exclusion principle and cannot occupy the same quantum state. They insist on social distancing. Bosons (such as photons and gravitons), in contrast, love to be in the same quantum state; a boson tends to go where other bosons are. Not only do bosons disobey the Pauli exclusion principle, they flaunt it and do the exact opposite. See figure 1.

Fermions are like certain prickly theoretical physicists who cannot stand to share an office with others while visiting institutes of theoretical physics.[7] Bosons, in contrast, are gregarious party animals and love to hang out together. They squeeze themselves into a single office.

Incidentally, that photons are bosons accounts for how lasers work.[8] A stupendous number of photons all occupying the same quantum state, that is, all having the same frequency, marching in step like a Roman legion, is what gives a laser its awesome power of coherence.

Bosons and fermions are said to obey Bose-Einstein statistics[9] and Fermi-Dirac statistics, respectively. Incidentally, the word "statistics" here has only a tenuous connection with its everyday use. Students taking a course on statistical physics learn to count the number of quantum states, not to crunch data.

Figure 2. Wolfgang Pauli and Niels Bohr studying a spinning top.
From Niels Bohr Archive, photograph by Erik Gustafson, courtesy of AIP Emilio Segrè
Visual Archives, Margrethe Bohr Collection.

How does a particle decide to be a fermion or a boson?

But who gets to decide whether a particle is a fermion or a boson? Neither genetically nor environmentally determined like humans, for sure. To tell you what physicists discovered, I have to first talk about angular momentum and spin (which after all started this book).

A child playing with a top[10] (see figure 2) can make it spin faster or slower by imparting it with more or less angular momentum. A skilled tennis player can hit the ball with more or less topspin, perhaps even backspin. The very rock we inhabit is spinning with an enormous amount of angular momentum determined by historical contingency. And so on and so forth. Examples abound. Evidently, the amount of angular momentum is a continuous variable.

By now, you have heard so much about the weirdness of the quantum world that you would not be surprised to learn that, in that world, angular momentum is quantized in units of Planck's constant $\hbar$: The amount of angular momentum can only be equal to $j\hbar$ with j an integer. Recall from chapter II.3 that $\hbar = 10^{-27}$ g cm^3/sec^2 and so in everyday life, even a puny object of mass of order a gram, of size of order a centimeter, and spinning on a time scale of a second would have a humongous amount of angular momentum of order $10^{27}\hbar$. Whether j is actually $10^{27} + 13$ or $10^{27} - 41$ would hardly matter or be noticed; in the everyday world, angular momentum would appear to be continuous.

But in that weird quantum world, oh so remote and yet so close to us, particles are spinning and have angular momentum $j\hbar$, with $j = 0, 1, 2, \ldots$.

An irrelevant[11] remark that might even throw some readers off: For the angular momentum associated with spin, the letter s is used instead of j. I will use both of these letters. The photon spins with $s = 1$, and the graviton $s = 2$, for instance.

Weird and weirder: spin $\frac{1}{2}$

Or at least that is what the early quantum pioneers believed, that j, or s, can only be a (non-negative) integer. Weird, but things soon get weirder!

Theoretical physicists were shocked to discover that the electron's spin is not equal to an integer times $\hbar$, but is given by $s = \frac{1}{2}$. (Yes, dear reader, this is related to Dirac's discovery that the gyromagnetic ratio $g = 2$ for the electron.) This many decades later, it is now understood that in the quantum world, j or s can take on both integer and half integer values: $j = 0, \frac{1}{2}, 1, \frac{3}{2}, 2, \ldots$.

Guess what? After getting over this shock, theoretical physicists realized that the particles with integer spin, such as the photon and a graviton, are bosons, while particles with half integer spin, such as the electron and the quarks,[12] are fermions. Spin determines statistics.

This profound realization, indeed one of the most profound and mysterious in theoretical physics, is known as the spin statistics connection. I will talk more about this in chapter VI.3. By the way, this peculiar connection between spin and statistics also explains a key observation in chapter VI.1: an even number of fermions together act like a boson, while an odd number (yes, including 1) of fermions together act like a fermion. This follows because spin angular momenta add.[13] A sum of odd numbers of half integers (such as $\frac{1}{2} + \frac{5}{2} + \frac{7}{2}$) can never equal an integer.

So, ta dah, this explains all that spooky stuff we saw while watching helium-4 and helium-3 nuclei scatter off each other. The helium-4 nucleus, containing 4 nucleons, is a boson, while, in sharp contrast, the helium-3 nucleus, containing 3 nucleons, is a fermion.

It appears to be one of the few places in physics where there is a rule which can be stated very simply, but for which no one has found a simple and easy explanation. The explanation is deep down in [quantum field theory] ...... This probably means that we do not have a complete understanding of the fundamental principle involved.
Richard Feynman[14]

I agree with Feynman. In my book on quantum field theory, I was not able to give a "simple and easy" explanation. What I did instead[15] was to show that if you were to choose the minus sign for a scalar field, we would end up violating a basic principle of special relativity. That exercise demonstrates the we need both quantum mechanics and special relativity to derive the spin statistics connection.

Early in life, I read in one of George Gamow's popular physics books that he could not explain quantum statistics—all he could manage for Fermi statistics was an analogy, invoking Greta Garbo's famous remark "I vont to be alone."— and that one would have to go to school to learn about it. Perhaps this spurs me, later in life, to write popular physics books also.[16]

Physics and mathematics mysteriously intertwine

The story told here about angular momentum, spin, and quantum statistics is to me the most mysterious and fantastic intersection between mathematics and physics in the long intertwined history of these two subjects. In the 19th century, mathematicians developed a topic in abstract algebra known as group theory,[17] as already mentioned in part V. Some mathematicians even crowed that they had finally invented something that physicists could not use. And indeed, group theory as such did not come into physics until quantum mechanics was developed, and even then, some old fogies resisted, saying that they didn't need it.[18]

Actually, in the series $j = 0, \frac{1}{2}, 1, \frac{3}{2}, 2, \ldots$, the integer values of j had already snuck into classical physics earlier in the guise of spherical harmonics, nowadays taught to undergraduates. But physicists were largely unaware of the half integer values. They make no sense in the classical world. I will now explain, or at least sketch, why.

As you may have intuited, angular momentum is more than intimately connected to rotation. Just like ordinary momentum is given by mass times velocity, angular momentum is given by the moment of inertia of the object in question times its angular velocity. And angular velocity simply measures how fast the object (for example, the spinning top that two greats of physics are studying intently in figure 2) is rotating. Thus, the value of j, be it integer or half integer, measures how the property of the object changes under rotation.

It is total common sense, so obvious that it is almost beyond common sense, that a rotation through 360° is equivalent to a rotation through 0°. Going around a circle through 360° brings you back to where you started. For the mathematical expressions corresponding to integer values of j, that is indeed the case: Rotations through 360° do not change them. This holds for everyday objects, such as the flag of Nepal (figure 3). Common sense reigns.

Figure 3. Consider rotating the flag of Nepal (this is in reference to the poem mentioned in chapter V.3). When rotated through 360°, it comes back to its original state.

Double covering of rotations

But almost beyond weird, the mathematical expressions corresponding to half integer values of j flip sign when rotated through 360°. They flip sign again when rotated through a further 360°. In other words, you have to rotate a mathematical expression corresponding to a half integer value of j through 720° to bring it back to its original state. Mathematicians call this peculiar phenomenon "double covering of rotations," and the relevant group theory was worked out by the French mathematician Élie Cartan among others.

You could almost hear some mathematicians rejoicing! Physicists will never be able to use something like this. Double covering of rotations, ha!

Indeed, I could imagine a 19th century physicist being shown this by a mathematician friend saying, "Interesting, but how could this possibly be relevant to physics? Unlike you, we do not deal with mental constructs, but with actual objects we could hold in our hands." Yes, even today, I would say that almost all physics students, when first exposed to this peculiar feature of rotations, have major difficulty wrapping their heads around it, so to speak.

I have labored long and hard to prepare you for this bizarre twist in physics, discussing identity, statistics, $\frac{1}{2}$ integer spin, and double covering. Along came quantum mechanics, and as you have learned, probability is given by the absolute square of the probability amplitude. The probability amplitude could flip sign without changing the probability. So physicists say, aha, in the quantum world we could possibly use this double covering mumbo jumbo, but still, it's only a mathematical possibility.

Well, you know full well I wouldn't be telling you about all this in a popular physics book if Mother Nature were not tickled by this curious possibility of a rotation through 360° producing a minus sign. She liked it so much that "half" of the quantum particles, the fermions, subscribe to this strangeness.

From counting with fingers to dealing with complex numbers

How can we incorporate fermions into the path integral formalism? Recall that in the simplest garden-variety quantum mechanics for a single particle, the path is described by the trajectory of the particle, specified by its location in space at any given time. Just three numbers varying with time. Similarly, the path followed by a field describing bosons (such as the photon and the graviton) is described by the configuration of the fields at any given time. These fields, the familiar electromagnetic field and gravitational field, are specified by a bunch of ordinary numbers. (For example, as we saw in chapter V.5, Einstein's gravitational field is specified by the metric of spacetime at any given point, just a bunch of ordinary numbers.)

But then how could we possibly describe something as bizarre as the electron field? To explain how, I have to remind you how ordinary numbers came about.

Starting with our fingers, humans invented the concept of positive integers. Then the Hindus added zero, eventually leading to the extension to negative integers. From there, kindly elementary school teachers showed us fractions, the numbers inhabiting the space between the integers. Fractions are also known as rational numbers, to distinguish them from irrational numbers,[19] such as the square root of 2. From irrational numbers we rose to confront transcendentals, such as $\pi = 3.14159\ldots$. This completes the real numbers.

The next great leap forward is to the imaginary numbers, based on defining $i \equiv \sqrt{-1}$, so that the equation $x^2 = -1$ can be solved. Numbers given by $a + ib$ with a and b two real numbers are known as complex numbers, which we encountered in chapter II.3 when discussing the path integral.

Strictly speaking, classical physics does not need complex numbers, except as useful mathematical tools in some situations. Quantum physics, in contrast, crucially requires complex numbers. As we saw in chapter II.3, the probability amplitude assigned to a path is given by a complex number with absolute square equal to 1.

So much for a lightning review of the system of numbers. For a long time, physicists were content to stop with complex numbers.[20]

Onward toward anticommuting numbers

Yet you feel deep down in your heart that a new kind of number would be needed for a quantum field describing fermions, such as the electrons. How do you describe a fermionic field in the path integral if interchanging the paths of two fermions is to produce a minus sign in the probability amplitude?

You learned in school that 3×7 is the same as 7×3. They both equal 21. Multiplication of real numbers is said to commute.[21] Incidentally, multiplication of complex numbers also commutes.[22]

So, theoretical physics needs a new type of number such that when we reverse the order of multiplication, the sign flips. It turns out that mathematicians had already studied what are now known as Grassmann or anticommuting numbers, which physicists denote by the "less used" Greek letters, such as $\xi, \zeta, \eta, \dots$ (pronounced xi, zeta, and eta, respectively).

As the name suggests, anticommuting numbers have the property that

$$\xi \times \zeta = -\zeta \times \xi$$

Henceforth, we will omit the multiplication sign, as is customary in algebra, and simply write $\xi\zeta = -\zeta\xi$. Setting ζ equal to ξ, we see that

$$\xi^2 = 0$$

Astonishingly,* anticommuting numbers square to 0.

So physicists went away happy ("Thank you, mathematicians!") and wrote the path integral for fermions in terms of anticommuting numbers. Anticommuting numbers muscle their way into physics via quantum field theory!

I now reveal a little secret to those readers who learned calculus: The calculus of anticommuting numbers is much much easier than the calculus of ordinary numbers. (Not so surprising,[23] eh, if the variables square to 0?)

	number system needed	key property
classical physics	real number	$1^2 = 1$
quantum mechanics	imaginary number	$i^2 = -1$
quantum field theory	anticommuting number	$\xi^2 = 0$

Spin, quantum statistics, and color

I started this chapter with atomic spectroscopy. In chapter V.2, I mentioned that Gell-Mann invented quarks, and then later, the decay of the neutral pion indicated that quarks come in three colors, which in turn led to quantum chromodynamics. You may be surprised to learn that there is a connection between these two topics.

Like atomic spectroscopy, a subject known as hadron spectroscopy studies how hadrons, that is, strongly interacting particles, are composed of quarks.

---

*I suppose that educators feel that the minds of school children are already blown by the imaginary number i, which squares to -1, so that there is no need to blow their minds further with Grassmann numbers ξ, which square to 0.

The proton, for example, consists of two up quarks and a down quark, as I had mentioned more than once in part V. When theorists studied how the quarks arrange themselves, they found that the Pauli exclusion principle was violated. For a while, this counted as another argument against Gell-Mann's quarks. Eventually, the introduction of color resolved this difficulty. The fact that quarks are fermions implies that the total quantum wave function has to be antisymmetric, but this now includes not only how the quarks move around, how they spin, and so on, but also how they are colored. The extra color "degree of freedom" saved the day: The Pauli exclusion principle holds for quarks also. As promised in chapter V.2, there is more than one reason for the necessity of color.

Notes

[1] This is the origin of the term "quantum mechanics."

[2] While a student in Cambridge, E. C. Stoner came to within a hair of stating the exclusion principle. Pauli himself only claimed to "summarize and generalize Stoner's idea" in his famous paper (*Zeitschrift für Physik* 31, 765, 1925). However, later in his Nobel Prize lecture, Pauli was characteristically ungenerous about Stoner's contribution. A detailed and fascinating history of the spin and statistics connection may be found in Duck and Sudarshan, *Pauli and Spin-Statistics Theorem*. The Matthew principle operates in gale force here.

[3] Duck and Sudarshan, page 21.

[4] When pairs of electrons in a metal form a boson, the metal can conduct electricity without any resistance.

[5] Note that this corresponds to the 4 and the 0, which were mentioned in a table in chapter VI.1.

[6] For readers who want more, I recommend McIntyre et al., *Quantum Mechanics* particularly chapter 13 and figures 13.3 and 13.9.

[7] Such as the one in Santa Barbara, California.

[8] The principle behind the laser is called "stimulated emission," as was first proposed by Einstein. The presence of a large number of photons of a specific frequency encourages atoms to emit more photons with that frequency.

[9] I refer those readers interested in history to S. Bergia's article "Who discovered the Bose-Einstein statistics?" in *Symmetries in Physics*

(1600–1980), edited by M. G. Doncel et al. In reading about history, I am always impressed by how muddy the actual developments were compared with the narrative in physics textbooks and in popular physics books. For instance, in 1911 (13 years before Bose), the obscure Polish physicist L. Natanson published a calculation of the number of ways one could put a specified number of "energy elements" into a specified number of "receptacles of energy," considering three cases: (1) both the receptacles and the elements are indistinguishable, (2) the receptacles are distinguishable but not the elements, and (3) both the receptacles and the elements are distinguishable. (Note that words like "photon" and "quantum states" were not yet used.) Bergia observed drily that the person who should be most interested in Natanson's work, namely, Planck, had no known reaction. As another example, the distinguished Dutch-American physicist P. Debye derived in 1910 the Planck distribution of photons without being aware that indistinguishability was an issue.

[10] Or, more likely, a fidget spinner, in 2017.

[11] One reason to insert this detail here is to forestall the jungle patrol from saying that I don't know the difference between j and s.

[12] When Murray Gell-Mann first presented the notion of quarks, he was met with considerable skepticism by senior theoretical physicists, except for Dirac. According to Gell-Mann, Dirac was pleased that they have spin $\frac{1}{2}$ and hence are described by the Dirac equation.

[13]Again, the mystery of quantum physics. Adding angular momenta a and b, you obtain, not just $a + b$, but also $a + b - 1$, $a + b - 2, \cdots$, $|a - b| + 1$, $|a - b|$. For example, $2 + 3 = 5$, 4, 3, 2, 1. To understand this kind of peculiar addition, you would have to look at a book on group theory, for example *Group Nut*, page 208. Undergrads in my course learn it with ease, so it is not too difficult.

[14]Feynman, *Feynman Lectures on Physics*, page 4–3.

[15]In *QFT Nut*, chapter II.4.

[16]See page x, Zee, *Einstein's Universe*.

[17]Évariste Galois (1811–1832) invented group theory the night before he was killed in a duel, a duel provoked possibly by his feelings for a young woman, but more likely caused by his Republican views. See *Group Nut*, page 45.

[18]For a brief history, see *Group Nut*, page 46.

[19]I could not resist showing you the simple proof of irrationality given by the Greeks. The proof is by contradiction. Assume that $\sqrt{2}$ equals a fraction: $\sqrt{2} = p/q$. We assume, of course, that we have already canceled off any common factors in p and q; it would be silly to write $282/202$, for instance, without canceling out the obvious common factor of 2, and write $282/202 = 141/101$ instead. Squaring $\sqrt{2} = p/q$ and multiplying by q^2, we obtain $p^2 = 2q^2$, which shows that p^2 is even. But we know that an odd number multiplied by an odd number is odd, and so we know that p is also even. So write $p = 2r$ with r an integer. Squaring and substituting, we obtain $q^2 = 2r^2$. Repeating

the argument we just went though, we see that q is also even. But the conclusion that p and q are both even contradicts our premise that we have already canceled out all the common factors in p and q. Hence, there is no way that $\sqrt{2}$ equals a fraction, no matter how large we make the numerator and denominator of such a fraction.

[20]Complex numbers could be further generalized to quaternions and octonions, but they are not needed for quantum field theory as presently formulated.

[21]It is amusing to see how the original sense of the word "commute," to change from one form to another (cf. "mute" as in "mutate" and "mutant"), was itself changed when it entered American English to mean moving from home to work and back. In the 19th century the word refered to a type of train ticket.

[22]In chapter II.3, I explain how to add two complex numbers. Recall that a complex number is characterized by its length and the angle it makes with a reference direction, such as the x-axis. To multiply two complex numbers, you multiply their lengths and add their angles to obtain a new complex number. This operation clearly commutes.

[23]An infinite variety of functions $f(x)$ exists if x is a real number, but there is essentially only one function $f(\xi) = a + b\xi$, with a, b two arbitrary real numbers, if ξ is an anticommuting number. Also, Taylor series (if you know what they are), instead of having an infinite number of terms, terminate after two terms.

Intellectual completeness ▬▬▬▬▬

When I give talks about quantum field theory to a general audience, I like to ask: "What is the purpose of studying physics?" And here is the slide I show (figure 1).

The purpose of physics is to understand Nature, which in this book means the physical universe. If so, then we should strive to understand Nature as fully as possible. It thus behooves me to explain what "fully" means.

The purpose of physics is to understand Nature

Figure 1. The purpose of studying physics.

Quantum mechanics is not intellectually complete

I like to introduce the notion of intellectual completeness to measure different areas of physics by. For example, fluid dynamics is an important and intellectually splendid area of physics, with many unsolved, or at least poorly understood, problems, such as turbulence. Fluid dynamics is also self sufficient in the sense that there is no need to explain what fluids are made out of in order to explain how fluids behave. To calculate how fast[1] a tsunami moves across the vast Pacific, we hardly have to invoke the fact that water consists of the peculiar molecule H_2O. (This relates back to the discussion of "more is different" and reductionism in chapter V.5.)

But intellectual splendor, or vitality if you prefer, and self sufficiency do not imply completeness. Fluid dynamics as developed in the 18th and 19th centuries would not be able to explain why water, contrary to most other fluids, expands when it freezes (thus allowing fish in frozen lakes in icy climes to survive winters[2]). For that, we have to understand how a multitude of the H_2O molecules come together and interlock with each other.

This is all simple enough. But now the main message of this chapter.

Quantum mechanics, no matter that it sounds profoundly puzzling, is intellectually incomplete! If you ever take a course on quantum mechanics, be sure to ask the professor where the Pauli exclusion principle comes from.[3]

Hopefully, you are totally convinced of the importance of the Pauli exclusion principle and the spin statistics connection for even a gross understanding of the world we inhabit. And yet, in quantum mechanics, it is just an arbitrary empirical rule of thumb that "fell from the sky," without any understanding to prop it up. Intellectually, quantum mechanics is egregiously incomplete. It cries out for quantum field theory.

Electrons are identical

How do you know that electrons are not manufactured in a factory somewhere? If so, there could be infinitesimal manufacturing defects that distinguish one electron from another, just like billiard balls, no matter how high the quality or how expensive.

The explanation offered by quantum field theory is so elegantly simple that it has the resounding ring of truth: Electrons are absolutely identical because they are all excitations in a single, the one and only, electron field. (See figure 2.)

How do you know that all the electrons are identical?

They could have been made in a factory somewhere...

We need QFT to explain this

Figure 2. Electrons are identical.

More examples could be given, such as why some forces are attractive and others repulsive. In a course on nonrelativistic quantum mechanics, we would only have the professor's say-so. But in a course on quantum field theory, this particular mystery could be deduced from the underlying principles, as we saw in chapter III.3. Whether a force is attractive or repulsive is determined by how the field responsible behaves under rotation. Hardly obvious, to say the least.

Why are the electric charges carried by the electron and by the proton exactly equal and opposite? The quantum electrodynamics of Schwinger and

Feynman is seriously incomplete. In the Dirac action, once we fix the coupling of the electromagnetic field to the electron field, we are free to multiply the coupling of the electromagnetic field to the proton field by any number we like, for example, $\sqrt{\pi/3}$ just for laughs. (See figure 3.)

Why is the electron charge exactly equal in magnitude to proton charge?

Quantum electrodynamics less complete than grand unified theory

---

Figure 3. That the electron and the proton have exactly equal and opposite electric charges is a total mystery in quantum electrodynamics but is forced on us by grand unification.

We need grand unified theory to explain this mysterious but striking, and strikingly precise, fact. Quantum electrodynamics is not as intellectually complete as grand unified theory.

A hierarchy of intellectual completeness

Quantum field theory can be written in any spacetime dimension you like. Recall that the action for quantum field theory, $S = \int d^4x \, \mathcal{L}$, is given by the Lagrangian density integrated over spacetime. You want to study quantum field theory in 7-dimensional spacetime? Just change the 4 to 7. Indeed, many quantum field theorists spend much of their careers looking at quantum field theory in different spacetime dimensions (but usually for dimension less than 4, since things generally become easier the smaller the dimension). (See figure 4.)

Why is spacetime (3 + 1) – dimensional?

Quantum field theory (can be written in any spacetime dimension)
Less complete than string theory

Figure 4. The dimension of spacetime.

In contrast, one intriguing feature of string theory is that it can be formulated consistently only in 10-dimensional spacetime. Too bad it is not 4, but 10! To get it down to the observed 4, string theorists have to shrink 6 of 10 dimensions. But still, that its logical structure fixes the dimension of spacetime is in itself interesting. In this sense, string theory is intellectually more complete than quantum field theory, even if the 4 has yet to be explained.

Quantum field theory is far too accommodating

Quantum field theory is far too accommodating, particularly concerning the weak interaction. As we saw in chapter V.3, quantum field theories were written down to respect P, C, CP, and T. But after the physics community was shocked successively by experimental results showing that these discrete symmetries were all violated by the weak interaction, theorists were able to construct quantum field theories to accommodate these violations. For instance, for parity violation, we simply let the weak interaction gauge bosons couple differently to the left handed components of the Dirac fields than they do to the right handed components.

Intriguingly, quantum field theories based on quantum principles and on special relativity cannot violate CPT. Thus, any experimental evidence of CPT violation would be serious business indeed and would call for a careful examination of the foundation of quantum field theory.

Quantum field theory is also accommodating in that it allows us, to a large extent, to put in any mass for the fields we would desire. If we want to describe a world with n noninteracting fermions, we could write, at the simplest level, $\mathcal{L} = \int d^4x \, \bar{\psi}_1 (i\gamma^\mu \partial_\mu - m_1)\psi_1 + \bar{\psi}_2 (i\gamma^\mu \partial_\mu - m_2)\psi_2 + \cdots + \bar{\psi}_n (i\gamma^\mu \partial_\mu - m_n)\psi_n$. The masses, $m_1, m_2, \ldots, m_n$, are just whatever you feel like putting in. You might have thought, from reading about the highly publicized Higgs mechanism, that theoretical physicists now understand where masses come from. Not really. With the Higgs mechanism, we simply replace m_a by $f_a\phi(x)$, with $\phi(x)$ being the Higgs field. When $\phi(x)$ assumes the value v (that is, when $\phi(x) = v$), the masses come out to be $m_a = f_a v$. That doesn't help much, does it? Where does f_a come from?

At present, our quantum field theory contains many parameters analogous to f_a, which remain unexplained. The hope, of course, is that the present day quantum field theory will be replaced by a more magnificent quantum field theory containing it, preferably one involving some as yet unknown physical principles, just as quantum field theory contains quantum mechanics.

Notes

[1] About as fast as a jet plane. This is in fact an easy calculation that the undergraduates in one of my courses could readily perform. See *FbN*, pages 282–283.

[2] The ice forms a layer on top, thus insulating the water beneath from the colder air. For fans of the anthropic principle (stating that the universe is the way it is because we exist), consider that the fish physicists could argue similarly why water must behave in this peculiar way. I prefer the explanation based on our understanding of molecular bonding as provided by quantum mechanics.

[3] People have tried to concoct arguments in terms of quantum mechanics, but most physicists find these concoctions unconvincing.

Recap of part VI

The issue of identity is fundamental to the quantum world. An electron is indistinguishable from all other electrons. When two quantum particles travel from some initial positions, say—just to fix our minds—one from Shanghai, the other from Tokyo, to some final positions, say, London and Paris, we have no way of knowing which of the two particles has arrived at a specific final position. Did the one from Shanghai arrive in London and the one from Tokyo in Paris, or did the one from Shanghai arrive in Paris, the other in London? There is no way, even in principle, to tell. In the classical world, it would be laughably easy to find out: Just hire two private eyes, one detective to tail the suspect leaving Shanghai, the other detective to follow the one leaving Tokyo. In classical physics, the position of the two objects under study could be determined at any given instant in time and reported back to us.

In the path integral, the amplitudes for the two possible histories could be added or subtracted, due to the deep mathematical insight that both $+1$ and -1 square to 1. Interestingly, Nature knows about this and uses both possibilities, so that fundamental particles are either bosons or fermions.

Quantum field theory tells us that particles with integer spin are necessarily bosons, and particles of half integer spin are necessarily fermions. This deep connection between spin and statistics, and its manifestation in the Pauli exclusion principle, is of crucial importance in determining the most consequential features of the physical world. In contrast, in ordinary non-relativistic quantum mechanics, this is an empirical rule of thumb without any explanation.

In this respect, and in many other respects, quantum mechanics is not as intellectually complete as quantum field theory. Similarly, quantum electrodynamics is not as complete as grand unified theory. We do not know the future of string theory, but if it turns out to be correct, we could say that quantum field theory is not as intellectually complete as string theory.

It is this glaring incompleteness that excites the minds of physicists. In the wild frontier between knowledge and ignorance, much will eventually flourish.

Parting comments and some unsolicited advice

Getting acquainted with quantum field theory

Dear reader, thank you for going on this exciting quest. Surely, it was a difficult journey, but you now know something about quantum field theory. Congratulations!

Quantum field theory grew out of the union of quantum mechanics and special relativity, and it furnishes essentially the only language and approach we have for exploring the realm of the very small and very fast, small and fast compared to the lumbering giants who we are. If we regard string theory as an intriguing but as yet unproven possibility, then quantum field theory is by far our most advanced and sophisticated subject in theoretical physics. At the very least, it is the most accurately tested.

The quantum action principle and the illusion of a classical world

Quantum field theory rests in part on a notion[1] originating in the 17th century in the minds of Fermat and others, that the bending of light could be accounted for by saying that light is in a hurry even as its speed changes from medium to medium. In the 18th century, Euler and Lagrange succeeded in formulating the analogous principle for material particles as an alternative to the differential formalism of Newton. Instead of asking for what happens in the next instant, this principle takes a global perspective on the many different paths that a particle could follow, picking out the one that extremizes the action.

One of the great, but unsung, triumphs of special relativity, from the philosophical point of view (or, as I prefer to say, from the point of view of the intellectual foundation of physics), is that it is able to unify these two great principles into one. Clearly, this would not have been possible before special relativity, since by definition, light travels at the speed of light, and special relativity is needed to describe particles traveling at speeds comparable to the speed of light.

Then came quantum mechanics, replacing certainty by probability, and not merely probability, but the more sophisticated concept of probability amplitude. In school, the unsuspecting young are taught the Schrödinger and Heisenberg formulations. But more fundamentally, the quantum world is described by the Dirac-Feynman path integral formulation.

All possible paths are examined, and their corresponding probability amplitudes are summed up. The key is that, unlike probability, the probability amplitude is not a number, but a 2-dimensional arrow of unit length. This multitude of arrows could add up to an enormously long arrow if everybody is pointing more or less in the same direction, or to a vanishingly small arrow if everybody is pointing in a random direction. The path whose neighboring paths have probability amplitudes all pointing in the same direction as its probability amplitude wins!

This path ends up being a classical path, followed by a classical particle in the classical world, thus giving the illusion of deterministic motion from instant to instant. In contrast, the path whose neighboring paths have probability amplitudes pointing each and every way loses.

The classical world is an illusion.

The path integral formulation lends itself naturally to quantum field theory. If the action we put in for the field is relativistic, then special relativity and quantum mechanics are automatically joined together. In contrast, quantum mechanics as usually taught in school manifestly privileges time: The Schrödinger equation tells us how the wave function evolves in time. If you insist on formulating quantum field theory using the Schrödinger approach, you could do it, but you would need to be a contortionist, with your calculation looking nonrelativistic at every step and then having to show that the final result is in fact relativistic. It can be done, but it's a bit clumsy.

After spectacular triumphs and a premature burial, a roaring return

An early triumph of quantum field theory is the explanation of how the fundamental forces in the physical universe come about. They are caused by the exchange of quantum particles. Whether a force is attractive or repulsive depends on the sign of the interaction energy. The delicately choreographed dance of the universe emerges. Before quantum field theory, there was no understanding of this scintillating interplay between the four fundamental interactions, some attractive, some repulsive, some short ranged, and some long ranged.

Another intellectual triumph is the explanation that all electrons are identical, simply because they are all excitations in the one and only electron field. How else would you explain this fundamental mystery?

After the postwar success of the perturbative approach to quantum electrodynamics, quantum field theory was left for dead because of its inability to tackle the strong interaction. The burial was attended with pomp and circumstances in view of its earlier triumphs. But then, in defiance of the "nattering nabobs of negativism," quantum field theory came roaring back, culminating in a victory parade that makes the spectators gasp with joy.

While a complete quantum field theory of gravity is not yet in sight, the perturbative approach, nailing down corrections to Newtonian gravity, could be, and has been, worked out. The spectacular grand unification of three of the four fundamental interactions was awesome to behold. It answered several previously unanswerable questions, for instance, the equal and opposite electric charges carried by the electron and by the proton.

Breaking the shackles of Feynman diagrams

In my textbook on quantum field theory, a chapter is titled "Breaking the shackles of Feynman diagrams." This process, by my reckoning, took off around 1974 with the realization that certain quantum field theories contain magnetic monopoles whose properties (such as its mass) are proportional to $1/e^2$. Thus, a perturbative series of the form $e^2 + e^4 + \cdots$ could never reveal the existence of magnetic monopoles. You could calculate Feynman diagrams until steam comes out of your ears, and you'll never see a single magnetic monopole.

This drive of going beyond Feynman diagrams already has enjoyed some spectacular successes in connection with the so-called twistor or helicity amplitude approach.[2] To give you a flavor of this impressive advance, I mention the amplitude in quantum chromodynamics for two gluons scattering into three gluons to lowest order of perturbation theory. Sounds easy, no? The task is merely to calculate to lowest order. Using Feynman diagrams, you would soon discover that the answer consists of almost 10,000 terms. Yes, you read that right. People used computers (equipped with advanced software capable of symbolic manipulations, in contrast to brute number crunching) to calculate these, and even a small part of the result when printed on a textbook page ends up looking like a black smudge.[3]

Amazingly, using modern twistor[4] technique, quantum field theorists can now write down the answer, not only for two gluons scattering into three gluons, but for two gluons scattering into n gluons for $n = 3, 4, \ldots$, any integer you like, in one line. One line!

Surely that convinced you that the traditional Feynman diagram method is not the way to go. Recent advances continue to reveal unexpected deep structures totally hidden from the perturbative approach.

Many parameters, or impressively few?

In an introductory physics course, we learned that the acceleration of falling objects is given by $g \simeq 9.8$ meter per second squared, fed to us as an empirical quantity. Physicists refer to quantities such as g as a parameter, that is, a constant coming from outside the theory under discussion. As we advanced, we were told that g is actually determined in terms of Newton's constant G and the earth's mass M and radius R. Clearly, M and R are only of local interest, hardly the concerns of fundamental physics, and certainly G is more fundamental and universal than g. For instance, if you were studying cosmology, g is hardly relevant, but G would appear everywhere in your equations. Physics is full of such empirically measured but unexplained parameters.

Parenthetically, since in natural units, G equals the inverse square of the Planck mass M_P, nowadays the puzzle of why G is so small amounts to asking why M_P is 10^{19} times larger than the proton mass m_p. Since our current understanding of the proton mass is based on our understanding of the strong interaction, this reduces to the crushing strength of the aptly named strong interaction compared to the gravitational interaction.

Now that you know what a parameter is, let it be told that at present, quantum field theories typically contain a bunch of parameters. Take the Dirac action for an electron, $S(\psi) = \int d^4x \, \bar{\psi}(i\gamma^\mu(\partial_\mu - ieA_\mu) - m)\psi$, for instance. It contains two parameters: the mass of the electron, denoted by m and measured to be, in human made units, $\simeq 0.911 \times 10^{-27}$ gm, and the coupling $e \simeq 0.303$. There is no understanding why m has this particular value and not some other (say, 1.97×10^{-26} gm just to be definite). Similarly for e. (Why not 30.3?)

Grand unified theories now contain, depending on how you count, around 20 parameters, including the masses[5] of the quarks and leptons, for example. These parameters are assigned the values they are empirically known to have, without any further understanding.

The hope for an ultimate theory is of course that these parameters will all be explained eventually. At this point, the personalities of various physicists come in. Some would say, how awful, so many parameters! These are the congenital negativists. The positivists would say, this is utterly amazing, a fantabulous achievement! In just a few decades, physicists have reduced the almost astronomical number of entries in the *Handbook of Chemistry and Physics*, a massive tome close to two thousand pages that has been updated annually for more than a century, to a mere 20 parameters or so. At least in principle, although certainly not in practice, all these entries can be calculated in terms of these parameters.

Independently of the advances in particle physics described in this book, a condensed matter theorist once exulted to me long ago that all the properties of matter, from, say, the conductivity of copper to the viscosity of water, could in principle be calculated in terms of two dimensionless numbers: the ratio of the proton to the electron mass $m_p/m_e \simeq 1836$ and and $e \simeq 0.303$. Before the mob of negativists turn into a lynching party coming after my condensed colleague and me, let me emphasize that, of course, the emphasis here is on "in principle," not "in practice." This goes back to the discussion about reductionism in chapter V.4 on grand unification. Each to his or her own taste and needs.

Incidentally, the reader should understand that many of the 20 or so parameters bedeviling quantum field theorists do not appear to have anything to do with the price of beans. For example, surely we could vary the mass of the bottom quark without affecting the viscosity of water.

Calculable versus incalculable

In the early 1970s, swept up during the victory parade mentioned earlier in this book, quantum field theorists became much more ambitious. Steve Weinberg and others pointed out that in a given gauge theory, some quantities are calculable while others are incalculable. Suppose that a quantity x equals zero in the action of some gauge theory. In general, quantum fluctuations would generate x. As Murray Gell-Mann once said,[6] in the quantum world, what is not explicitly forbidden would always happen. The fluctuations run through all possibilities, however unlikely.

However, you cannot simply set x to zero by hand. In that case, x would indeed be generated, but to an arbitrary value, said to be incalculable. For x to be calculable, there has to be some built-in symmetry, some internal logic in the theory, which sets x to zero in the first place, but such that the symmetry is not respected by quantum fluctuations. This sounds a bit convoluted, but in fact it reflects the beauty of gauge theories that such a possibility could arise.

I suggested to my graduate student Steve Barr that for his PhD thesis, he take up this challenge of constructing theories in which the electron mass m_e is calculable in the sense defined by Weinberg. He and I managed to find such theories. But unfortunately m_e comes out in terms of unknown quantities. Barr got his degree. Decades have passed, and he has retired from his professorship. Yet nobody has managed to overcome this challenge, that is, to calculate the electron mass in terms of known quantities (such as the muon mass). Indeed, as I wrote this, Weinberg himself had once again taken up this challenge.[7] Incidentally, it is also possible to construct theories in which the masses of the three neutrinos are generated by quantum fluctuations and determined in terms of quantities that are in principle measurable.

Here is a rather rough and imperfect analogy. In the 19th century, the viscosity of water was a parameter in the equations of fluid dynamics. Suppose someone proposes a highly speculative theory in which water consists of a huge number of hypothetical "molecules" somehow built out of positively charged and negatively charged particles interacting via the electric force. The viscosity could then be calculated in principle, but only in terms of the unknown masses of these hypothetical particles. The theory is useful only when these hypothetical charged particles are shown to exist and their properties measured.

The disappointing conclusion at present is that it is possible to construct theories in which the electron mass, and the neutrino masses, are calculable, but only in terms of other quantities (which may or may not exist in the real world). Thus, this exercise is more or less in the nature of an existence proof in mathematics or the construction of a model proof of concept in engineering.

Identity and indistinguishability: loners versus party animals

More than any other result in theoretical physics, the mysterious connection between spin and quantum statistics shapes the microscopic world into what it is. How much a quantum particle spins, whether that amount is an integer or half an integer as measured in natural units, somehow fixes its personality. It may want to be alone like Greta Garbo, or, to the contrary, it may want to hang out with others of its kind. If a fellow student of physics had told me this when I was an undergraduate, my response would have been both idiomatic and American, "Get out of here, you've got to be kidding me!" What does personality have to do with spin?

Nevertheless, this connection is as real as pecan pie à la mode. Furthermore, it underlies many technological devices having major impacts on our lives, such as lasers. Indeed, were this connection not true, the world would be unrecognizable. Atoms would be much smaller than they are, and everything would be far more compact.

Indeed, the heavens are shaped by this deep connection between spin and statistics. When stars exhaust their fuel, they turn into white dwarves[8] held up by the electrons' absolute refusal to share space with other electrons. Quite astonishing to me that such gigantic astrophysical dwarves owe their existence to the inner logic of quantum field theory.

Something as important as the spin statistics connection was totally unexplained and unexplainable by theoretical physics until the advent of quantum field theory. To students taking an introductory course in quantum mechanics: Try asking the professor why the Pauli exclusion principle holds, assuming that he or she is a quantum mechanic but not a quantum field theorist. The professor could have no answer. Just a rule that fell out of the sky.

The general public sometimes assumes that theoretical physics can explain everything. I mean everything in the physical world of course; physicists, excluding the charlatans, cannot explain whether or not there is life after death, for example. That is clear. And of course, physicists still cannot explain fundamental concepts, such as time. But still, it was a revelation, at least to me when I was an undergraduate, that people could not explain how a laser truly works without quantum field theory.

Quantum field theory as presently known is intellectually incomplete

The hope is that quantum field theory will one day expand into an intellectually more complete theory, infusing quantum field theory with presently unknown physical principles, perhaps analogous to the way quantum mechanics expanded into quantum field theory when infused with Einstein's special relativity. Or perhaps analogous to the way that field theory was rejuvenated by the Yang-Mills construction. This future theory may or may not be a quantum field theory in its present form.

Indeed, many string devotees would say that that theory is already here. While a quantum field is defined locally at a spacetime point, string theory is formulated in terms of a 1-dimensional string. Thus it is to be expected that if the distance scales of the phenomena we are interested in is much larger than the string length, string theory should reduce to quantum field theory. In this sense, quantum field theory is here to stay, regardless of whether string theory is here to stay, just as quantum mechanics is here to stay even with the advent of quantum field theory. Indeed, Newtonian physics is very much alive and well, thank you.

Trapped by harmony

The reader should distinguish the content of a theory from the calculational techniques that could be applied to the theory. Quantum field theorists are frustrated by how they are imprisoned in perturbation theory.[9]

We are stuck in a harmonic paradigm. Let us recall how this comes about. Although when expressed in equations, Maxwell's theory may look complicated to the uninitiated, it is actually extremely simple when formulated in terms of an action: It is quadratic and hence harmonic, with only one cubic term describing the coupling of the electromagnetic field to a charged Dirac field. Because this coupling happens to be small, quantum electrodynamics could be solved perturbatively.[10]

Yang-Mills theory, in some sense the natural generalization of Maxwell theory, is only moderately more complicated. It contains cubic and quartic terms.

Once again, Yang-Mills theory looks impossible to solve until the realization that its couplings strength goes to zero at high energies or short distances. The perturbative approach comes to the rescue yet again.

Einstein's theory of gravity is another order of magnitude more complicated, with an infinite number of terms beyond the quartic.[11] All the mysterious results that enchant physicists and the general public alike, such as warped spacetime and black holes, can be treated by classical or semiclassical methods. Quantum effects are either completely turned off or just treated to lowest order.

Once again, the perturbative approach of keeping only the quadratic terms suffices to study the propagation of gravity waves, just as Maxwell, keeping only the quadratic terms, revealed that light is secretly an electromagnetic wave. (In the regime where the perturbative approach fails, such as in the vicinity of two merging black holes, numerical techniques on high powered computers can take over.)

Symmetries

Over the decades, quantum field theorists have struggled to break out of this harmonic jail. One approach is to inject more and more symmetries into the action, never mind whether the real world exhibits these symmetries or not. Understandably, more symmetric actions are easier to deal with. In school, we learned that it is easier to calculate the area of a circle than the area of an ellipse.

Consider quantum electrodynamics, with a bosonic photon field interacting with fermonic fields carrying electric charge, such as the electron field. Similarly, in quantum chromodynamics, we have the bosonic gluon fields interacting with fermonic quark fields. In the real world, there does not exist a symmetry transforming the photon field and the electron field into each other, nor is there a symmetry transforming the gluon fields and the quark fields into each other. Nevertheless, for the past several decades, an entire generation of theoretical physicists have devoted their lives to toying with quantum field theories in which the bosonic and fermonic fields could be transformed into each other, theories known as supersymmetric quantum field theories, seemingly named by some Hollywood agent wannabe. The intent of these individuals, at least initially, was to play with such theories with the hope of discovering some features that might shed light on more realistic quantum field theories, in the same way that playing with circles might give us some hint of how to deal with ellipses.

But eventually, people became more speculative and ambitious (who wouldn't be after building their lives around this possibility?), arguing that there may be traces of supersymmetry left in the real world and thus advocating experiments to search for them. Thus far, the results[12] can only be described as disappointing.

For me and many others, perhaps the most alluring prospect is that although there is no hope, at least at the moment, of solving Yang-Mills theory, there are hints that the supersymmetric version of Yang-Mills theory may be solvable in some limit.[13]

Nonperturbative approaches

Even in this relatively small (in terms of the number of practitioners) area dedicated to unraveling the mysteries of the universe, theoretical physics operates on many fronts. In the present context, we should distinguish between constructing a theory to explain some new and hitherto unknown phenomena, notably dark energy and dark matter, and extracting the physical content of an established quantum field theory, such as quantum chromodynamics. On these two fronts, we run into different kinds of troubles.

Quantum field theory is too accommodating, as I remarked when discussing the weak interaction. Contrary to the impression that you may have gotten from the popular press, it is actually rather easy to incorporate dark energy and dark matter into quantum field theory. As a result, perhaps tens of theories of dark matter exist, if not hundreds. All you have to do is to introduce a field that participates extremely weakly, or even not at all, in the three non-gravitational interactions, so that its effects barely show up except in cosmological settings. What is lacking is a compelling argument favoring one theory over all others in this multitude. Yes, quantum field theory is too accommodating. Dark energy also can be readily accommodated in the Einstein-Hilbert action but without any rhyme or reason for its observed magnitude.

In contrast, the difficulty of extracting the content of any given quantum field theory beyond the harmonic paradigm may be conveyed by this somewhat imperfect analogy. Imagine that after Maxwell had distilled all that was known about electromagnetic phenomena into a handful of equations, the mathematical techniques for solving these (partial differential) equations were missing, so that he could deduce the consequences of these equations only for some especially simple situations.

Readers with some exposure to elementary physics probably know that the first physics examples that are taught involve particles restricted to move in 1-dimensional space, or perhaps even 2-dimensional space. Evidently, such motions can be solved more easily than in the full splendor of the 3-dimensional space we live in. Similarly, some quantum field theories can be solved nonperturbatively in (1+1)-dimensional spacetime, that is, in a world with one time dimension in one space dimension.

Some readers with an engineering background may know about conformal transformation of the 2-dimensional plane, namely, stretching and modifying various geometrical shapes. Conformal field theories in (1+1)-dimensional spacetime are constructed or designed to have actions invariant under these conformal transformations, which allowed them to be solved, yielding surprising results[14] that could not be found using the traditional perturbative approach.

To a large extent, these theories are all cleverly designed to be solvable, but regrettably, they are not the theories that Nature actually favors and that we like to solve in (3+1)-dimensional spacetime. There is a well known parable about looking for a lost object at night beneath a lamppost, because that is the only place we can look. In some sense, physicists have taken this parable further by expressly constructing lampposts where the looking is "easy," rather than where they would really like to look. The difficult task is to invent a flashlight so that they could look where there are no lampposts.

Many questions remain

Many questions remain, and many of the brightest students in theoretical physics continue to flock to these fundamental puzzles. Neither recruitment nor advertisement is necessary; the fascination is natural. Why are there three families of quarks and leptons? Where do the masses of the fundamental particles come from? Perhaps such questions must wait until we better understand gravity, but perhaps not. Conceivably, gravity cannot be quantized unless we understand why quarks and leptons come in three copies, but it seems to me unlikely.

Why is the cosmological constant so much smaller than what quantum field theorists have estimated? This question appears to be related to gravity, but even that is not clear. Perhaps it, and related questions, can only be answered after gravity is understood more deeply, but again, perhaps not. At this stage, theoretical physicists can only take betting odds.

We don't know how to quantize gravity. Theorists are not only arguing over the proper approach to take; some even feel that gravity should not be quantized. But those who favor quantizing gravity invariably assume that the principle of quantum physics as we know it holds all the way up to the Planck energy.

Personally, I would prefer to see the quantum principle as embodied in the path integral formulation modified at some energy scale far below the Planck energy. Perhaps the way we sum the probability amplitudes is not entirely appropriate. Or perhaps paths close to each other on the Planck scale should not be counted as distinct. The possibilities are only limited by the imagination of the young minds coming up.

Some unsolicited advice

I am under the impression that many individuals out there are into reading popular books on physics, but they lack the courage or motivation to tackle a textbook. I now take the liberty of offering such well intentioned but no doubt frustrated people some unsolicited advice.

As many, perhaps all, physicists have said, the universe is a book written in the language of mathematics. That is a fact. I believe that you could read a popular book about evolutionary biology or brain science and get fairly much the gist of the main ideas of the field. But theoretical physics, starting with quantum mechanics, is exploring a domain so remote from everyday experience that, almost by definition, it cannot be described by human language. And quantum field theory is even more abstract and remote! Probably 99% of the physics students who ever got through a course on quantum mechanics did not move on to quantum field theory.

The readers of popular physics books are frustrated, and the authors (excepting those hacks out to make a few bucks by slapping some words together, of course) are also frustrated. Many obvious analogies come to mind. Trying to understand physics by reading popular books is like reading about music without ever listening to it. Nature is singing to us, and generations of people far more intelligent than most of us have discovered the language, however imperfect, that enables us to listen. Perhaps it is not high fidelity, but more like a scratchy early record, but still, it conveys far more than any popular book written in human language could possibly convey.

Just like frustrated puzzle doers who couldn't wait to have the answer revealed to them, many, or perhaps even all, theoretical physicists long for that blissful moment right after death when the Creator reveals how the universe is put together. (Surely, even some of the self-proclaimed atheists whom I know might occasionally sneak into the closet late at night to wish for the answer to this ultimate homework problem.) My favorite such image has Pauli saying to the Creator "It's not even wrong!"

But also surely, the Creator, when revealing the answer, will use a language as far beyond mathematics as the most abstruse mathematics that theoretical physicists have barely heard of is beyond clunky human languages! It will be like explaining quantum field theory to ants!

Coming back into the real world, don't you think that it is high time for you to tackle a real textbook on quantum field theory, now that you have made it this far? I realize that I'm speaking to a wide variety of readers, but I would like to address this especially to curious and ambitious high school and college students. Now that you have an overview of quantum field theory, what do you need to start my textbook *QFT Nut*?

We will not let our ambition run wild. Let me be specific and easy going. Your goal: Get through the first two parts of my textbook. By the end of that, you would be able to calculate the Feynman diagrams for electron electron scattering or electron photon scattering, to mention just two examples. Sure, people like Feynman and Schwinger and Dirac did it 70 or 80 years ago, but still, think how satisfying that would be!

Just to calibrate yourself, two juniors took a reading course based on my book last year, and they were able to get to the point I mentioned above in 3 or 4 months. So let's get started. What do you need to know before embarking on this project?

In mathematics, you have to be comfortable with partial derivatives and with the complex exponential. You have to know what a matrix is. You certainly don't need any of the fancy schmancy mathematical mumbo jumbo that some people use to "frighten children" with, as my mentor Murph Goldberger used to say to me.[15]

In physics, you have to know some quantum mechanics, but that does not necessarily mean having to get through the entire undergraduate curriculum on the subject. You are not preparing to take a final exam requiring you to calculate this and that. Instead, you have to know some basic concepts, such as state, operator, wave function, Hamiltonian, and a few other similar items. As I explained, quantum field theory is the child resulting from the marriage of quantum mechanics and special relativity. So you need to know some special relativity, especially the notion of Minkowski spacetime. You have to be comfortable with the 4-vector notation. But again, this does not mean that you have to be a whiz in figuring out those headache inducing paradoxes and puzzles in special relativity.

I should also advise you that, if you come across something that you totally do not understand, it may be better to move on and come back to it later. More often than not, with the knowledge you gained, you would find what earlier seemed opaque becomes more transparent[16] later.

But of course I should not make any false promises. If you have totally forgotten your high school algebra and you have never encountered calculus, then what I said is simply not for you, period. The last thing I want to do is to increase the amount of frustration in the world. On the flip side, I would really like some of you to have the pleasant surprise of discovering that a clearly written textbook on quantum field theory is not as forbidding as you might think.[17]

Notes

[1]I refrain from mentioning the self-evident, such as the notion of field.

[2]See the book *Scattering Amplitudes in Gauge Theory and Gravity* by Henriette Elvang and Yu-tin Huang.

[3]If you would like to see the fabled smudge, look at the figure on page 484 of *QFT Nut*.

[4]Invented by R. Penrose, who by the way, wrote the preface to my book *Fearful*.

[5]The popular media might have given you the notion that these masses are now understood due to the Higgs mechanism. This is rather misleading, as the mass of a quark or a lepton is given by the coupling of the Higgs field to that quark or a lepton and its vacuum expectation value.

The Higgs mechanism has in fact introduced more parameters.

[6]This happens to be one of the many MGM dicta that I learned as a student.

[7]S. Weinberg, "Models of lepton and quark masses," *Physical Review* D 101, 035020, 2020.

[8]See, for example, *FbN*, page 228.

[9]As explained in chapter V.2.

[10]As described in chapters IV.2 and IV.3.

[11]As explained in chapter V.5.

[12]For me, the search for supersymmetry at attainable accelerator energy seems more driven by sociology than physics. A generation of physicists learned supersymmetry in graduate school and in turn trained their students in it.

[13]As explained in chapter V.2, quarks come in 3 colors. One favorite limit studied by quantum field theorists is to replace 3 by ∞.

[14]These have important applications in condensed matter physics.

[15]Or, more vulgarly, "bottles of piss water," as Feynman once said to me. I am saddened by how many autodidacts have been frightened and misled by such purveyors of "piss water."

[16]Not to mention that nowadays there are online forums offering help, not available to poor me during the hot humid New Jersey summer long, long ago.

[17]See endnote 11 in chapter III.1.

Timeline

René Descartes (1596–1650), 54
Pierre de Fermat (1607–1665), 58
Isaac Newton (1642–1726/27), 85
Leonhard Euler (1707–1783), 76
Charles-Augustin de Coulomb (1736–1806), 70
Joseph-Louis Lagrange (1736–1813), 77
Pierre-Simon, marquis de Laplace (1749–1827), 78
Jean-Baptiste Joseph Fourier (1768–1830), 62
Thomas Young (1773–1829), 56
Johann Carl Friedrich Gauss (1777–1855), 78
Michael Faraday (1791–1867), 76
Niels Henrik Abel (1802–1829), 27
Hermann Günther Grassmann (1809–1877), 68
Georg Friedrich Bernhard Riemann (1826–1866), 40
James Clerk Maxwell (1831–1879), 48
Josiah Willard Gibbs (1839–1903), 64
Ludwig Eduard Boltzmann (1844–1906), 62
Oliver Heaviside (1850–1925), 75
Joseph John Thomson (1856–1940), 84
Heinrich Rudolf Hertz (1857–1894), 37
Max Karl Ernst Ludwig Planck (1858–1947), 89
Hermann Minkowski (1864–1909), 45
Élie Cartan, (1869–1951), 82
Albert Einstein (1879–1955), 76
Theodor Kaluza (1885–1954), 69
Hermann Weyl (1885–1955), 70
Niels Henrik David Bohr (1885–1962), 77
Erwin Rudolf Josef Alexander Schrödinger (1887–1961), 74
Louis Victor Pierre Raymond de Broglie (1892–1987), 95
Satyendra Nath Bose (1894–1974), 80
Oskar Benjamin Klein (1894–1977), 83

Edmund Clifton Stoner (1899–1968), 69
Wolfgang Ernst Pauli (1900–1958), 58
Enrico Fermi (1901–1954), 53
Werner Karl Heisenberg (1901–1976), 75
Ernst Pascual Jordan (1902–1980), 78
Paul Adrien Maurice Dirac (1902–1984), 82
Ettore Majorana (1906–1938? probably no later than 1959), 32
Shin'ichiro Tomonaga (1906–1979), 73
Hideki Yukawa (1907–1981), 74
David Joseph Bohm (1917–1992), 75
Yoichiro Nambu (1921–2015), 94
Julian Schwinger (1918–1994), 76
Murph Goldberger (1922–2014), 92
Richard Feynman (1918–1988), 70
Freeman John Dyson (1923–2020), 97
Philip Warren Anderson (1923–2020), 97
Geoffrey Foucar Chew (1924–2019), 95
Yuval Ne'eman (1925–2006), 81
Abdus Salam (1926–1996), 70
Robert Laurence Mills (1927–1999), 72
Robert Brout (1928–2011), 82
Murray Gell-Mann (1929–2019), 90
Tom Kibble (1932–2016), 84
Jun John Sakurai (1933–1982), 49
Steven Weinberg (1933–2021), 88
Nicola Cabibbo (1935–2010), 75
Ken Wilson (1936–2013), 77
Gerald Guralnik (1936–2014), 78
Sidney Coleman (1937–2007), 70

Note: The age at death does not take into account the months of birth and of death. Thus, for example, Hertz actually died when he was 36.

A short list of mathematical symbols ■

$\vec{x} = (x, y, z)$: the 3 Cartesian coordinates of the 3-dimensional space we live in. Using the index notation, we often write this as $\vec{x} = (x^1, x^2, x^3)$, often referred to just as x^i, with i ranging over $1, 2, 3$.

$(t, x, y, z) = (x^0, x^1, x^2, x^3)$: the coordinates of the (3+1)-dimensional Minkowskian spacetime we live in, often referred to just as x^μ, with μ ranging over $0, 1, 2, 3$. In particular, $x^0 = t$.

x: sometimes just the first of the 3 Cartesian coordinates (x, y, z), but often used in quantum field theory to refer to the 4 coordinates of $(t, x, y, z) = (x^0, x^1, x^2, x^3)$ spacetime. (Usually, the context clears up any possible confusion.)

———

$\frac{df}{dx}$: the derivative of f with respect to x measures the variation of the function $f(x)$ as x varies. You can think of it as the infinitesimal change of f, called df, with d denoting "difference" divided by the infinitesimal change of x, called dx.

$\frac{\partial f}{\partial x}$: the partial derivative of f with respect to x is used when we want to emphasize that f may depend on a whole lot of other variables, called y, $z, \ldots$, and that these are to be fixed when x varies.

$\frac{\partial f}{\partial x^\mu}$: the partial derivative of f with respect to x^μ for a specific μ, that is, with x^μ varying but x^ν held fixed for $\nu \neq \mu$.

$\partial_\mu f$: $\frac{\partial f}{\partial x^\mu}$ is usually abbreviated to $\partial_\mu f$.

$\partial_\mu = \frac{\partial}{\partial x^\mu}$

$\partial_0 = \frac{\partial}{\partial t}$

$\partial_i = \frac{\partial}{\partial x^i}$

———

$\omega = 2\pi f$: frequency as used in quantum field theory is related to the "everyday frequency" f by a multiplicative factor of 2π. The use of ω instead of f eliminates the many pesky factors of 2π (which bedevil school children).

$k = 2\pi/\lambda$: the "wave number" of a wave used in more advanced physics is related to the commonly used wavelength λ, defined as the distance between two crests, by a multiplicative factor of 2π.

The wave vector $\vec{k}$ points in the direction the wave propagates and has magnitude k. It is a more useful concept than wavelength. The frequency ω and the wave vector $\vec{k}$ together characterize a wave.

———

$\vec{E}$: the electric field.
$\vec{B}$: the magnetic field.
$A_\mu(x)$: often called the gauge field here for simplicity, but strictly speaking, it is the gauge potential.

———

S: the action S is given in general by $S = \int dt\, L(t)$, namely, the integral of the Lagrangian over time.

$\mathcal{L}$: the Lagrangian density. In quantum field theory, the word "density" is more often than not omitted. In a field theory, classical or quantum, the Lagrangian $L(t)$ is given by the integral of the Lagrangian density over space: $L(t) = \int d^3x\, \mathcal{L}(t, \vec{x})$. Thus, in such theories, the action S is given by the integral of the Lagrangian density over spacetime: $S = \int d^4x\, \mathcal{L}(x) = \int d^4x\, \mathcal{L}$, with the x dependence of $\mathcal{L}$ often suppressed.

———

The integration sign $\int$ is allegedly a distortion of the capital letter S denoting summation. For the purposes of this book, you may think of integration as a fancy kind of summation, if you wish.

$\int d^3x$ and $\int d^4x$: the d^3x and d^4x simply reminds us that we are integrating over 3-dimensional space and 4-dimensional spacetime, respectively.

———

There are two e's in this book! One comes from the word "exponential," the other from "electromagnetism." Do not confuse the two.

Euler's number, $e = 2.71828\ldots$, is an exact mathematical number which form the basis of the exponential function. It appears frequently in this book, for example, in the path integral defining quantum physics.

The electromagnetic coupling, $e \simeq 0.303$, is measured experimentally to indicate how strongly the electron and the photon fields are coupled together. In contrast, Euler's e was determined by pure thought.

———

α: the coupling strength of electromagnetism, defined by $\alpha \equiv e^2/4\pi \simeq 1/137.036\ldots \simeq 0.007\ldots$. That it is so small was crucial for the development of quantum field theory.

$\hbar = h/(2\pi)$: we will refer to this as Planck's constant. (Some people call h Planck's constant and $\hbar$ the reduced Planck's constant. We won't bother with this distinction in this book.) By common convention among quantum field theorists, $\hbar$ is usually set to 1.

In quantum physics, the energy of a particle is given by $E = \hbar\omega$, with ω the frequency of the wave "associated" with the particle. Its momentum is given by $\vec{p} = \hbar\vec{k}$, with $\vec{k}$ the wave vector of the wave "associated" with the particle. The de Broglie wavelength λ of a particle is the inverse of the magnitude of the wave vector $\vec{k}$. Thus, $\lambda \sim \hbar/p$. For a particle at rest, λ is usually set to $\lambda \sim \hbar/m$.

Two ways of representing a complex number:
$z = x + iy$, with $i = \sqrt{-1}$ and x, y two real numbers, defines a complex number.
$z = re^{i\theta}$: We can think of a complex number as an arrow with length r making an angle of θ with the x-axis in a 2-dimensional plane. (This is sometimes called the "polar representation.")
Square of a complex number: $z^2 = (x + iy)^2 = x^2 - y^2 + 2ixy$; or in the polar representation, $z^2 = (re^{i\theta})^2 = r^2 e^{2i\theta}$.
Absolute square of a complex number: $|z|^2 = |x + iy|^2 = x^2 + y^2$; or in the polar representation, $|z|^2 = |re^{i\theta}|^2 = r^2$, which does not depend on the angle θ.

In quantum physics, the probability amplitude A is a complex number. The concept of probability amplitude does not exist in everyday life.
The probability corresponding to the probability amplitude A is given by $P = |A|^2$. The concept of probability has the same meaning as in everyday life: P is a real number between 0 and 1, and if we repeat an experiment N times, with N large, then the number of times the event associated with P is expected to occur is given by PN.
Note that the absolute square, which is what we use in quantum physics to obtain probability from probability amplitude, is considerably simpler than the garden variety everyday square. Loosely speaking, quantum interference is due to the phase angle θ of the probability amplitude. Taking the absolute square of a complex number z, we lose information about its phase angle.

Einstein repeated index summation convention: $a^i b^i$ means $\sum_i a^i b^i$ for two quantities a and b carrying the same index i. Einstein simply proposed dropping the summation symbol $\sum_i$.

In the Einstein convention, the scalar dot product between two 3-dimensional vectors $\vec{u}$ and $\vec{v}$ is written as $\vec{u} \cdot \vec{v} = u^i v^i = u^1 v^1 + u^2 v^2 + u^3 v^3$.

The scalar dot product between two 4-dimensional vectors A and B is written as $A_\mu B^\mu = \sum_\mu A_\mu B^\mu \equiv A_0 B^0 + A_1 B^1 + A_2 B^2 + A_3 B^3$. The index μ is often suppressed, so that we write simply AB.

––––––

e^{iS}: in quantum physics, the probability amplitude of the path or history with classical action S. If we do not set $\hbar$ equal to 1, as we often do in this book, then the probability amplitude is $e^{iS/\hbar}$.

Stationary phase: an approximation used in evaluating the path integral. Instead of summing or integrating over all possible path, we pick out the path whose neighbors all have essentially the same phase, that is, whose "arrows" all more or less point in the same direction.

––––––

Under an electromagnetic gauge transformation, a charged field $\psi \to e^{i\Lambda(x)}\psi$.

Under a Yang-Mills gauge transformation, a charged field $\psi \to e^{i\theta^a(x)T^a}\psi$, where T^a is a set of matrices determined by the group. The number of values the index a runs over is also determined by the group.

––––––

$S(\psi) = \int d^4x \; \bar\psi (i\gamma^\mu \partial_\mu - m)\psi$: the Dirac action.

$\quad (i\gamma^\mu \partial_\mu - m)\psi = 0$: the Dirac equation.

$\quad \gamma^\mu$: the 4 Dirac gamma matrices.

––––––

$\eta_{\mu\nu}$: the Minkowski metric of flat spacetime (defined by $\eta_{00} = -1$, $\eta_{11} = +1$, $\eta_{22} = +1$, $\eta_{33} = +1$ with all other components equal to 0).

$\quad g_{\mu\nu}$: the metric of spacetime.

The distance ds between two neighboring points in spacetime differing in their coordinates by dx^μ is given by $ds^2 = g_{\mu\nu}(x)dx^\mu dx^\nu$.

$\quad g$: a quantity formed out of $g_{\mu\nu}$ known as the determinant. It measures the volume of spacetime.

$\quad R$: the scalar curvature, a quantity formed out of the Riemann curvature tensor, which measures the curvature of spacetime.

$\quad \int d^4x \; \sqrt{g}R/G$: the Einstein Hilbert action.

$\quad \Lambda$: the cosmological constant introduced by Einstein.

Bibliography

Books

R. Baierlein, *Newton to Einstein: The Trail of Light: An Excursion to the Wave-Particle Duality and the Special Theory of Relativity,* 2001.

J. Bjorken and S. Drell, *Relativistic Quantum Mechanics,* 1964.

J. Bjorken and S. Drell, *Relativistic Quantum Fields,* 1964.

B. R. Brown, *Planck: Driven by Vision, Broken by War,* Oxford University Press, 2015.

I. Duck and E.C.G. Sudarshan, *Pauli and the Spin-Statistics Theorem,* World Scientific Publishing, 1998.

G. Farmelo, *The Strangest Man: The Hidden Life of Paul Dirac, Mystic of the Atom,* Basic Books, 2011.

R. P. Feynman, *QED: The Strange Theory of Light and Matter,* with a preface by A. Zee, Princeton University Press, 2014.

R. P. Feynman et al., *The Feynman Lectures on Physics,* vol. 3, Addison Wesley, 1971.

R. P. Feynman and A. R. Hibbs, *Quantum Mechanics and Path Integrals,* McGraw-Hill, 1965.

S. Gasiorowicz, *Elementary Particle Physics,* Wiley, 1966.

D. McIntyre, C. Manogue, and J. Tate, *Quantum Mechanics: A Paradigms Approach,* Pearson, 2012.

A. Pais, *Subtle Is the Lord,* Oxford University Press, 1982.

J. J. Sakurai, *Invariance Principles and Elementary Particles,* Princeton University Press, 1964, 2016.

A. D. Stone, *Einstein and the Quantum: The Quest of the Valiant Swabian,* 2015.

S. Weinberg, *The Discovery of Subatomic Particles,* W. H. Freeman, 1983.

H. Yukawa, *Tabibito—The Traveler,* translated by L. Brown and R. Yoshida, World Scientific, 1982.

Books by the author

In addition to the books listed above, I refer quite often, naturally, to the textbooks and popular books I have written as listed below according to the following abbreviations: *QFT Nut, GNut, Group Nut, Fearful, Toy, Unity, G,* and *FbN.* All are published by Princeton University Press, except as noted.

Quantum Field Theory in a Nutshell, 2003, 2010.
Einstein Gravity in a Nutshell, 2013.
Group Theory in a Nutshell for Physicists, 2016.

Fearful Symmetry: The Search for Beauty in Modern Physics, Macmillan, 1986; Princeton University Press, 2016.

An Old Man's Toy: Gravity at Work and Play in Einstein's Universe, Macmillan, 1989; retitled as *Einstein's Universe: Gravity at Work and Play,* Oxford University Press, 2001.

Unity of Forces in the Universe, World Scientific, 1982.

On Gravity: A Brief Tour of a Weighty Subject, 2018.

Fly by Night Physics, 2020.

Index

acceleration: F=ma, 4–5, 55n8, 150–51; gravity and, 174, 334; motion and, 4, 21n3, 75, 80, 82, 150, 174, 222–23, 334; Newton and, 4–5, 67, 80, 82, 150, 222

accelerators: electromagnetism and, 106–7, 109, 120; Fermi National Accelerator Laboratory and, 240; grand unification and, 257; Large Hadron Collider and, 13n27; particles and, 12, 106–7, 109, 120, 217, 240, 257, 309, 342; weak interaction and, 217

action: at a distance, 22–25, 65, 75, 93, 118, 128, 151, 216, 283, 289; choice of history and, 74–78; classical physics and, 72, 331–32; Dirac, 153–54, 158, 160, 170–78, 326, 334; Einstein-Hilbert, 271–72, 275, 280, 289, 291, 339; energy and, 74–76; Euler and, 72, 78, 79n9; Euler-Lagrange formulation and, 78, 82, 90–91, 93, 127, 173, 275, 331; finding extremum and, 76–78; global/local formulation and, 79n8; grand unification and, 244–45; kinetic energy and, 51–52, 74–75, 81–83, 90, 104, 127, 151, 160, 173–74, 178n1; Lagrange and, 72–75, 78, 79n9; least time principle and, 69–72, 83n2; light and, 69–74, 78n2; Maxwell and, 78n3; Minkowski and, 72; motion and, 75; Newton and, 71–78; path integral and, 97, 153, 158–59, 162, 175–76, 181, 279, 332; Planck's constant and, 6, 33n17, 59, 90, 95, 119, 312, 316; potential energy and, 74–75, 81–82, 90, 104, 127–28, 151, 160, 173–74; quadratic scaling and, 150–51, 157–58, 161n8, 181; scattering and, 71; simultaneity and, 37–39; spacetime and, 71–74, 77–78, 79n8, 99; special relativity and, 331–32; speed of light and, 69–70

action principle: as external, 73–74; Feynman and, 80; gauge theory and, 174; global/local formulation and, 80, 82–83; gravity and, 272, 278n29; harmony and, 160; illusion of classical world and, 331–32; probability and, 93–94; toy example and, 78

Aharonov, Yakir, 176–77

algebra, 342; antimatter and, 190, 191n8; fields and, 163; harmony and, 161n8; quantum statistics and, 318, 321; spacetime and, 15, 39–42, 45n14, 51; special relativity and, 15, 39–42, 45n14, 51; strong interaction and, 199; weak interaction and, 218

Anderson, Carl, 114, 233, 237n25

Anderson, Phil, 247

angular momentum, 2, 316–18

anharmonic motion, 104, 149–50, 159, 161n5

anticommuting numbers, 320–21, 323n23

antimatter: antineutrinos, 212–18, 222, 225, 228–29, 235, 250–51, 258; antiprotons, 187, 193; antiquarks, 203, 207, 251; Dirac and, xvi, 187–88, 191n1, 191nn4–5, 239; Einstein and, 188–90, 191n8; electrons and, 187–90, 191n5, 191n8; energy and, 191n4, 191n8; Feynman diagrams and, 187–88, *189*, 191n7; grand unification and, 239–40, 249; Heisenberg and, 189, 191n8; Lagrangians and, 191n5; light and, 188–89; Lorentz transformation and, 190; momentum and, 191nn7–8; photons and, 187–88, *189*, 191n7; positrons, 11–12, 130, 144, 154–57, 165–66, 187–88, 190, 191n5, 191n8, 193, 205, 209n36, 222, 228, 232, 249, 252–55, 290; protons and, 187; quantum mechanics and, 187–88, 190, 295;